D Smith

Semiconductor diode

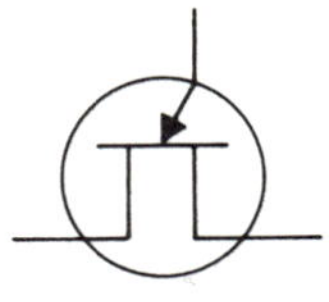
Unijunction transistor

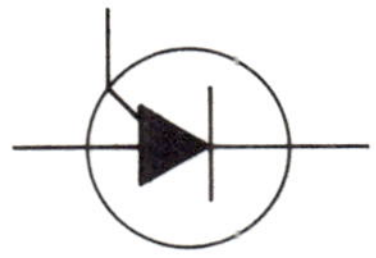
PUT

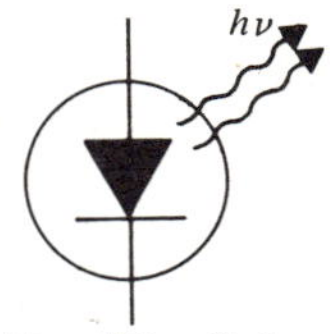

Light-emitting diode

Zener or breakdown diode

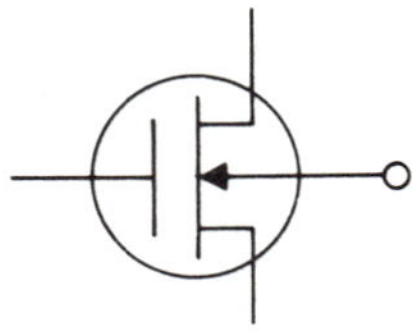
n-channel MOSFET

Diac

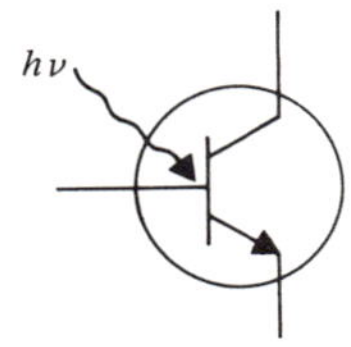

Phototransistor

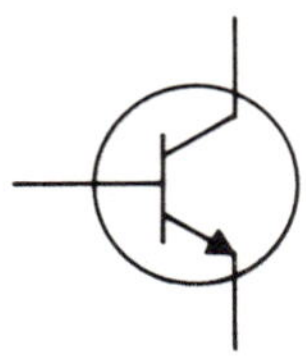
npn transistor

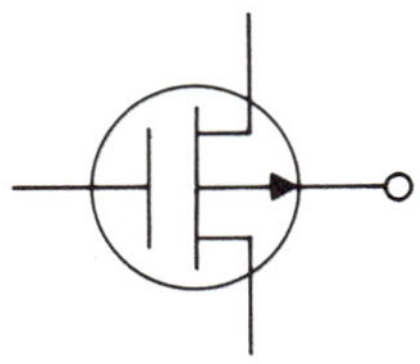
p-channel MOSFET

Silicon bilateral switch (SBS)

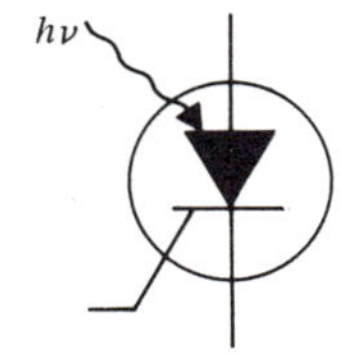

Light-activated SCR

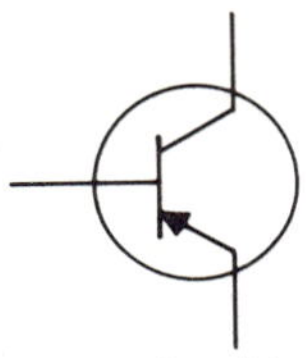
pnp transistor

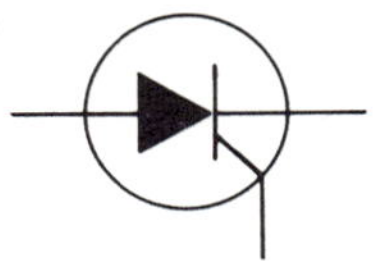
SCR and GTO

Triac

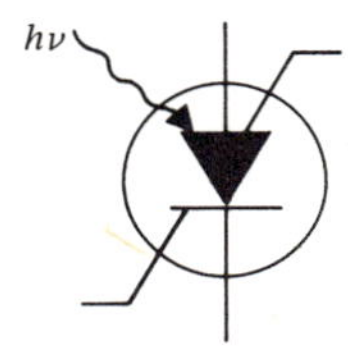

LASCS

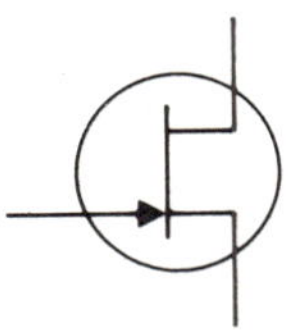
n-channel JFET

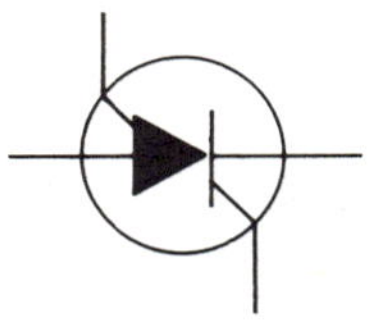
SCS

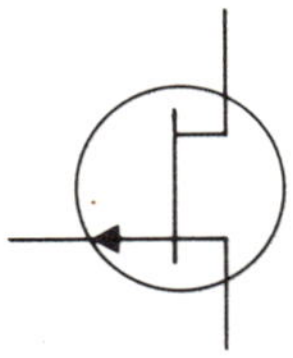
p-channel JFET

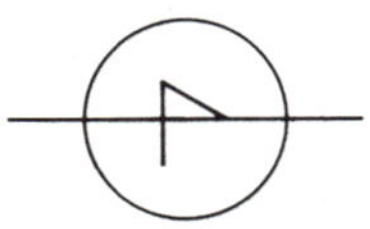
Four-layer diode

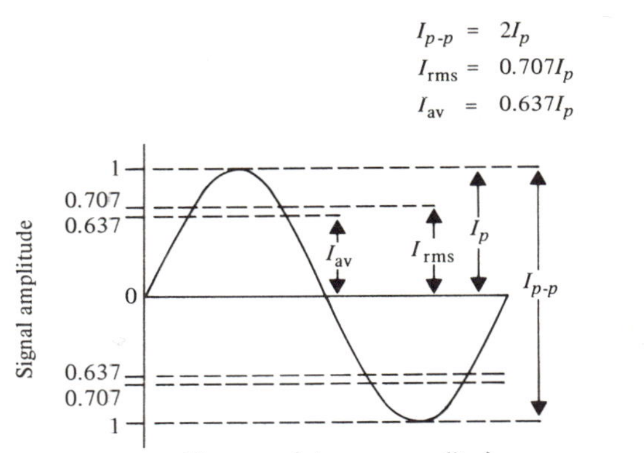

Measures of sine wave amplitude

3 – 35
45 – 61
66 – 72
89 – 130 Last Test
124 – 143
154 – 178
180 – 184
231 – 241

Instrumentation
For Scientists Series

CONTROL OF ELECTRICAL QUANTITIES IN INSTRUMENTATION

Module 2

Text with Experiments

HOWARD V. MALMSTADT
University of Illinois

CHRISTIE G. ENKE
Michigan State University

STANLEY R. CROUCH
Michigan State University

W. A. BENJAMIN, INC.
Menlo Park, California · Reading, Massachusetts
London · Amsterdam · Don Mills, Ontario · Sydney

THE MALMSTADT-ENKE INSTRUMENTATION FOR SCIENTISTS SERIES

 Printed in the United States of America. Published simultaneously in Canada. Library of Congress Catalog Card No. 73–7109.

ISBN 0-8053-6904-X
ABCDEFGHIJ-AL-79876543

Preface

The integration of basic electronic principles and experiments with major measurement and control concepts and instruments has been a major objective in this second module, as it also was in the first module in the Malmstadt-Enke Instrumentation for Scientists Series. Module 1, *Electronic Analog Measurements and Transducers,* was developed with the assumption that the reader had little or no background in electronics. This second module, *Control of Electrical Quantities in Instrumentation,* assumes a familiarity with the electrical quantities of charge, current, and voltage, their measurement, and the concepts of domain conversions wherein physical and chemical quantities are converted by transducers to measurable electrical quantities.

Module 2 begins with a study of the basic electrical properties of resistors, capacitors, and inductors, which provide control of electrical quantities, and then moves systematically to the semiconductor control elements that have revolutionized modern scientific instrumentation. The concepts and principles of electrical switching are developed in relation to specific solid state switching devices. In this manner it is possible to present first the basic principles and characteristics of semiconductor devices with their relatively simple and intuitive ON-OFF control applications. Then after the basic device characteristics have been introduced, the more advanced concepts of continuous control are presented through a study of basic amplifiers, including voltage, current, operational, and instrumentation amplifiers. This module ends with the development of the elegant automatic feedback control concepts and systems which are widespread in modern instrumentation systems.

Throughout this module the important applications of the devices and circuits are integrated with concepts and principles. The intimate interplay of fundamentals and applications unifies the study of electrical control and can provide new ideas on how to solve measurement and control problems. This approach also stimulates the student in his study of the principles because he can see the immediate "payoff" in important applications. The reader is referred to the preface of Module 1, *Electronic Analog Measure-*

ments and Transducers, for a more detailed discussion of the philosophy of our approach.

The order of presentation of the material in Modules 1 and 2 is unique and has been repeatedly refined through class testing and many discussions with students and practicing scientists. The presentation of analog measurements and transducers first in Module 1 provides an overall perspective in which the study of circuit elements, semiconductor devices, and control systems in Module 2 becomes both meaningful and exciting. Thus, by studying both Modules 1 and 2, the student can gain a thorough background in dc analog electronics, including dc analog circuits, transducers, measurements, operational amplifier circuits, semiconductor devices, and servo control concepts and systems. The material in Modules 1 and 2 is presented so that it can be used as a first course in college electronics or measurement instrumentation, or as the first part of a more advanced course in Electronics for Scientists.

In Module 3, *Digital and Analog Data Conversions,* measurement systems are described and analyzed as a sequence of interdomain data converters. It is shown how data can be encoded in three classes of electrical data domains—analog, time, and digital—and the methods of converting between electrical domains are developed. The presentation of these domain conversion concepts and circuits includes the material that would be included in modern introductory courses in operational amplifier circuits and digital electronics. Therefore, the combination of Modules 1, 2, and 3 includes the material for a complete basic course in analog and digital electronics, while also introducing transducers and the major concepts of scientific measurements.

For upper-level undergraduate or graduate courses in electronics or instrumentation, it is important to provide a thorough analysis of instrument limitations and to present the latest concepts and techniques for extracting useful data from noisy signals. Therefore, in Module 4, *Optimization of Electronic Measurements,* the components of electrical signals are analyzed and the modern methods and instruments for optimizing signal-to-noise ratios are presented. The combination of Modules 1–4 therefore provides a firm foundation in modern analog and digital electronics with emphasis on the concepts, implementation, and optimization of scientific measurements. With this background it should be possible to launch a systematic investigation into the most complex digital and analog instruments, computer interfacing, and automated systems. Through this modular book system, we have attempted to provide the individual with the means to select the level and depth of study most appropriate for his needs or curriculum.

The same text material is available in three forms: text-experiment modules, text modules, and the combination text materials from Modules 1–4 under one cover with the title *Electronic Measurements for Scientists.* Each text-experiment module contains detailed circuits and instructions for the experiments, and assumes implementation with specific types of instruments. The text material in each text module is identical to that in the corresponding text-experiment module, but the experiments are presented as summaries that state the experimental objectives and general approach; all detailed instructions and circuits for the experiments are omitted. The different methods of packaging the text material are intended to fulfill the requirements of different groups. Some want to take advantage of the many thousands of hours that have gone into writing and testing specific experiments that are uniquely designed to illustrate the important measurement principles and provide firsthand experience with state-of-the-art instrumentation. Others want to use the text materials contained in one or more modules as reference materials for many types of courses in physics, engineering, chemistry, and the life sciences. Others want a basic text for a lecture course or a series of courses that includes all the principles and concepts covered in the combined text *Electronic Measurements for Scientists,* and either do not have an interest in laboratory experiments or have a separate lab course that uses experiments and systems quite different from those in the text-experiment modules.

The authors would like to acknowledge the assistance of many others in the preparation of this module. The art work was executed by E. C. Smith Associates under the direction of Stan Parnell. Peri Warstler, Parna Westfall, and Donna Abernathy did the typing. Terry Woodruff and Scott Goode read the entire text and made many valuable suggestions and helped in the development of some of the experiments. Kelly O'Keefe, Dave Rothman, Keith Caserta, and Brian Hahn aided in the development of some of the experiments and problems.

July 1973

H.V.M.
C.G.E.
S.R.C.

Contents

Introduction

Many types of components are used in electronic circuits, but none more frequently than resistive and conductive elements. The movement of electrical charge in circuits is directed by conductors, and the rate of flow of charge can be controlled by resistors. In the first section of this module the characteristics of conducting materials used in electronic instrumentation are introduced through a basic examination of their structures, and practical resistors and conductors and their limitations are discussed. Transducers that convert temperature, light intensity, and strain into related resistance values, and the techniques for resistance measurement and network analysis, are then presented.

Capacitors and inductors play a major role in many instruments for the control of electrical charge and voltage. When they are combined in various ways with other elements, the resulting circuits can provide a variety of important functions. The characteristics of practical capacitors and inductors and their applications in basic *RC* and *RL* circuits are described, and the types of transducers that convert various physical quantities into related values of capacitance and inductance are considered. The important operational amplifier integrators and differentiators that utilize capacitors for charge storage, and their use in several applications, are described in the second section.

The development of inexpensive high speed switches is basic to the modern instrumentation/automation revolution. Also, new types of power switches have greatly facilitated the design of reliable and compact instruments. In the third section of this module the important types of diodes, transistors, FET's, optoelectronic devices, SCR's, and other types of switching elements that control electrical quantities are introduced. The application of program-controlled switches for waveform generation is illustrated.

All measurement and control instruments require various types of power supplies in order to operate. Most modern instruments operate from the ac power line, but generally require regulated dc voltages such as $+5$ V and ± 15 V. The conversion from ac to the required dc voltages is described in the fourth section. There is also a discussion of electrochemical and solar batteries that are used in portable equipment.

It is often necessary to amplify without distortion the signals from input transducers in order to provide reliable measurements or control. Various elegant amplification concepts were introduced in Module 1, *Electronic Analog Measurements and Transducers,* and several practical linear amplifiers were described on a functional block basis. In the final two sections of this module the internal operation of the amplifier functional block is considered in more detail and some of the major concepts and basic relationships for linear amplification and control are discussed. Very accurate types of current and voltage regulators with linear feedback control are also presented.

Resistive Devices for Measurement and Control

Section 2–1

The electrical properties of a material or device determine the way it responds to the electrical quantities of charge, potential difference, current, and power. All electrical and electronic devices possess in some degree the properties of resistance, capacitance, and inductance, and all device characteristics can be explained in terms of these three properties.

When a device is designed to have a single property (e.g., a resistor to have only the property of resistance), the presence of the other electrical properties is considered an imperfection in that device. The magnitude of the electrical property of an ideal "linear" device is independent of the magnitude of the electrical quantities applied. Thus the resistance of a linear resistor is independent of the applied voltage or current; the capacitance of a linear capacitor is independent of the quantity or sign of stored charge; and so on. The conditions under which deviations from the expected linear behavior of devices are encountered constitute possible limitations on the device's applicability. For example, the maximum power dissipation rating of a resistor limits the constant current-voltage product that can be applied to it without risking overheating or destruction. On the other hand, many devices such as diodes are intentionally made to be very nonlinear. The properties of nonlinear devices can be varied in response to applied electrical quantities.

The magnitudes of the electrical properties of devices are often affected by environmental conditions such as temperature, humidity, pressure, radiation, and magnetic field strength. In some cases the sensitivity of a device to these factors limits the scope of its applications. In other cases, devices for which some electrical property is intentionally sensitive to a particular physical parameter are used to convert the magnitude of the physical parameter into an electrical property that can be measured electrically. Thus the measured resistance of a temperature-dependent resistor is related to the temperature of the resistor.

No electrical device has complete freedom from all unwanted electrical characteristics, or complete independence from age or ambient conditions. Various materials and methods of construction result in various combinations of the desired qualities and limitations. This explains why there are so many different types of electrical components even though there are only three basic electrical properties. In this section, practical conductors and resistors will be described; the measurement of resistance will be discussed; and the characteristics of circuits containing combinations of resistors will be investigated.

2–1.1 CONDUCTANCE AND RESISTANCE

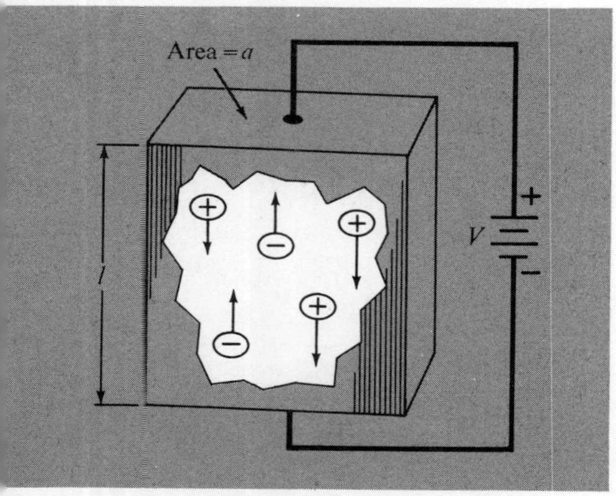

Fig. 2–1 Motion of charge carriers in an applied electric field.

When an electrical voltage is applied across a piece of material, a net motion of the mobile charge carriers in that material is generated in the direction of the electric field created by the applied voltage. This is illustrated in Fig. 2–1 for a material with cross-sectional area a cm^2 and length l cm.

The net motion of charge carriers, or electrical current, is measured by the net charge crossing a plane (which is perpendicular to the motion) each second. The current, in coulombs per second, is given by

$$I = \frac{aVQ_e}{l}(n_1\mu_1 + n_2\mu_2 + \ldots + n_p\mu_p), \tag{2–1}$$

where there are p different kinds of charge carriers designated 1, 2, 3, and so on up to p. For the kind of charge carrier designated 1, n_1 is the density in carriers/cm^3, Q_e is the electron charge in coulombs, and μ_1 is the mobility of the carrier in cm^2/volt-sec. It is assumed in Eq. (2–1) that the charge carriers each have a charge equal to Q_e.

The conductance G in mhos of the piece of material as connected in Fig. 2–1 is thus

$$G = \frac{I}{V} = \frac{aQ_e}{l}(n_1\mu_1 + n_2\mu_2 + \ldots + n_p\mu_p). \tag{2–2}$$

The conductivity σ, characteristic of the material, is the conductance between opposite faces of a 1 cm tube with an applied voltage of 1 volt (or the current per cm^2 with an applied electric field strength of 1 volt per cm). Therefore

$$\sigma = Q_e(n_1\mu_1 + n_2\mu_1 + \ldots + n_p\mu_p), \tag{2–3}$$

where σ has the units amperes/volt-cm. Resistance R in ohms is defined as the reciprocal of conductance. Therefore

$$R = \frac{V}{I} = \frac{l}{a\sigma} \tag{2–4}$$

and the resistivity ρ in volt-cm/ampere is $1/\sigma$. Thus the resistivity of any material depends on the population density, charge, and mobility of the mobile charge carriers in the material; the resistance of a device made of that material depends on the dimensions of the material and its resistivity.

The work required to move charge through a conducting material is equal to the product of the total charge moved and the potential difference across the conductor:

$$W = QV. \tag{2–5}$$

When the units of Q and V are coulombs (C) and volts (V) respectively, the work W is in joules (J). This is work done in the conductor which produces heat, light, or force. If a current of I amperes exists in a conductor for t sec, the total charge moved is $Q = It$ coulombs, and the work performed on (and dissipated by) the conductor is

$$W = ItV. \tag{2–6}$$

Power (P) is defined as the rate of doing work, or W/t. A work rate of 1 J/sec is equal to 1 watt (W) of power. Therefore,

$$P = W/t = ItV/t = IV. \tag{2–7}$$

By combining Ohm's Law (Eq. 2–4) with Eq. (2–7), the power dissipated in a conductor can be related to the resistance of the conductor and the current through it (or the voltage across it). Thus

$$P = I^2R = V^2/R, \tag{2–8}$$

which shows that the power dissipated in a resistor increases as the square of the applied current or voltage.

2–1.2 MECHANISMS OF CONDUCTION

It was shown in the previous section that the conductivity of any material depends on the density of mobile charge carriers, n, and their mobility, μ. If n and μ were constant quantities like Q_e, that is all we would need to know to characterize resistive devices. However, n and μ (and thus σ and R) can change with changes in temperature, humidity, light intensity, voltage, and other factors. Furthermore, the way in which the population density and mobility of the charge carriers vary with these environmental parameters depends on the mechanisms by which the charge carriers are created and transported. Materials used for electronic devices are classed according to their mechanism of charge carrier generation and motion. They generally fall into three main categories: metallic conductors, semiconductors, and insulators. In the next section the basic mechanisms of conduction will be described, and it is from these that the characteristics of the common types of conducting materials will be developed.

Since matter is generally composed of electrically neutral atoms, a mechanism must exist for the separation of electrons from some of the atoms in conducting materials to form free negatively charged electrons and positively charged ions. The differences in the mechanisms of **ionization** among each of the types of conducting material contribute greatly to the differences in their conductive properties. Free electrons are the principal charge carriers for many conducting materials but the positive ions also often play a significant, and sometimes dominant, role in the conduction process.

The electrons in an isolated atom are confined to specific orbitals or energy levels. Each energy level can be populated by two electrons of opposite spin. There are many more energy levels than are needed to accommodate the atom's normal (neutral) complement of electrons. In its lowest energy state the electrons in the atom populate the lowest possible (innermost) energy levels. An electron may be promoted to a higher energy level by absorbing energy from a photon or by a collision with another particle. The atom is now said to be in an **excited state.** Excited state atoms generally return to the ground state or lowest energy state by losing energy through collisions and/or by emission of radiation. If the energy source which excites an atom is sufficiently energetic, it can provide enough energy for an electron to escape the atom completely. The atom is then said to be **ionized,** and, in the process, two charged particles are created. In the presence of an electric field, the electrons will be accelerated in the direction of the positive charge and the positive ions will be accelerated in the direction of the negative charge.

Metals

The outermost and most easily removed electrons of an atom are frequently called the **valence** electrons. It is the interaction of the valence electrons of atoms that is primarily responsible for the formation of crystals and molecules. When similar atoms combine to form a pure crystalline solid, the discrete energy levels of the valence electrons broaden into bands of allowed energy levels. The maximum electron population of each energy band is twice the number of atoms in the solid, but the electrons may have any energy within the band limits. One such band is the **valence** or **bonding**

band. In this band the valence electron(s) of each atom share orbitals with the electrons of the nearest neighbor atoms to form **covalent bonds** between the atoms. If there is an odd number of valence electrons in each atom and there are two available energy states per atom in each band, one of the energy bands will be only half filled. The unoccupied energy states in the energy band allow the electrons in that band to be accelerated by an electric field. Thus electrons in a partially filled band are **conductive electrons** and a partially filled band is a **conduction band.** Monovalent metals such as lithium, sodium, and potassium are observed to have one free conducting electron per atom as expected. The metal atoms in this case are positively charged ions and would also contribute to the metal's conductivity if they were reasonably free to move.

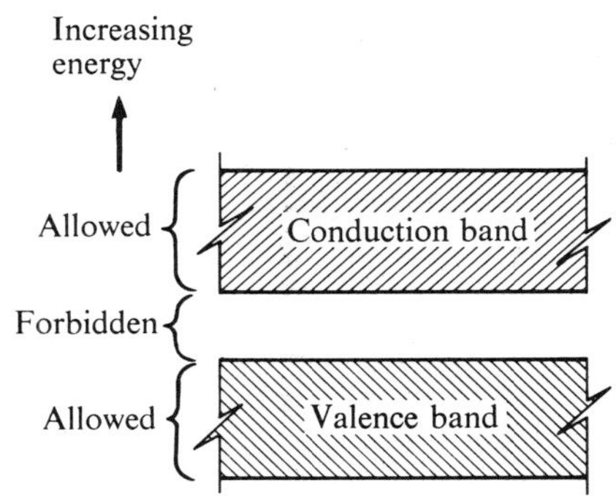

Fig. 2–2 Electron energy bands in a solid.

In materials with an even number of valence electrons in the lowest energy state, the valence bands are filled and therefore the electrons in them are not free to move. An allowed energy band higher than the highest valence band will be empty. This is illustrated in Fig. 2–2. If some of the valence electrons can be energized (promoted) into the next higher band, they will be free to move in this partially filled band. This next higher band is thus the conduction band in this case. The magnitude of the energy gap between the filled valence band and the conduction band determines the conductance type of a given material. If the energy gap is zero, it is a **metallic conductor;** if it is small, it is a **semiconductor;** and if it is large, it is an **insulator.**

In any solid material, there is an equilibrium distance between the atoms. This is the distance for which the energy of combination is a minimum. Variations in this equilibrium distance with temperature are responsible for variations in bulk dimensions with temperature. The closer atoms are to each other, the more their electron orbitals interact. This causes the energy range of the allowed bands to increase and that of the forbidden bands to decrease. This is shown in Fig. 2–3. In many metals the interatomic distance is so small (as evidenced by the high density of such materials) that the allowed valence and conduction bands overlap. When this occurs, the combined allowed energy band is unfilled and the valence electrons can conduct. This explains the high conductivity of bivalent metals such as magnesium, calcium, and strontium even at temperatures near absolute zero (minimum energy state). If the atoms have an even number of valence electrons *and the valence and conduction bands do not overlap,* the material must be a semiconductor or an insulator.

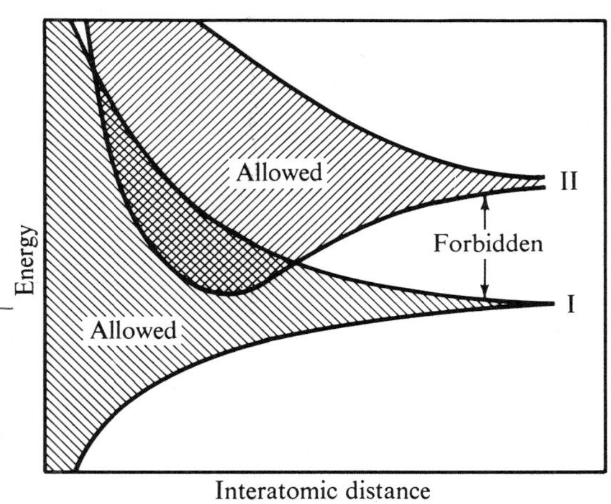

Fig. 2–3 Energy bands vs. distance.

The electrical properties of most metals are consistent with a model in which the metal atoms are positive ions in a "sea" of mobile electrons. The resistance of the metal results from the collision of the mobile electrons with the metal atoms. At very low temperatures (near absolute zero) the metal structure is almost perfectly periodic so that the electron motion,

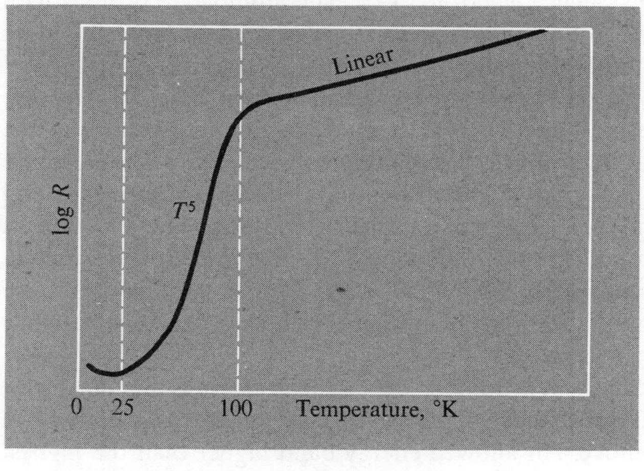

Fig. 2–4 Effect of temperature on the resistance of a metal.

described as a wave function, meets little (in a superconductor, zero) resistance. As the temperature increases from 0°K, random vibrations of the metal atoms cause a rapid increase in the collision rate. The resistivity in the low temperature region actually increases as the fifth power of T for many metals, as shown in Fig. 2–4. At higher temperatures (still much lower than room temperature) the resistivity increases less rapidly and eventually increases approximately linearly with temperature. The resistivity of most metals increases from 3 to 6 parts per thousand per degree centigrade in the linear region. Such changes can cause difficulty in precision circuits but they are turned to advantage in the platinum resistance thermometer.

Semiconductors

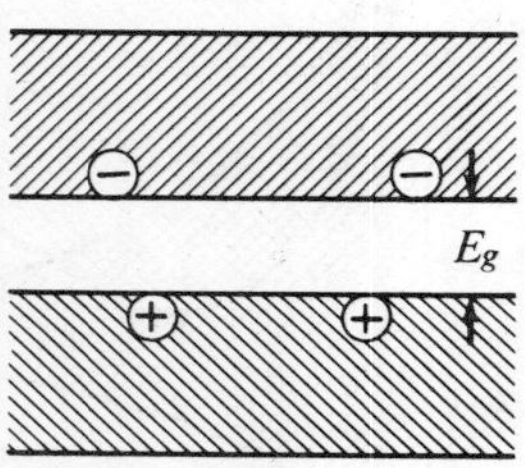

Fig. 2–5 Band structure of semiconductor or insulator with valence electrons excited to the conduction band.

A material with an even number of valence electrons and a forbidden energy gap between the valence and conduction bands cannot conduct electricity at 0°K. At absolute zero temperature the valence band is filled and the conduction band is empty. A source of energy is required to promote valence electrons to the conduction band. This energy might be provided by quanta of radiation (photons), resulting in photoconduction, or by thermal excitation at higher temperatures. An energy source at least equal in energy to the forbidden gap is obviously required for excitation to conduction. The result of promoting an occasional valence electron to the conduction band is shown in Fig. 2–5. For each electron in the conduction band, there is a lack of one electron in the valence band. The point where a valence or bonding electron is missing is called a **hole.** It must be kept in mind that while the conduction band electrons are free to move throughout the material (and do), the valence electrons are limited to interactions

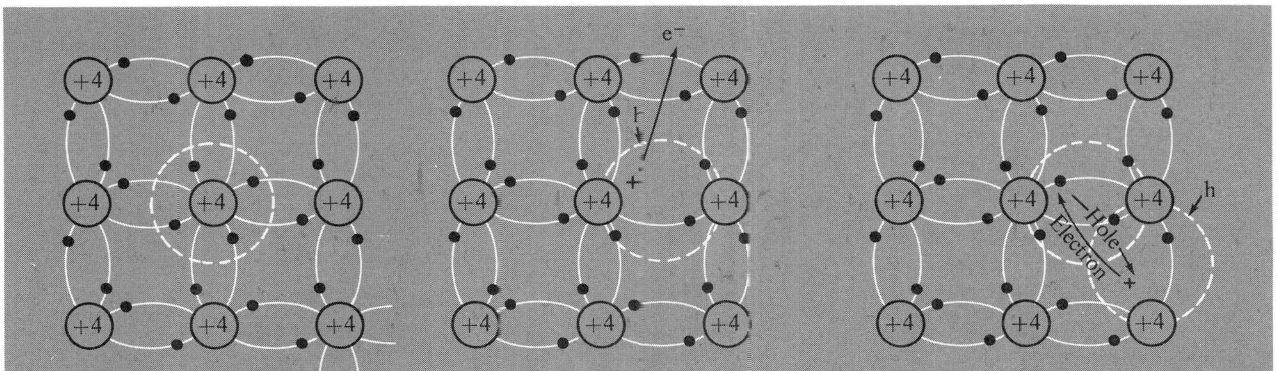

(a) Covalent bonding with no bonds broken

(b) Consequences of a broken covalent bond

(c) An oversimplified picture of hole motion through the motion of bound valence electrons

Fig. 2–6 Motion of conduction electron and hole in semiconductor or insulator.

with near neighbor atoms. Thus the area of a missing valence electron (the hole) is a localized region of positive charge. When an electric field is applied to a material with excited electrons, the conduction electrons will drift in the direction of increasingly positive potential. Valence electrons can only move in response to the electric field if they are adjacent to a hole. The net result of the valence electron motion is a drift of the positive hole in the direction of increasingly negative potential. This action is illustrated in Fig. 2–6.

The population density of electrons n_e in the conduction band at any given temperature is greater for materials with lower forbidden gap energies, E_g, and increases with increasing temperature for any material. The relationship between n_e, T, and E_g is given approximately by the following equation:

$$n_e = 2(2\pi m_e kT/h^2)^{3/2} e^{-E_g/2kT}, \tag{2–9}$$

where m_e is the mass of the electron, k is Boltzmann's constant, and h is Planck's constant. For every conduction electron there is also a hole which acts as a positive charge carrier. Thus $n_h = n_e$ and from Eq. (2–3) the total conductivity is

$$\sigma = Q_e(n_e\mu_e + n_h\mu_h) = n_e Q_e(\mu_e + \mu_h). \tag{2–10}$$

The energy gap and mobilities for several pure metals and semiconductor materials are given in Table 2–1.

From Eqs. (2–9) and (2–10) it can be seen that if μ_e and μ_h are not strongly dependent on temperature, then the conductivity will increase approximately exponentially with increasing temperature, as is illustrated in

Table 2–1 Conducting characteristics of several pure materials at 25°C

Material	Energy gap, eV	Mobility at 25°C, cm^2/V-sec: Electrons	Holes	Carrier density, n	Resistivity ρ, Ω-cm
Elements					
Copper	—	35	—	$\cong 10^{23}$	$\cong 1.5 \times 10^{-6}$
Silver	—	56	—	$\cong 10^{23}$	$\cong 1.5 \times 10^{-6}$
Diamond	5.47	1,800	1600	—	—
Germanium	0.80	3,900	1900	2.5×10^{12}	8.9×10^4
Silicon	1.12	1,500	600	1.6×10^{10}	2.5×10^5
III-V compounds					
Ga Sb	0.67	4,000	1400		
Ga As	1.43	8,500	400	1.1×10^7	
In Sb	0.16	78,000	750		
In·As	0.33	33,000	460		
II-VI compounds					
CdS	2.42	300	50		
CdSe	1.7	800			
ZnO	3.2	200			

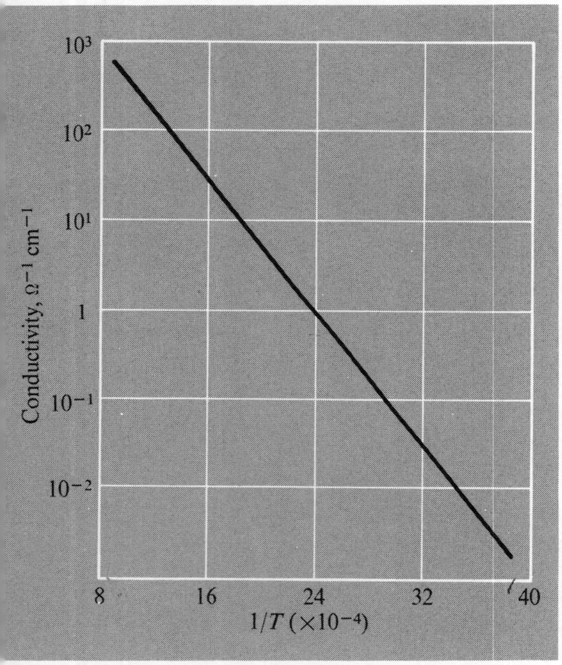

Fig. 2–7 The electrical conductivity vs. reciprocal absolute temperature for the intrinsic semiconduction region in germanium.

Fig. 2–7. This thermal dependence is the basis of the temperature-dependent resistor (thermistor).

A semiconductor which conducts by equal numbers of holes and electrons created by the excitation of valence electrons is called an **intrinsic semiconductor.** All *pure* semiconductors and insulators are intrinsic semiconductors. Whether a material is a semiconductor or an insulator depends on the magnitude of the energy gap E_g compared with the operating temperature or energy of excitation. At low enough exciting energies, all such materials are insulators, and at high enough exciting energies all such materials are conductors. Table 2–1 shows how markedly different the resistivity can be for such materials at room temperature.

Extrinsic Semiconductors. The majority of semiconductors used in electronic materials are not pure, or intrinsic, semiconductors as described above, but are impure, **doped or extrinsic semiconductors.** The intentional impurity (or dopant) is an extremely small amount of an element with 3 valence electrons per atom, such as boron or indium, or 5 valence electrons per atom, such as arsenic or antimony. The impurity level used is generally of the order of 1 part in 10^8 so that the structure, dimensions, and chemical characteristics of the semiconductor material are essentially unaffected by doping. The impurity atoms simply occupy scattered sites in the normal semiconductor crystal lattice.

An atom with 5 valence electrons in a crystal structure which uses 4 valence electrons per atom for bonding, has one valence electron more than it needs. The energy level of this additional electron is only a few hundredths of an electron-volt from the conduction band for the semiconductor. Thus at normal temperatures, essentially all the impurity atoms are ionized and have contributed an electron to the conduction band. No holes are produced in this process since each atom still has 4 bonding electrons. The valence band is still filled and thus holes are only produced by intrinsic type excitation. In germanium there are about 10^{23} atoms per cm^3. A concentration of arsenic dopant of only 1 part in 10^8 will produce 10^{15} conducting electrons per cm^3. This charge carrier density is roughly 100 times larger than the intrinsic charge carrier density of germanium at room temperature. Thus a trace impurity can have an enormous effect on the electrical properties.

The 5-valent impurity is called a **donor** impurity because it donates electrons to the conduction band. The resulting material is called an **n-type semiconductor** because the majority of charge carriers are negatively charged. In any semiconductor, the product of n_e, the conducting electron density, and n_h, the hole density, is a constant at any given temperature. Thus

$$n_e \cdot n_h = n_i^2(T), \tag{2–11}$$

where $n_i^2(T)$, the charge density product, is a function of temperature *but not of impurity concentration.* For intrinsic germanium at room temperature, $n_i^2(T) \approx (10^{13})^2 \approx 10^{26}$. Since this constant is independent of n_e or n_h, for the example above where $n_e = 10^{15}$ electrons/cm^3, $n_h = 10^{26}/10^{15} = 10^{11}$ holes/cm^3. Thus there are 10^4 more electron charge carriers than hole charge carriers in the doped germanium in this example. The electrons are truly the majority of charge carriers and are aptly called **majority carriers.**

The addition of 3-valent impurity atoms to a semiconductor material has a complementary effect to that just described. When 3-valent atoms take their place in the lattice, one valence electron is missing per atom. This creates, in effect, a hole in the valence band which greatly reduces the energy required to ionize a nearby semiconductor atom. Just as was shown with the 5-valent impurity, the required ionization energy is small compared to the thermal energy (kT) at room temperature. An impurity level of 10^{15} atoms of indium per cm^3 of germanium would thus produce essentially 10^{15} holes per cm^3 at room temperature. Since by Eq. (2–7), n_e would be 10^{11} electrons per cm^3, the holes are the majority carriers. The 3-valent

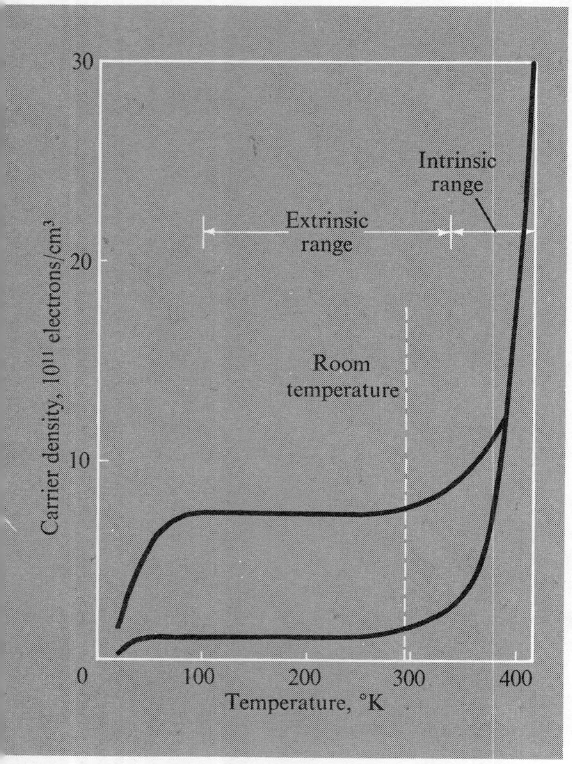

Fig. 2–8 Carrier density vs. temperature in two n-doped germanium crystals.

dopant is called an **acceptor** and the acceptor-doped semiconductor is called a **p-type semiconductor.**

If both donor and acceptor impurities are present, they tend to neutralize each other. The effective impurity material is the one present in excess and its effective concentration is equal to its excess concentration.

The temperature dependence of the carrier concentration of a doped semiconductor is illustrated in Fig. 2–8. At low temperatures the number of charge carriers n_i contributed by the dopant increases as kT becomes significant compared to E_i, the impurity ionization energy. At temperatures between 100° and 300°K, essentially all the impurity atoms are ionized and the charge density is quite constant. Essentially all the charge carriers are being created by the extrinsic or impurity mechanism. Above room temperature, kT becomes significant with respect to E_g, the intrinsic ionization energy gap, and the charge carrier density increases by the intrinsic mechanism. The operation of extrinsic semiconductors in the intrinsic temperature range poses a threat to the component and is generally avoided. In the intrinsic range, an increase in temperature can cause an increase in current through the device which can cause further heating of the material, and so on. For germanium devices, the maximum safe operating temperature is about 85°C, while silicon devices, with a higher E_g, can be safely operated up to 200°C.

2–1.3 RESISTORS

A large variety of electronic components are made for the sole purpose of introducing resistance in an electric circuit. Such components are called **resistors.** They are used to limit the current in a circuit, to convert current to voltage, and to divide voltages and currents. Resistors are made from a variety of materials and come in many shapes, sizes, and values. They are characterized primarily by the nominal resistance value, the accuracy with which the actual resistance agrees with the nominal value, the power dissipation, the stability of the resistance with respect to time, temperature, humidity, etc. The characteristics of several of the most common resistor types will be discussed in this section.

Wire

Metallic wire is an essential part of all electronic systems. Its low resistance can provide a nearly ideal electrical connection between components. It is also used in the fabrication of many devices such as inductors, transformers, and wire-wound resistors. The usual electrical hookup wire is made of copper, a metal which provides very high conductivity and good flexibility at moderate cost. The copper wire is often plated with a thin layer of another

Table 2–2 Copper wire characteristics

A.W.G.*	Number of strands	Diameter per strand, mm	Resistance, Ω/m	Current capacity, A
30	1	0.255	0.346	0.144
24	1	0.511	0.0804	0.577
24	7	0.022	0.0804	
22	1	0.644	0.0501	0.918 (5)
22	7	0.025	0.053	
20	1	0.812	0.0316	1.45 (7.5)
20	7	0.032	0.0332	
18	1	1.024	0.0198	2.32 (10)
18	7	0.040	0.0203	
16	1	1.291	0.0125	3.69 (13)
16	19	0.029	0.0130	
14	1	1.628	0.0078	5.87 (17)
12	1	2.053	0.0049	9.33 (23)

* American Wire Gauge (A.W.G.) is a means of specifying relative wire diameter. The lower the A.W.G. number, the larger the diameter and the lower the resistance per unit length.

metal such as silver or tin to make it easier to solder other components to it,[1] and is then covered with an insulator, usually plastic. Copper wire is available in many diameters, called **gauges.** From Eq. (2–4), it is clear that the larger the diameter, the lower the resistance per unit length. Table 2–2 gives the resistance of several sizes of copper wire. A conductor made of several collected small wires is called **stranded wire**. It offers improved flexibility and is less likely to break under repeated flexing.

Note 1. Tinned Wire.
A copper wire that has been coated with a layer of tin or solder to prevent corrosion and to simplify soldering.

To provide some perspective, 22 gauge wire is the normal hookup wire size, 18 gauge is used for household lamp cords, and 12 and 14 gauge are used for house wiring. The current values given in parentheses in Table 2–2 would bring the temperature of the wire to 100°C if the wire were bundled or enclosed and the ambient temperature were 57°C (135°F).

Carbon Resistors

The most common type of resistor is made of hot-pressed carbon granules **(composition resistor)** or a thin film of finely divided carbon granules deposited on an insulating substrate **(carbon film resistor).** These are the familiar resistors shown in Fig. 2–9a. They have the advantages of being inexpensive, remarkably free of capacitance and inductance, and available over a very wide range of values (from 2.7 Ω to 2.2 × 10^7 Ω and from $\frac{1}{8}$ to 2 W). On the other hand, they are only moderately accurate (tolerances are ±5% and ±10% from the nominal value for carbon composition resistors and ±1% and ±2% for carbon film resistors), they introduce more electrical noise into circuits than do other resistor types, and they have a relatively high **temperature coefficient of resistance** (variation of resistance with temperature).

Color Code. The nominal resistance value of most carbon resistors is marked on them by three colored bands painted around the body of the resistor. These three colors can be translated into a number with two significant figures by using the standard resistor color code. This is illustrated in Fig. 2–9b. Each number from 0 to 9 has been assigned to a color as shown in Table 2–3. Note that from 2 to 7 the colors follow the spectrum.

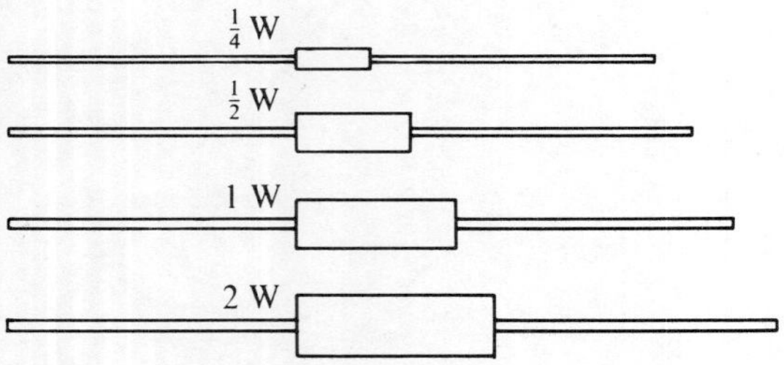

(a)

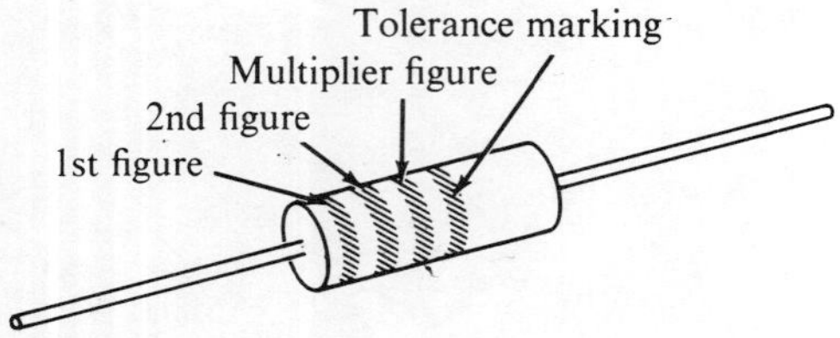

(b)

Fig. 2–9 Carbon resistors: (a) physical size for given power dissipation; (b) color code.

Table 2–3 Resistor color code

Black	0	Green	5
Brown	1	Blue	6
Red	2	Violet	7
Orange	3	Gray	8
Yellow	4	White	9

The color of the band closest to the end of the resistor represents the first figure of the resistance; the second band, the second figure; and the third band, the number of zeros to add to the first two figures to get the total resistance. Thus a resistor which is coded with yellow, violet, and red is 4700 ohms or 4.7 kΩ. Blue, gray, green is 6,800,000 Ω, or 6.8 MΩ; and green, blue, black is 56 Ω. For resistance between 1 and 10 Ω, gold is used for the third band. Thus orange, white, gold is 3.9 Ω.

The Accuracy of the Nominal Resistance. The limit of accuracy assured by the manufacturer of the resistor is called the **tolerance.** This is expressed in percent of the nominal value. For instance, a 100 Ω resistor with a tolerance of 10% may have any value from 90 to 110 Ω. In color-coded resistors the tolerance is indicated by the color of the fourth band, silver being 10%, gold being 5%. Closer tolerances are printed on the resistor body.

The 10% tolerance requirement of most common resistor applications has resulted in the apparently strange values of the resistors manufactured. The standard values of resistances available in 10% tolerance between 1 and 10 kΩ are 1, 1.2, 1.5, 1.8, 2.2, 2.7, 3.3, 3.9, 4.7, 5.6, 6.8, 8.2, and 10 kΩ. Note that there is a value within 10% of any required resistance.

The use of "round" numbers for resistance values would result in an unnecessarily large number of values required to cover the same resistance range. Where 5% tolerance is specified, 5% resistors are supplied in the values 1.1, 1.3, 1.6, 2.0, 2.4, 3.0, 3.6, 4.3, 5.1, 6.2, 7.5, and 9.1 kΩ, in addition to all the standard 10% values. In all other decades, the 5 and 10% values are the same as those listed above.

Many factors affect the accuracy of a resistor. For resistors of very low values (less than 10 Ω) the resistance of the connection to the resistor can become a significant part of the total resistance of the circuit. For resistors of very high values (greater than 10 MΩ) leakage of current between the contacts on the surface of the resistor can significantly decrease the resistance of the circuit. This effect, which will depend on the ambient humidity, can be reduced through the use of special insulating materials. In applications where stability is important it is desirable to use a resistor with a much higher power rating than the circuit requires. This will reduce temperature changes in the resistor.

Wire-Wound Resistors

Resistors are made of wire for three different applications: low resistance, high precision, and high power dissipation. All are made by winding a coil of a relatively high resistivity metal alloy wire onto a spool (often ceramic) and making contact to each end of the wire as shown in Fig. 2–10a. Low resistance wire-wound resistors require little wire, are inexpensive, and are more stable than composition resistors. Resistors of low resistance are often called upon to dissipate relatively large amounts of power. For dissipation greater than 2 W, **power resistors** are used. For this application, rather low gauge resistance wire is wound on a ceramic form which will withstand high temperatures. Power resistors should be mounted so that they can readily dissipate heat to the air or the instrument case.

When a low temperature coefficient alloy wire is used, very precise and stable resistors can be made. Accuracy and stability of the order of ±1% to ±0.001% are obtainable in this way. Wire-wound resistors become less desirable as the nominal resistance increases beyond 1 kΩ. The longer wire lengths required increase the cost and the inductance of the resistor.

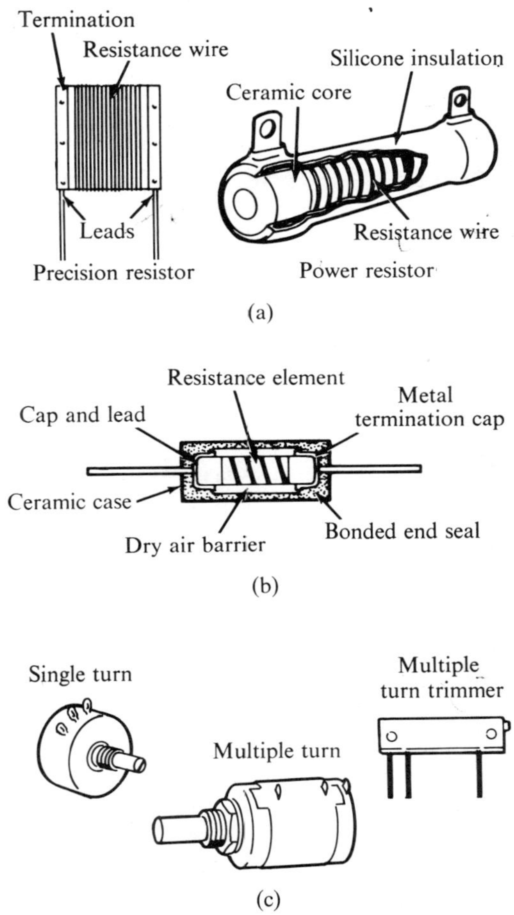

Fig. 2–10 Practical resistors: (a) wire wound; (b) metal film; (c) variable resistors (potentiometers).

Metal Film Resistors

The improvement in the techniques of metal film deposition and the development of better substrate materials have provided a valuable addition to resistor types. Metal film resistors are made by vacuum-depositing a thin metal layer on a low thermal expansion substrate. The resistance of the film is increased and adjusted by etching or grinding a pattern through the deposited film as illustrated in Fig. 2–10b. Accuracy and stability approaching the better wire-wound resistors are now available. Metal film resistors are much easier to fabricate in high resistance values than are wire-wound resistors, are essentially noninductive, and introduce a minimum of *electrical noise*. Noise characteristics of resistors are discussed in Module 4.

Variable Resistors

Some resistors are made to have an adjustable or variable resistance. This is accomplished by exposing the resistive element so that contact can be made at any point along the resistor by a movable contact, or "wiper." Such a device is called a **rheostat** or a **potentiometer.** The value of the total or maximum resistance is usually stamped on the case. The resistive element is usually either a coil of resistance wire or a strip of resistive film. Several types of potentiometers are shown in Fig. 2–10c.

2–1.4 RESISTIVE TRANSDUCERS

If the resistance of a device is a known and reproducible function of temperature, light, position, strain, voltage, magnetic field, or another physical or chemical parameter, the device can be used as an interdomain converter which converts from the domain of the functional dependence to the domain of resistance. The functional relationship constitutes the transfer function for the conversion. The resistance is then measured by one of the techniques to be described in Section 2–1.5, and the magnitude of the temperature, light, or strain is deduced from the transfer function. In this section, a number of common resistive transducers and their characteristics are described.

Temperature-Dependent Resistors

As illustrated in Section 2–1.2, the conductance of all conducting materials varies with temperature. If the resistance-temperature relationship for a particular conducting device is characterized and reproducible, the measurement of that device's resistance can be used to determine the temperature of the device. In this application both metallic and semiconductor

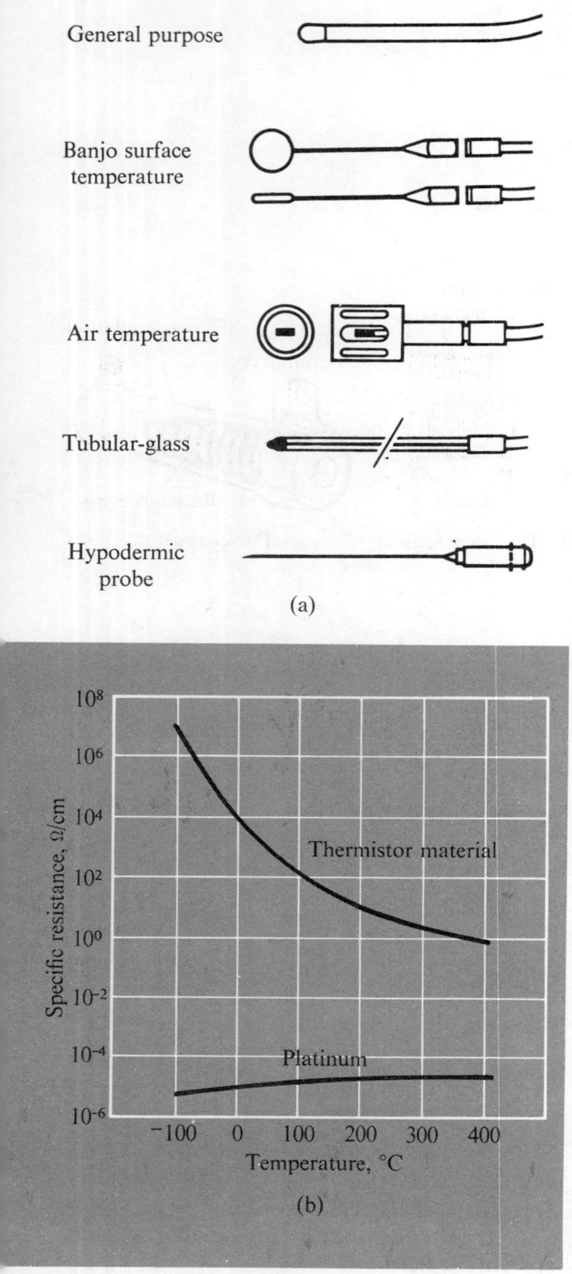

Fig. 2–11 Resistance-temperature response of typical thermistor material, compared with platinum and several practical thermistor probes.

materials are used. The most common metallic thermal resistance sensor is the platinum resistance thermometer. Temperature-dependent semiconductor resistors are called **thermistors.**

Platinum Resistance Thermometer. The sensing element of a platinum resistance thermometer is a high-purity platinum wire (99.999% pure) wound on a ceramic core and placed in the tip of a probe. Copper contacts are made to the ends of the sensing element wire and brought out through the other end of the probe shaft. The resistance of a platinum resistance thermometer is generally 100 to 500 Ω at 0°C. As in the case of all common metallic conductors, the resistance increases with increasing temperature. The resistance-temperature relationship for pure platinum is essentially linear over the range from −270°C to 1100°C with a resistance change of 0.392% per °C. Platinum resistance thermometers are extremely accurate, sensitive, and stable.

Thermistor. The thermistor is made from an intrinsic semiconductor. As we have seen, the number of charge carriers in intrinsic semiconducting material increases linearly with $1/T$. Thus the resistance of the thermistor decreases sharply with increasing temperature. Typical resistance-temperature curves are shown in Fig. 2–11 together with some typical configurations of thermistor probes. The relatively large resistance change per degree makes the thermistor a useful device in temperature measurement or control devices requiring high accuracy or resolution. Temperature changes as small as 5×10^{-4} °C can be detected. Thermistors are used in the temperature range of −100°C to +300°C. Any current in a resistance does work which appears as heat. Care must be taken that the amount of heat produced by the current in the thermistor or platinum resistance thermometer does not affect the temperature of the measured system.

Light-Dependent Resistors

When light falls on a semiconductor, the photon energy may be sufficient to move valence electrons into the conduction band. The resulting increased density of charge carriers due to the light makes the material more conducting. Thus the resistivity decreases as the light intensity increases. The semiconducting sulfide, selenide, and telluride salts of cadmium are most often used for this purpose. Some practical photoconductive cells and the spectral response curves of such devices are shown in Fig. 2–12. For each material to photoconduct, the incident light must have a short enough wavelength (high enough energy) for the photons to have enough energy to move electrons from the valence to conduction states. As the energy of the light increases beyond the minimum required for promotion of valence electrons to the conduction band, the absorption of the

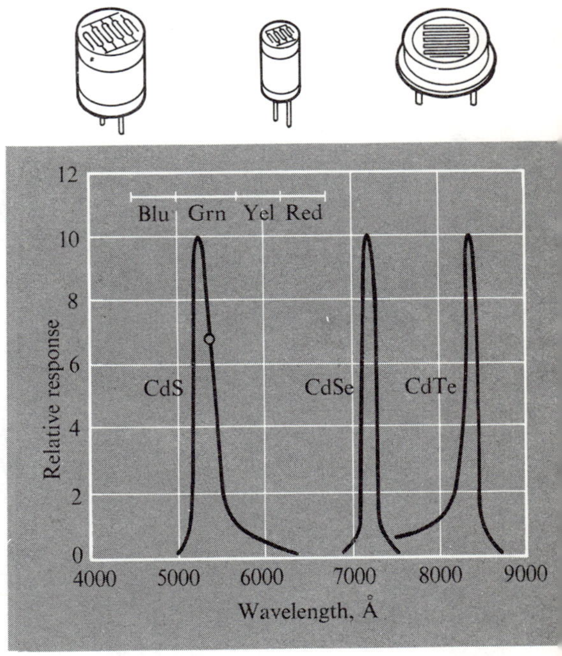

Fig. 2–12 Practical photoconductive cells and the photoconductivity of CdS, CdSe, and CdTe vs. wavelength.

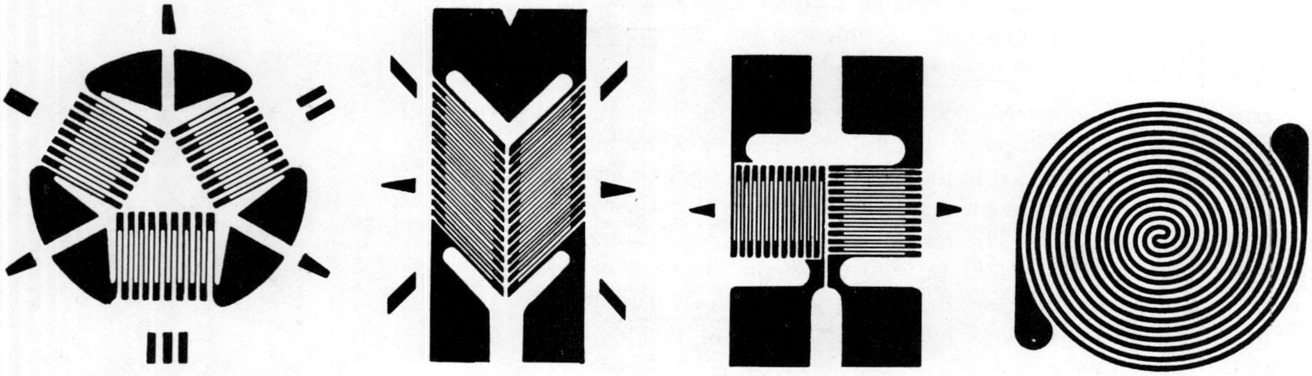

Three-element rosette 60° planar (foil) | Two-element rosette 90° shear planar (foil) | Two-element rosette 90° stacked (foil) | Single-element spiral (foil)

Fig. 2–13 Practical strain gauges.

light by the material increases rapidly. This results in a marked decrease in the fraction of the light which penetrates to the active region of the device.

Strain-Dependent Resistors

A resistor which is made of a fine wire has a resistance $R = \rho l/a$, where ρ is the resistivity of the metal and l and a are the wire length and cross-section respectively. If the wire is distorted, i.e., stretched, l increases and a decreases. Actually, ρ may also change. The result is a resistor for which the resistance is a function of the strain. Such a device is called a **strain gauge.** Strain gauges have extensive applications in mechanical and biological measurements. Often the fine wire is bonded to a flexible insulating substrate so that the wire dimensions change as the substrate is flexed. Some strain gauges are made by depositing a thin metal film pattern on the substrate material. Several practical strain gauges are shown in Fig. 2–13. Since the resistance change with strain is a very small fraction of the total resistance, very sensitive resistance measuring techniques are required. Another difficulty is separating the changes in resistance due to temperature from those due to strain.

2–1.5 RESISTANCE MEASUREMENT

Resistance measurements in the laboratory tend to fall into two categories: testing or troubleshooting, and measurements using resistive transducers. The ohmmeter function provided in the common multimeters and FET voltmeters fulfill the testing requirement well. They are useful for checking the continuity of a circuit, searching for short circuits, testing components for catastrophic breakdown, and rough resistance measurements. The common ohmmeter circuits are described in this section. However, because of their generally poor accuracy and nonlinear scales, few ohmmeters are useful for laboratory measurements of resistive transducers. Improved accuracy can be obtained from the resistance-to-voltage converter circuit based on the current follower, which can provide an output linear in conductance or resistance. The laboratory standard for resistance measurement is the Wheatstone bridge, which is shown to be a true null resistance comparison technique.

Voltage-Current Ratio

By Ohm's Law the resistance of a device can be determined when there is a current through it by measuring the ratio of the *IR* drop (voltage) across it to the current through it. This is often the most convenient way to measure the resistance of a device in an operating circuit and under operating conditions. The resistance of an isolated device is often measured by placing it in a circuit with a battery, current meter, and other resistors. The measured current is related to the device resistance. Such a circuit which has the meter face calibrated in ohms is called an ohmmeter. For a linear conversion of resistance or conductance to voltage, an operational amplifier current follower circuit is generally used.

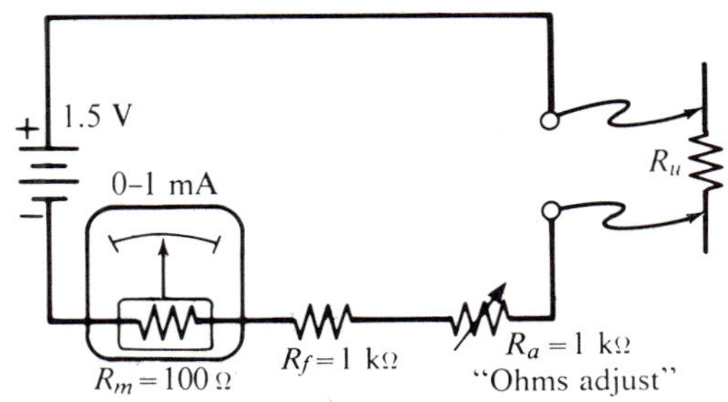

Fig. 2–14 An ohmmeter of simple series type.

The Ohmmeter. A series-type ohmmeter circuit is shown in Fig. 2–14. With the test probes short-circuited ($R_u = 0$), the "ohms adjust" control is turned so that the current I_1 through the total circuit resistance $R = R_m + R_f + R_a$ deflects the meter exactly full scale. Now, by connecting the test probes across the unknown resistance R_u, the current is decreased to a value I_2 which depends on the value of R_u. The 1.5 V battery is across the total resistance in the two cases; so

$$I_1 R = 1.5\ \text{V} \quad \text{and} \quad I_2 R + I_2 R_u = 1.5\ \text{V},$$

from which it follows that

$$R_u = \left(\frac{I_1}{I_2} - 1\right) R \tag{2–12}$$

and

$$I_2 = I_1 \left(\frac{R}{R_u + R}\right). \tag{2–13}$$

Since a 1 mA meter movement is shown in Fig. 2–14, and the battery is 1.5 V, the resistance $R = R_m + R_f + R_a$ will have to be set to 1500 Ω for full-scale deflection; so, from Eq. (2–12),

$$R_u = \left(\frac{I_1}{I_2} - 1\right)1500.$$

With $I_2 = (1/2)I_1$ (midscale deflection), the unknown resistance $R_u =$ 1500 Ω; with $I_2 = (1/3)I_1$, $R_u = 3000$ Ω, etc. It is apparent from Eqs. (2–12) and (2–13) that the transfer function between R_u and I_2 or scale position is not linear, since R_u goes from ∞ to 0 as I_2 goes from 0 to I_1. Values of R_u much higher than 1500 Ω become crowded on the "infinite ohms" end of the scale, and values much lower become indistinguishable from zero. The resistance at midscale could be decreased tenfold to 150 Ω by shunting the meter to make it 10 mA full scale and decreasing the series resistance (with test probes short-circuited) to 150 Ω. The resistance at midscale could be increased to 15 kΩ by increasing the circuit resistance R to 15,000 Ω and then using a 15 V battery or a 100 μA current meter. Clearly the more sensitive the current meter, the higher the value of resistance that can be measured accurately.

A more often used variation on the circuit described above is illustrated in Fig. 2–15. This ohmmeter is referred to as the "voltmeter type" because the voltage is measured across a resistance R_s that is in series with the unknown resistance R_u. The voltmeter branch has a total resistance R_v that is in parallel with R_s, so as to give an effective resistance $R_p = R_sR_v/(R_s + R_v)$, which is in series with R_r and any unknown resistance R_u. The battery voltage V is divided across the voltmeter, R_r, and R_u.

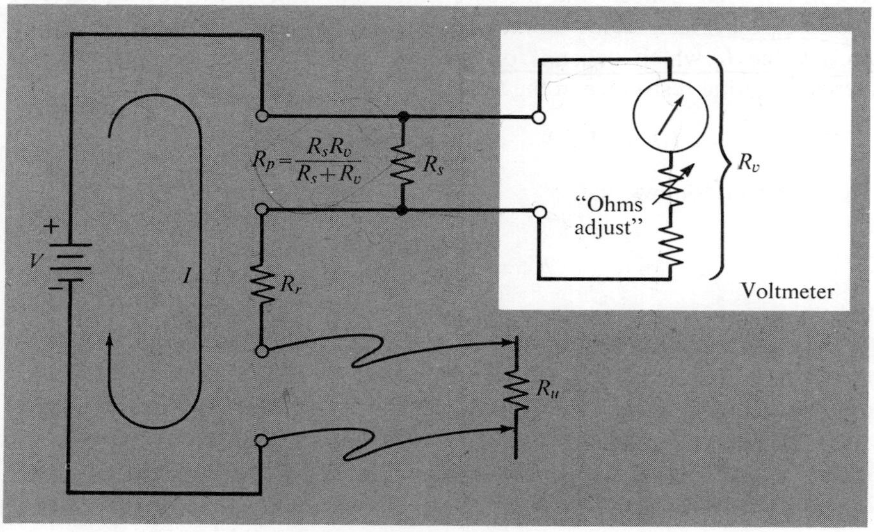

Fig. 2–15 An ohmmeter of voltmeter type.

When the test leads are shorted, the measured voltage is

$$v_1 = V\left(\frac{R_p}{R_p + R_r}\right)$$

and when the test leads are connected to R_u,

$$v_2 = V\left(\frac{R_p}{R_p + R_r + R_u}\right).$$

The relationship among v_1,v_2, and R_u is found by combining the above two equations,

$$v_2 = v_1\left(\frac{R_p + R_r}{R_u + R_p + R_r}\right) \tag{2–14}$$

and

$$R_u = (R_p + R_r)\left(\frac{v_1}{v_2} - 1\right). \tag{2–15}$$

If the "ohms adjust" resistance is varied until the meter reads full scale with the test leads short-circuited, the meter will have a midscale reading ($v_2 = v_1/2$) when $R_u = R_p + R_r$ and the same nonlinear scale as the meter of Fig. 2–14.

If $R_p = 11.5\ \Omega$ and $R_r = 1.0\ \Omega$, then $R_u = 12.5\ \Omega$ for a midscale meter reading. The meter could be calibrated to read ohms directly. If $R_p + R_r$ is increased to 1250 Ω, the resistance value will equal the meter reading × 100. In this case $R_u = 1250\ \Omega$ for a midscale reading if the "ohms adjust" was previously varied so that the meter current was full scale with the test leads short-circuited.

As the circuit resistance, $R_p + R_r$, is increased for measuring larger values of R_u, the circuit current decreases. Consequently the fraction of circuit current sent through the meter branch must be increased in order for the meter to read full scale. The range switch, therefore, not only switches the absolute value of $R_p + R_r$ to the desired resistance but also changes relative values of R_s and R_v and, for a very high resistance range, it may switch in a higher battery voltage, in order to get sufficient current through the meter.

When the ohmmeter function is part of a very high input resistance voltmeter such as a vacuum tube or field-effect transistor voltmeter, the simpler circuit of Fig. 2–16 is often used. The voltmeter is used to measure the fraction of V which appears across the unknown resistor R_u. When the test probes are shorted, the measured voltage is zero. The ohms adjust is a gain control on the amplifier which is adjusted to give full-scale de-

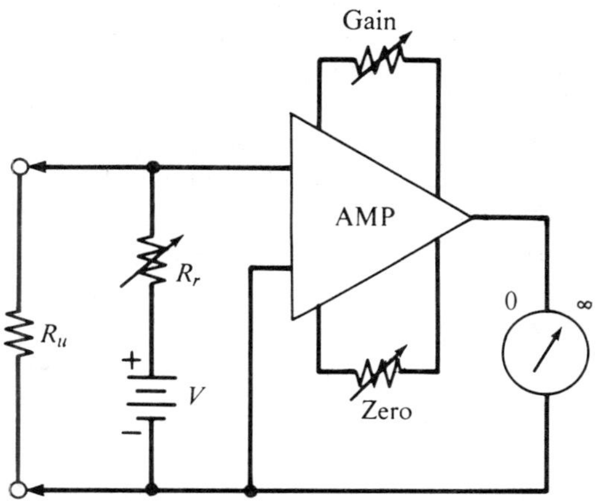

Fig. 2–16 VTVM or FET VM type of ohmmeter.

flection *when the test probes are open* ($R_u = \infty\ \Omega$). Note that the scale on this type of meter is reversed from the scales of Figs. 2–14 and 2–15. When an unknown resistance R_u is connected between the test probes, the relation between the measured voltage v_2 and R_u is

$$v_2 = V\left(\frac{R_u}{R_r + R_u}\right) \tag{2–16}$$

and

$$R_u = R_r\left(\frac{v_2/V}{1 - v_2/V}\right). \tag{2–17}$$

As Eqs. (2–16) and (2–17) show, the scale is still nonlinear with a range from 0 to $\infty\ \Omega$ and a half-scale value of R_r. The ohmmeter range is changed simply by switching in various values for R_r up to a maximum value of 0.1% or 1% of the amplifier input resistance.

An ohmmeter is never used while the circuit is in operation,[2] and thus there is no circuit distortion introduced by the measurement. For resistances that depend on circuit conditions, the only solution is to establish normal operating conditions, measure the voltage across the resistance, measure the current through the resistance, and calculate the resistance value from Ohm's Law.

The resistance of devices that might be damaged by moderate currents cannot be measured with an ordinary ohmmeter. Such devices include meter movements and some fuses, lights, relays, tube filaments, diodes, etc. When the danger of damage exists, some other means must be devised to make the measurement.

Note 2. In-Circuit Resistance Measurements.

Even when the power supply for an electronic circuit is off, the circuit and power supply capacitors might remain charged and discharge only slowly as the resistors in the circuit are checked with an ohmmeter. These capacitor voltage sources can cause erroneous resistance measurements or damage to the ohmmeter. Therefore, after the instrument is turned off, all capacitors should be discharged before ohmmeter measurements are attempted. Also, discharged capacitors in a circuit might become charged when the ohmmeter probes are connected across a resistor. This can cause the measured value to drift until the voltages in the circuit reach a steady state.

Linear Resistance-to-Voltage Converters. The nonlinear relationship between the unknown resistance and the current or voltage measured in the above ohmmeter circuit is inconvenient for many types of resistive transducer measurements. From Ohm's Law, $R = V/I$, the voltage across a resistor is proportional to the resistance for a constant value of current. Similarly, the conductance, $G = I/V$, is proportional to the current through a resistor for a constant applied voltage. To take advantage of these linear relationships, a constant current source is required in the first case, and a current measurement with negligible input voltage in the second case. The current follower circuit introduced in Module 1 provides a convenient and accurate means to achieve either of these requirements. In Fig. 2–17a R_u is connected to the voltage source V, and the result is a current $i = V/R_u$ which is proportional to the conductance of R_u. The operational amplifier maintains an output voltage such that point S is at the common potential.

Therefore, the input current, i, produces a voltage $v_o = -iR_r$ at the current follower output. Also, the connection of the G-to-I converter circuit to point S is equivalent to a connection to common; i.e., essentially no additional voltage or current is added to the input circuit. Since $i = V/R_u$ and $v_o = -iR_r$,

$$v_o = -VR_r/R_u = -VR_rG_u, \tag{2-18}$$

resulting in an output voltage directly proportional to the conductance of R_u within the output voltage and current capabilities of the operational amplifier.

In the circuit of Fig. 2–17b, a fixed resistance, R_r, is used in the input current generating circuit. Since this same current passes through R_u and the operational amplifier (OA) maintains point S at the common potential, the output voltage, $v_o = -iR_u$, is proportional to R_u. From $i = V/R_r$ and $v_o = -iR_u$,

$$v_o = -\frac{V}{R_r}R_u. \tag{2-19}$$

Of course, both circuits are simply the OA inverting amplifier introduced in Section 1–4.2, for which $v_o = -v_{\text{in}}R_f/R_{\text{in}}$ with $v_{\text{in}} = V$ and R_u taking the place of R_{in} in the first case and R_f in the second.

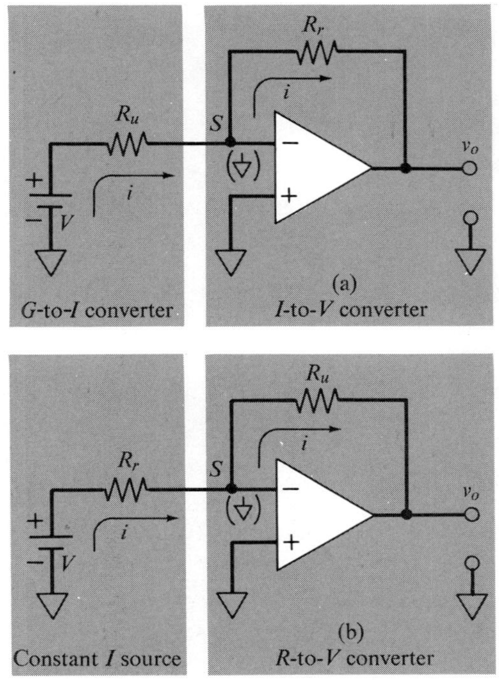

Fig. 2–17 Current follower circuit for resistance measurement: (a) $v_o = -VR_r/R_u$; (b) $v_o = -VR_u/R_r$.

Experiment 2–1 *Light-Sensitive and Temperature-Sensitive Resistive Devices*

A photoconductive cell is used to illustrate how the resistance of a device can change as a function of the illumination, and how the measurement of current through the resistive cell can be related to the illuminance. A thermistor is then used to show the relationship between resistance and temperature for a very useful temperature transducer, again using a current meter to obtain the resistance information.

Experiment 2–2 *Measurement of Thermistor Resistance with a Voltmeter*

In this experiment a voltmeter is used to measure the voltage drop across a thermistor through which a constant current is applied. The temperature is varied and the voltmeter readings are related to resistance values, from which temperature values are calculated.

Experiment 2–3 *Measurement of Thermistor Resistance with an Ohmmeter*

Temperature information was obtained in Experiments 2–1 and 2–2 by measuring current and voltage, respectively, which were then related to resistance values, from which temperature was calculated. It is shown in this experiment that an ohmmeter can be used directly to provide the resistance measurements that are related to temperature. Thus the first three experiments demonstrate the use of three types of conventional meters (current meters, voltmeters, and ohmmeters) for obtaining similar information from a resistive transducer.

Experiment 2–4 *Operational Amplifier Measurement of Thermistor Resistance*

Temperature information is obtained in this experiment from a thermistor whose resistance is calculated from measurements of the output voltage of an operational amplifier circuit containing the thermistor. A constant current is applied to the thermistor by placing it in the feedback loop of an OA inverting amplifier to which a known input current is applied. Output voltage measurements are made and related to the thermistor resistance and thus to temperature. Alternatively the thermistor is used as the input resistor for the OA and a constant input voltage is applied. The thermistor resistance determines the input current, which is then measured by a current comparison method and related back to the temperature of the thermistor.

Null Comparison Measurement

In a null comparison measurement of resistance, the effect of an unknown resistance must be compared with the effect of a variable standard resistance under as near identical conditions as possible. Since a resistance affects the electrical quantities in a circuit, the unknown and standard resistances are placed in identical circuits in such a way that the affected electrical quantity (generally voltage or current) in each circuit can be compared. Several methods for performing this comparison have been devised, of which the Wheatstone bridge is by far the most common. Comparison methods for resistance measurement offer great accuracy, resolution, and relative independence of the electronic sources and detectors used.

The Wheatstone Bridge. The Wheatstone bridge shown in Fig. 2–18 provides the most direct and best-known circuit for comparison of un-

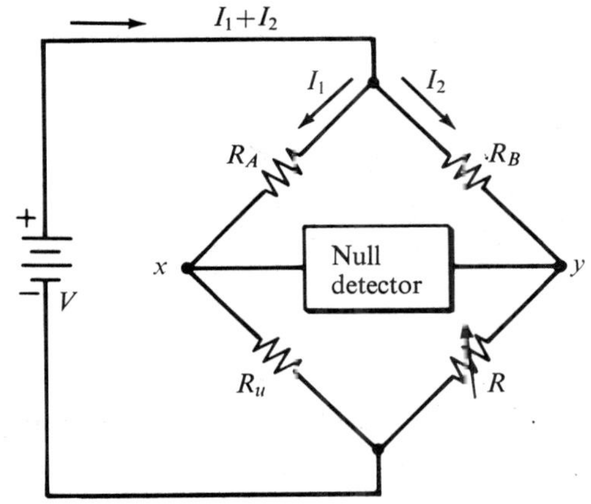

Fig. 2–18 Wheatstone bridge.

known resistances against standard resistances. Resistances R_A, R_B, and R are standard resistance values which are used in the measurement of the unknown resistance R_u. Resistance R is made variable and is adjusted until the null detector indicates that the bridge is balanced.

When the circuit is at balance, there is no current through the null detector and no potential difference between terminals x and y. At balance, four significant conditions exist:

1. The current through R_A and R_u is I_1.
2. The current through R_B and R is I_2.
3. $I_1R_u = I_2R$.
4. $I_1R_A = I_2R_B$.

Therefore, $R_u/R = R_A/R_B$, and

$$R_u = R\frac{R_A}{R_B}. \tag{2–20}$$

It can be seen from Eq. (2–20) that the unknown resistance R_u is determined from the values of the three standard resistances: R, R_A, and R_B. It is common practice to make the ratio R_A/R_B some exact fraction or multiple such as 0.01, 0.1, 1, 10, 100, etc., and to refer to it as the "multiplier." Resistance R is made variable in small increments or continuously, so that the dial reading of R times the multiplier setting equals the unknown resistance R_u.

The main sources of error in a Wheatstone bridge are the inaccuracies of the three standard resistances R, R_A, and R_B. However, these can be made with errors of only about 0.001%. Other factors limiting the accuracy are errors in establishing the null point, thermal emf's, and changes in resistance values due to heating (too high currents). It is important to

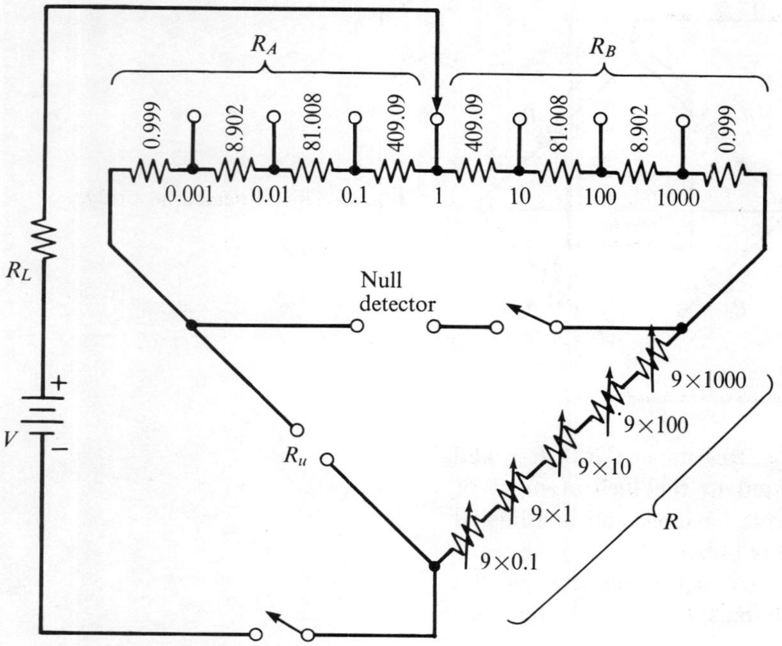

Fig. 2–19 General-purpose Wheatstone bridge.

keep the I^2R loss in the resistors at a safe value, because excessive heating may cause a permanent shift in resistance values. The maximum values specified by the manufacturer should never be exceeded in a bridge measurement.

A schematic of a typical general-purpose Wheatstone bridge is shown in Fig. 2–19. The multiplier R_A/R_B varies from 0.001 to 1000 in seven ranges. The total resistance $R_A + R_B = 1000\ \Omega$. The resistance R is a decade box varying in steps of 0.1 Ω up to a total resistance of 9999.9 Ω.

The magnitude of the off-balance indication of the null detector of a Wheatstone bridge is sometimes used as a measure of the change in resistance of the unknown resistor R_u. For a derivation of the sensitivity and linearity of the bridge as a resistance-to-voltage converter, the reader is referred to problem 16 at the end of this section. In this application, R_u is frequently a light-, temperature-, or strain-dependent resistor. When matched devices are used for R and R_u, or R_A and R_u, an off-balance output is obtained which is related to the *difference* in the resistance of the two devices and thus the difference in the measured quantity at the two devices. This "differential" resistance measurement with matched devices can tend to cancel the effects of resistance changes due to environmental factors other than the measured quantity.

Experiment 2–5 *Accurate Measurement of Thermistor Resistance with a Wheatstone Bridge*

Unknown resistances are most accurately measured by a comparison measurement, which is conveniently accomplished with the Wheatstone bridge. The resistance bridge is constructed and then used to measure the resistance of a thermistor. Precision resistors and a decade resistance box are used in making the bridge, and a VOM is used on its most sensitive current scale as the null detector. The sensitivity, precision, and accuracy of the resistance (and temperature) measurements with the Wheatstone bridge are compared with those of the methods used in the first three experiments. Some "feel" for the applicability and limitations of several techniques for obtaining resistance information should have been obtained in the first four experiments. It is important to note that the conversion of the resistance information into related electrical signals is necessary so that these signals can be manipulated in the electrical domain so as to provide direct display of temperature, light, or other physical information, or for feedback control information that can be used to hold these parameters constant.

2–1.6 RESISTANCE NETWORK ANALYSIS

Practical circuits are often complex combinations or **networks** of components connected in series and parallel. The methods of analyzing resistance networks to determine the total network resistance or the currents and voltages in various branches of the network will be introduced in this section. The principles involved in these methods are very useful tools, not only for complete network analysis, but also for greatly simplifying the problem of the action of a circuit on a single component or the effect of varying that component on the circuit. This last problem is a matter of concern whenever two devices, circuits, or instruments are connected together.

Successive Reduction

To analyze a complex network of resistances, simple series and parallel branches of the network can often be reduced to single equivalent resistances until only one resistance remains; this single resistance is equivalent to the entire network. If current and voltage values are desired for various network components, they can be calculated as the real network resistances are reconstructed from their equivalent values.

The process of successive reductions is illustrated in Fig. 2–20. The first step is to reduce the simple series and parallel combinations to equivalent values. Thus the parallel combination of R_1 and R_2 is equivalent to a

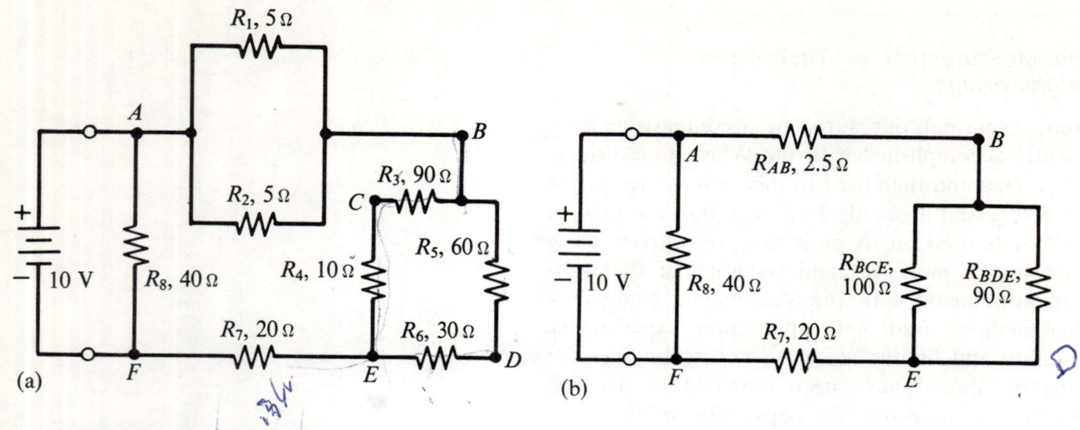

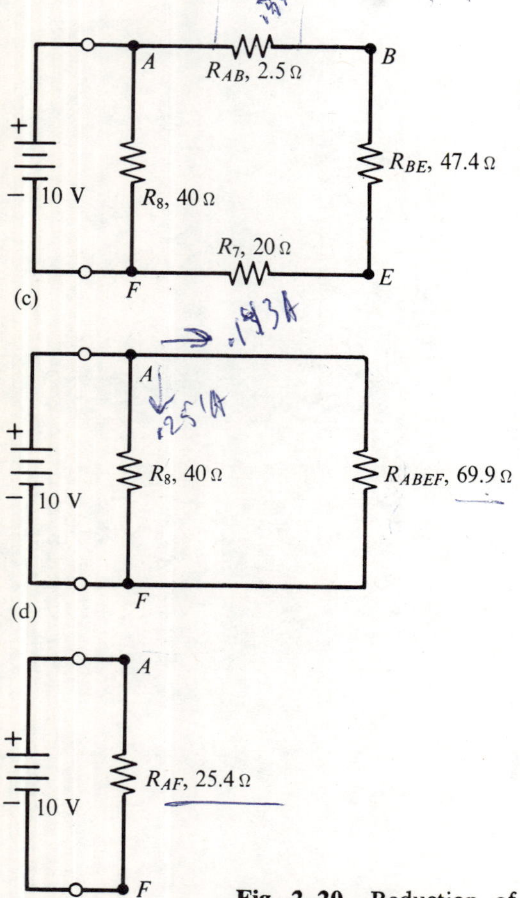

Fig. 2–20 Reduction of series-parallel network.

single 2.5 Ω resistor which we shall call R_{AB}. Similarly, the series combination of R_3 and R_4 is equivalent to a single 100 Ω resistor R_{BCE}, and the combination of R_5 and R_6 is equivalent to a single 90 Ω resistor R_{BDE} as shown in Fig. 2–20b. These steps and subsequent similar steps in the complete reduction are as follows:

From Fig. 2–20a to Fig. 2–20b,

$$R_{AB} = R_1R_2/(R_1 + R_2) = 2.5\ \Omega,$$

$$R_{BCE} = R_3 + R_4 = 100\ \Omega,$$

$$R_{BDE} = R_5 + R_6 = 90\ \Omega.$$

From Fig. 2–20b to Fig. 2–20c,

$$R_{BE} = R_{BCE}R_{BDE}/(R_{BCE} + R_{BDE}) = 47.4\ \Omega.$$

From Fig. 2–20c to Fig. 2–20d,

$$R_{ABEF} = R_{AB} + R_{BE} + R_7 = 69.9\ \Omega.$$

From Fig. 2–20d to Fig. 2–20e,

$$R_{AF} = R_8R_{ABEF}/(R_8 + R_{ABEF}) = 25.4\ \Omega.$$

If the current and voltage values for each resistor are desired, they can now be calculated by "reconstructing" the network. From Fig. 2–20e, for R_{AF}, $V = 10$ V and $I = 0.39$ A. Now from Fig. 2–20d, the current-splitting equation can be used to calculate the current through R_8 and R_{ABEF}.

These steps and the remaining steps in the complete analysis are tabulated below. Solutions for actual components are indicated by an asterisk (*).

From Fig. 2–20e:

For R_{AF}, $V = 10$ V, $I = 10$ V/25.4 Ω $= 0.394$ A.

From Fig. 2–20d:

*For R_8, $V = 10$ V, $I = (0.394\text{ A})(R_{ABEF})/(R_8 + R_{ABEF}) = 0.251$ A.

For R_{ABEF}, $V = 10$ V, $I = 0.394\text{ A} - 0.252\text{ A} = 0.143$ A.

From Fig. 2–20c:

For R_{AB}, $I = 0.143$ A, $V = (0.143\text{ A})R_{AB} = 0.36$ V.

For R_{BE}, $I = 0.143$ A, $V = (0.143\text{ A})R_{BE} = 6.77$ V.

*For R_7, $I = 0.143$ A, $V = (0.143\text{ A})R_7 = 2.86$ V.

From Fig. 2–20b:

For R_{BDE}, $V = 6.77$ V, $I = (0.143\text{ A})R_{BCE}/(R_{BDE} + R_{BCE}) = 0.075$ A.

For R_{BCE}, $V = 6.77$ V, $I = 0.143\text{ A} - 0.075\text{ A} = 0.068$ A.

From Fig. 2–20a:

*For R_1 and R_2, $V = 0.36$ V, $I = 0.143/2 = 0.072$ A.

*For R_3, $I = 0.068$ A, $V = (0.068\text{ A})R_3 = 6.10$ V.

*For R_4, $I = 0.068$ A, $V = (0.068\text{ A})R_4 = 0.68$ V.

*For R_5, $I = 0.075$ A, $V = (0.075\text{ A})R_5 = 4.50$ V.

*For R_6, $I = 0.075$ A, $V = (0.075\text{ A})R_6 = 2.25$ V.

Kirchhoff's Laws

Occasionally circuits are encountered where it is not possible to reduce the resistances to simpler circuits by the method of successive reductions illustrated above. To analyze such networks, two simple but very important laws are used. They are called Kirchhoff's Laws, and are often stated as follows:

1) The algebraic sum of all the currents flowing toward a junction is zero. (A junction is a common connection to three or more components.)

2) The algebraic sum of all the source voltages is equal to the algebraic sum of all the IR drops in any loop. (A loop is any closed conducting path in the network.)

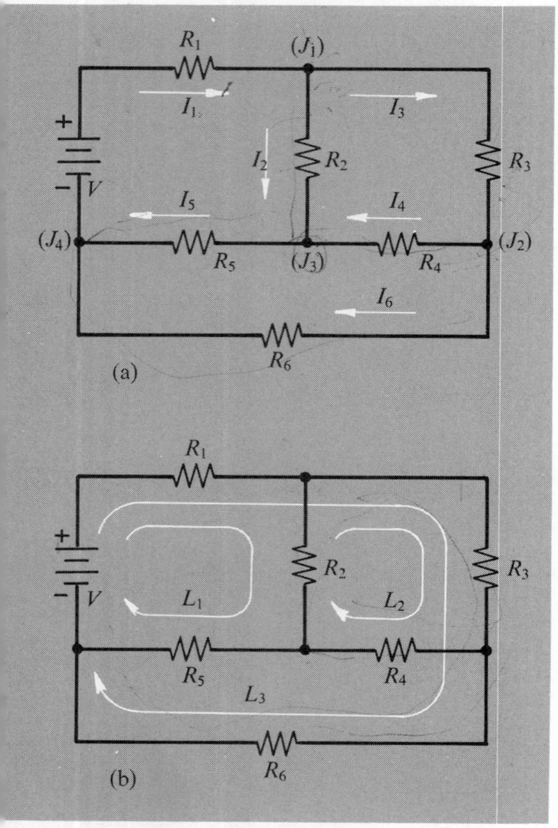

Fig. 2–21 Illustration of Kirchhoff's laws: (a) junction law; (b) loop law.

Kirchhoff's junction law is illustrated in Fig. 2–21a. This network cannot be solved by successive reductions. There are four junctions identified by J's in the figure. A separate equation can be written for each junction as follows:

$$(J_1) \quad I_1 = I_2 + I_3$$

$$(J_2) \quad I_3 = I_4 + I_6$$

$$(J_3) \quad I_4 + I_2 = I_5$$

$$(J_4) \quad I_6 + I_5 = I_1$$

For j junctions there are $j - 1$ independent equations; hence one of the above equations is redundant. Since there are six unknown currents, more equations are needed. These are obtained by applying Kirchhoff's loop law.

The application of Kirchhoff's loop law is illustrated in Fig. 2–21b. Loops are drawn to include every component as shown. Many other combinations could be drawn, but it is seen that the minimum number is three in this case. Summing the voltages around each loop (using the currents and directions from Fig. 2–21a) yields three equations as follows:

$$(L_1) \quad V = I_1R_1 + I_2R_2 + I_5R_5$$

$$(L_2) \quad 0 = I_3R_3 + I_4R_4 - I_2R_2$$

$$(L_3) \quad V = I_1R_1 + I_3R_3 + I_6R_6$$

Thus from Kirchhoff's Laws, six equations have been obtained for this circuit which can be used to solve for six unknown quantities. In general, there will always be enough Kirchhoff's Law equations to solve for all the currents when the voltages and resistances are known.

The simultaneous equations are then solved by successive substitution or by determinants. The Kirchhoff Law method of analysis is thorough, but for complex networks the tedious and mechanical nature of the method invites errors. For this reason, other analysis techniques which are simpler for partial solutions are often used instead. Kirchhoff's Laws, however, are very useful in quickly determining the relationship among currents at a single junction or potentials in a single loop.

Wye-Delta Transformation

A "Y" (wye) connection is the common point of three components such as shown in Fig. 2–22a. The other terminal of each component connected

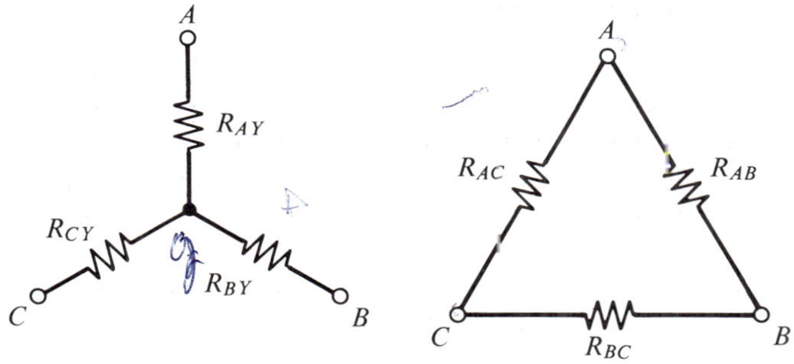

$$R_{AY}=\frac{R_{AB}R_{AC}}{R_{AB}+R_{AC}+R_{BC}}$$

$$R_{BY}=\frac{R_{AB}R_{BC}}{R_{AB}+R_{AC}+R_{BC}}$$

$$R_{CY}=\frac{R_{AC}R_{BC}}{R_{AB}+R_{AC}+R_{BC}}$$

(a)

$$R_{AB}=R_{AY}+R_{BY}+\frac{R_{AY}R_{BY}}{R_{CY}}$$

$$R_{AC}=R_{AY}+R_{CY}+\frac{R_{AY}R_{CY}}{R_{BY}}$$

$$R_{BC}=R_{BY}+R_{CY}+\frac{R_{BY}R_{CY}}{R_{AY}}$$

(b)

Fig. 2–22 Wye-delta transformation: (a) delta to Y; (b) Y to delta.

to the Y is labeled *A*, *B*, and *C* respectively. The Y-connected resistors can be shown to be equivalent to a circuit with a resistor between each of the *A*, *B*, and *C* terminals as shown in Fig. 2–22b. This is called the "Δ" (delta) form. Since equivalence relationships exist, a portion of a circuit with Δ-connected components can be transformed into an equivalent Y arrangement of components and vice-versa. The equivalence relationships are given in Fig. 2–22 and are derived in Appendix A.

The complex circuit of Fig. 2–23 can be solved by the use of these transformations and relationships. In the equivalence shown Δ*ABC* and Δ*CDE* have been transformed into Y's. The circuit can now be analyzed by successive reduction and the voltages at *A*, *B*, *C*, *D*, and *E* calculated. Since these voltages must also hold for the equivalent points in the original circuit, the voltage across and the current through any component can be easily calculated. The Y-Δ transformation could also have been used to solve the circuit of Fig. 2–23.

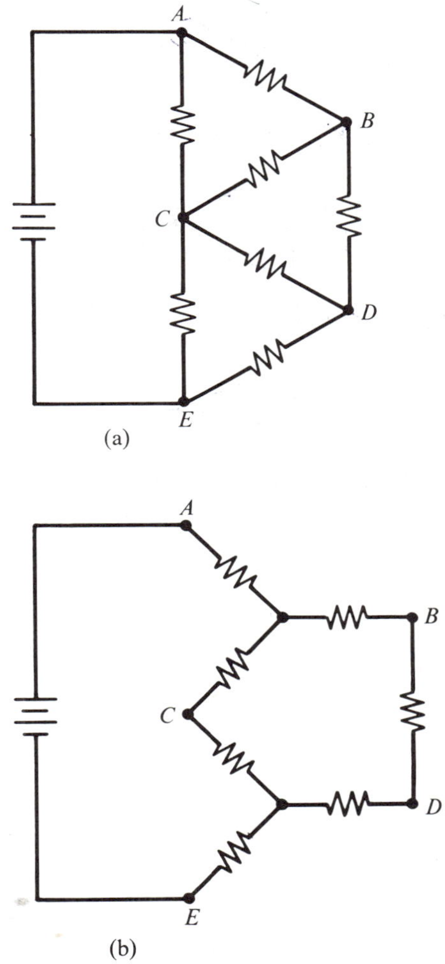

Fig. 2–23 Application of the Δ-Y transformation.

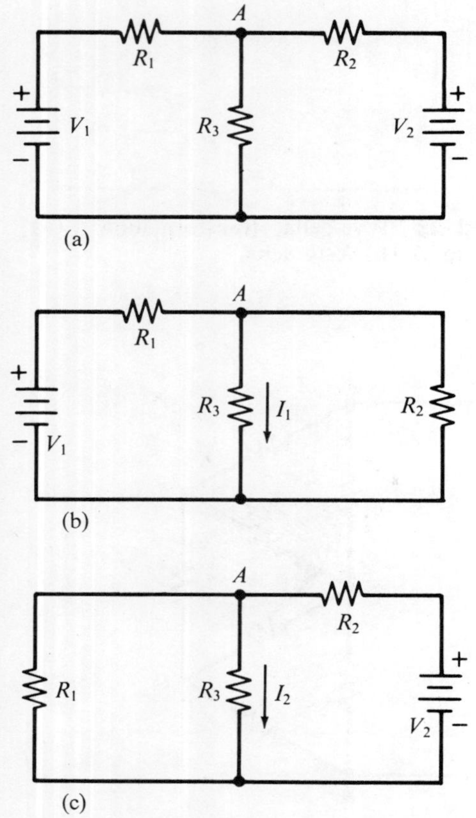

Fig. 2–24 Superposition theorem application.

Superposition Theorem

The superposition theorem states that each voltage source in a network produces current in any given branch independently of any other voltage sources in the network. Thus the current in any branch will be the sum of the currents from all of the voltage sources taken one at a time. This theorem is especially useful in analyzing circuits with several voltage sources. A common example is the half-bridge shown in Fig. 2–24a. The desired quantity is the potential difference across R_3. When the currents from source V_1 are being considered, V_2 must be replaced by a short circuit as in Fig. 2–24b. The current through R_3 due to V_1 can be obtained by multiplying the total current from V_1 by the current splitting fraction to obtain

$$I_1 = \frac{V_1}{R_1 + \dfrac{R_2 R_3}{R_2 + R_3}} \left(\frac{R_2}{R_2 + R_3} \right) = \frac{V_1 R_2}{R_1 R_2 + R_1 R_3 + R_2 R_3}.$$

From Fig. 2–24c,

$$I_2 = \frac{V_2 R_1}{R_1 R_2 + R_1 R_3 + R_2 R_3},$$

so that

$$I = I_1 + I_2 = \frac{V_1 R_2 + V_2 R_1}{R_1 R_2 + R_1 R_3 + R_2 R_3}$$

and the voltage across R_3 is

$$IR_3 = \frac{R_3 (V_1 R_2 + V_2 R_1)}{R_1 R_2 + R_1 R_3 + R_2 R_3}.$$

Thévenin's Theorem

One of the most useful theorems in circuit analysis is Thévenin's theorem. It states that *any voltage source which has only two output terminals and is composed of resistors and batteries can be represented by a series combination of a resistor R_{Th} and a battery V_{Th}* as in Fig. 2–25. From the Thévenin equivalent circuit it can be seen that V_{Th} is the open-circuit potential difference at the two terminals of the network and R is the resistance between the output terminals when the battery V_{Th} is short-circuited. Thus R_{Th} is the effective source resistance of the network acting as a voltage source of V_{Th} volts.

The network given as an example in Fig. 2–25a is a voltage divider.

The open-circuit voltage V_{Th} is

$$V_{Th} = V\frac{R_2}{R_1 + R_2} = bV, \tag{2–21}$$

where b is the fraction of the total divider resistance across which V_o is taken; this equation shows the proportional relationship between the divider output voltage and the divider fraction. When the battery is replaced by a short circuit, the resistance between the output terminals is the parallel combination of R_1 and R_2:

$$R_{Th} = \frac{R_1 R_2}{R_1 + R_2}. \tag{2–22}$$

If the total divider resistance $(R_1 + R_2)$ is R, $R_2 = bR$ and $R_1 = (1 - b)R$. Substituting these relations in Eq. (2–22), one obtains

$$R_{Th} = \frac{bR(1 - b)R}{bR + (1 - b)R} = b(1 - b)R. \tag{2–23}$$

This equation shows that the output resistance of the divider is a maximum when $b = \frac{1}{2}$, and zero when $b = 1$ or 0.

If $V = 10$ V, $R_1 = 120\ \Omega$, and $R_2 = 80\ \Omega$, $V_{Th} = 4.0$ V and $R_{Th} = 48\ \Omega$. This particular voltage divider is thus electrically equivalent to a battery of 4.0 V in series with a resistor of 48 Ω. The actual output of the divider for any load resistance connected to the output terminals can be readily calculated. In fact, through the use of this theorem, the calculation of the output voltage, under load, of *any* voltage source is equally simple. Thévenin's theorem is frequently used to determine the effect of interconnecting two circuits or devices.

Thévenin's theorem is particularly useful when one wishes to calculate the effects of changing or adding one branch in a network. Consider the half-bridge of Fig. 2–26a, which was previously analyzed by superposition. The component for which the voltage and current were sought (R_3) is temporarily removed, and the Thévenin equivalent values are calculated from the remaining network. These calculations are now very simple:

$$V_{Th} = V_1 + R_1(V_2 - V_1)/(R_1 + R_2) \quad \text{and} \quad R_{Th} = R_1 R_2/(R_1 + R_2).$$

When R_3 is connected to the equivalent terminals A and B in Fig. 2–26b, the current through R_3 is simply $V_{Th}/(R_{Th} + R_3)$.

Thévenin's theorem can also be used to remove a component that makes simple reduction of a network impossible and then add it back to the network after the equivalent circuit has been developed. This technique is illustrated for the Wheatstone bridge of Fig. 2–27a. The 10 Ω resistor

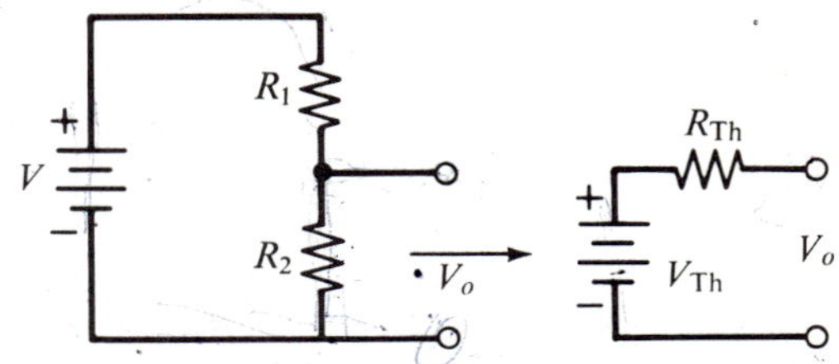

Fig. 2–25 Thévenin's theorem applied to the voltage divider.

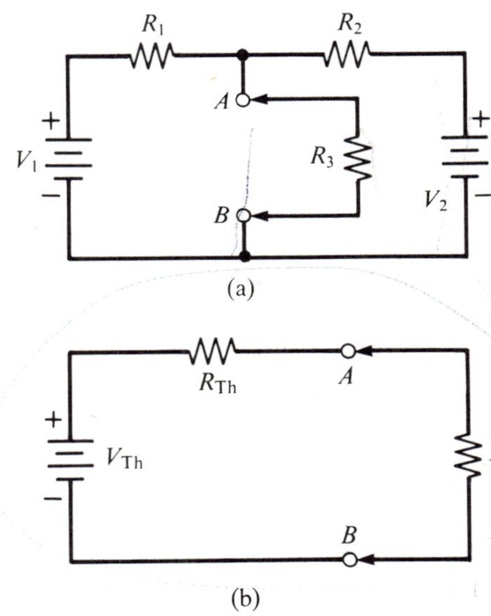

Fig. 2–26 Example of Thévenin's theorem: (a) half-bridge; (b) Thévenin equivalent circuit.

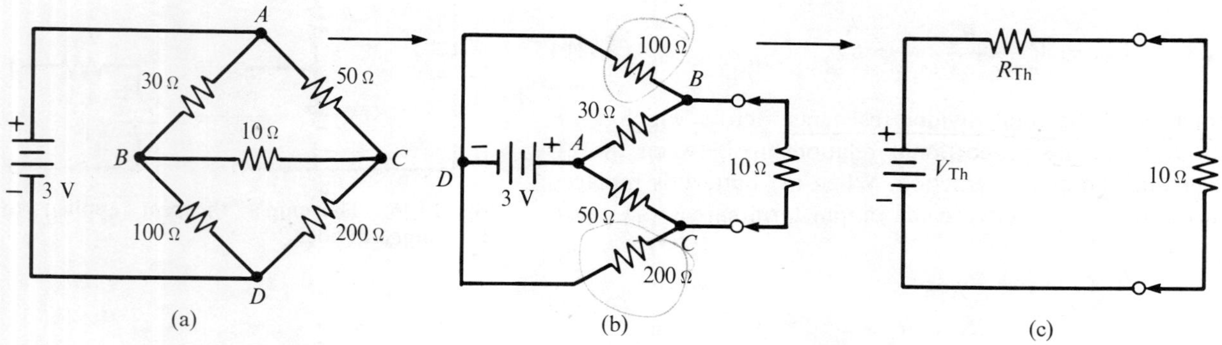

Fig. 2–27 Thévenin's theorem applied to the Wheatstone bridge: (a) original bridge; (b) bridge redrawn; (c) Thévenin equivalent circuit.

from B to C prevents simple analysis of the circuit. (Also, it is the current through this element that is sought in bridge analysis.) The 10 Ω resistor is removed temporarily and the bridge circuit redrawn for clarity in Fig. 2–27b. The Thévenin equivalent voltage is the difference between the voltage at B from the divider ABD and the voltage at C from the divider ACD. Thus

$$V_{\mathrm{Th}} = 3\ \mathrm{V}\,\frac{30}{100+30} - 3\ \mathrm{V}\,\frac{50}{200+50} = 0.093\ \mathrm{V}.$$

The Thévenin equivalent resistance, R_{Th} (taken with V shorted), is the parallel combination of the 30 and 100 Ω resistors in series with the parallel combination of the 50 and 200 Ω resistors:

$$R_{\mathrm{Th}} = \frac{50 \times 200}{200+50} + \frac{30 \times 100}{100+30} = 63.1\ \Omega.$$

When reconnected, the current through the 10 Ω resistor is thus 0.093 V/(63.1 + 10) = 1.26 mA. The Δ-Y transformation would also have been useful in analyzing the Wheatstone bridge.

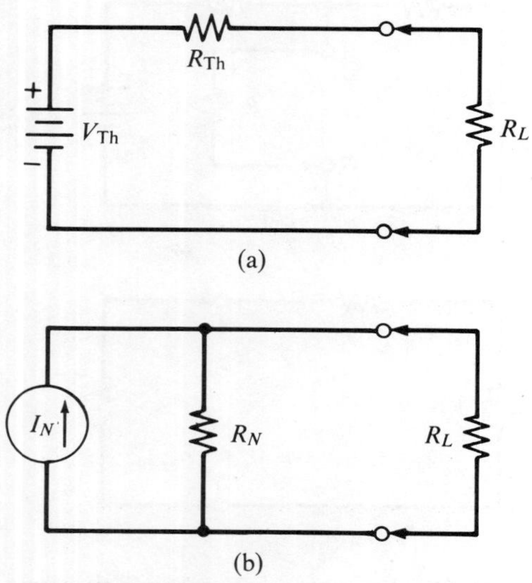

Fig. 2–28 (a) Thévenin and (b) Norton equivalent circuits.

Norton's Theorem

A corollary of Thévenin's voltage source theorem is Norton's current source theorem. Norton's theorem states that any two-terminal network of resistors and voltage or current sources is equivalent to a current source with a parallel resistance. The Thévenin and Norton equivalent circuits are shown in Fig. 2–28. In the Norton equivalent circuit, I_{N} is the current through the terminals with the output shorted. With the half-bridge example of Fig. 2–26a, $I_{\mathrm{N}} = V_1/R_1 + V_2/R_2$. Since an ideal current source has infinite resistance, the equivalent source resistances (R_{N} and R_{Th}) are the same in both cases. Thus $R_{\mathrm{N}} = R_1R_2/(R_1 + R_2)$ for the half-bridge.

When the load resistor (R_3 in this case) is connected, the current I_N is split between R_N and the load. Thus for the half-bridge,

$$I = I_N R_N / (R_3 + R_N).$$

All three solutions are thus seen to give identical results.

Since Norton and Thévenin circuits are both equivalent to any two-terminal network, they must be equivalent to each other. The relationships between the equivalent circuit values for the two cases are easily seen from Fig. 2–28. The open-circuit voltage of Thévenin's circuit is V_{Th} and that of Norton's circuit is $I_N R_N$. Therefore,

$$V_{Th} = I_N R_N. \tag{2–24}$$

The short-circuit current of Thévenin's circuit is V_{Th}/R_{Th} and of Norton's circuit is I_N. Now,

$$I_N = V_{Th}/R_{Th}. \tag{2–25}$$

From Eqs. (2–24) and (2–25) it follows that

$$R_N = R_{Th}. \tag{2–26}$$

The choice of circuit analysis technique is largely a matter of personal preference and convenience. However, from the above discussions it is clear that the theorems underlying each method provide important insights into component and circuit interactions. These circuit analysis techniques will be used throughout this and other modules to aid in understanding how circuits perform.

Experiment 2–6 *Linearization of a Transducer Transfer Function*

The relationship between certain desired information in the physical/chemical domain and the output from a specific transducer is often a nonlinear function. In order to make it convenient to display this information on linear current or voltage measurement devices (DVM's, strip chart recorders), it is important to introduce linearization circuits. In this experiment a network is constructed that provides a linear voltage output directly related to temperature even though the resistance values obtained from the thermistor transducer are a nonlinear function of temperature.

PROBLEMS

1. The resistance of a thermistor is to be measured by applying 5 V to a circuit consisting of a thermistor and an ammeter in series. A current of 0.99 mA is measured. (a) If the ammeter resistance is neglected, what is the thermistor resistance? (b) If it is known that the meter has a resistance of 5 kΩ on the 50 μA scale, what resistance does it have on the 1 mA scale used to measure the current through the thermistor? (c) What is the true resistance of the thermistor and what is the relative perturbation error (RPE) caused by inserting the 1 mA full scale meter into the circuit?

2. It is desired to measure the resistance of a thermistor at a temperature near 25°C where the thermistor has a resistance of 10 kΩ. Measurement of the current through the thermistor and of the voltage drop across it are both considered. Only a 1 V source is available. Two meters are available: a current meter with a 50 μA basic sensitivity 5 kΩ resistance and a 100 μA scale; and a voltmeter whose input resistance is 1 MΩ on the 1 V scale. Which of the two meters will cause less loading error (perturbation error)?

3. An ohmmeter of the voltmeter type, similar to that of Fig. 2–15, is used to measure resistance. The voltage source V is a 1.5 V battery, $R_p = 12.5\ \Omega$, and $R_r = 1.5\ \Omega$. If full scale is set to 0 Ω with the "ohms adjust" and a reading of two-thirds of full scale is obtained with the unknown resistance R_u, what is the value of R_u?

4. The Wheatstone bridge of Fig. 2–18 is balanced with $R_A = 909.09\ \Omega$ and $R_B = 90.909\ \Omega$, and the variable standard resistance $R = 628\ \Omega$. What is the unknown resistance R_u?

5. The circuit shown in Fig. P-5 is constructed. What is the current I_7 through resistor R_7?

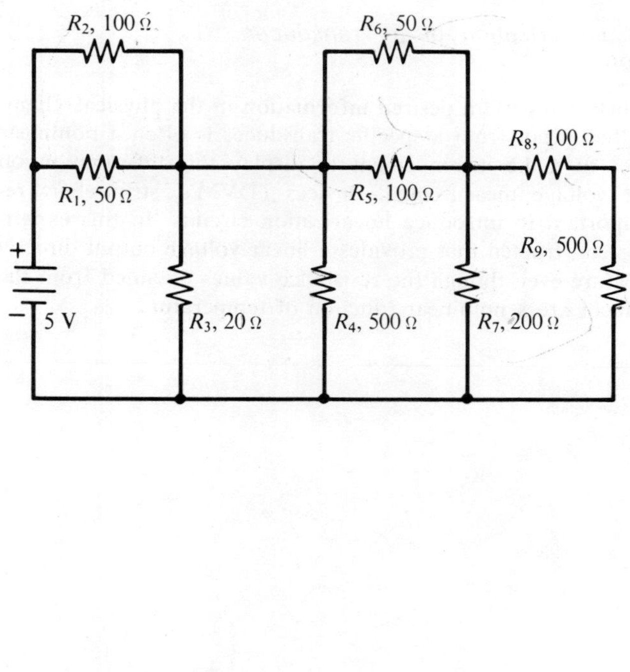

Fig. P–5

6. In the circuit of Fig. P–5 what value of source voltage V is needed so that the current through R_3 is 1 mA?

7. The circuit of Fig. 2–21 is constructed in the laboratory; $V = 10$ V, $R_1 = 1$ kΩ, $R_2 = 500$ Ω, $R_3 = 750$ Ω, $R_4 = 1$ kΩ, $R_5 = 2$ kΩ, and $R_6 = 500$ Ω. Find the current through each resistor (currents I_1 through I_6).

8. What is the minimum input resistance required of a voltmeter to measure the IR drop between points A and B in Fig. P–8 with an accuracy of at least 0.01%? Use Thévenin's theorem.

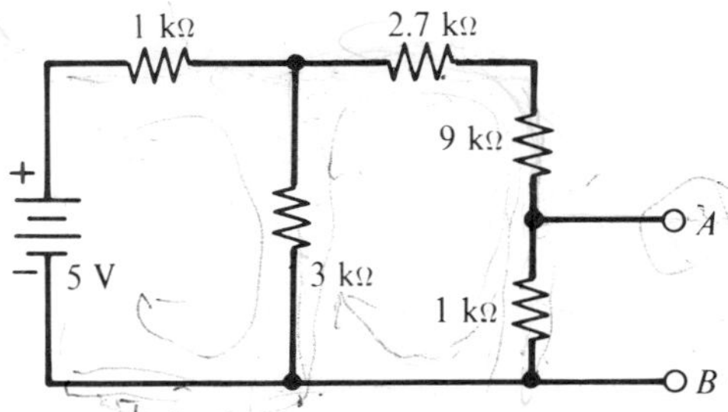

Fig. P–8

9. In the half-bridge of Fig. 2–26, $V_1 = 10$ V, $V_2 = 15$ V, $R_1 = 1$ kΩ, $R_2 = 5$ kΩ, and $R_3 = 2.5$ kΩ. What is the current through each resistor?

10. In the half-bridge of Fig. 2–26, suppose that the voltage V_2 is variable while V_1 is a fixed voltage. Given that V_2 is changed until the current through $R_3 = 0$ A, develop an equation relating V_2 to V_1 and the resistors R_1 and R_2. Draw an operational amplifier circuit which will give the same transfer function and operates on the current null comparison principle of the half-bridge.

11. In the half-bridge of Fig. 2–26, $V_1 = 14$ V, $V_2 = 9$ V, $R_1 = 20$ Ω, and $R_2 = 5$ Ω. Determine the value of R_3 which will absorb the greatest amount of power and determine the maximum power dissipated.

12. The Wheatstone bridge of Fig. P–12 is to be used to measure resistance. If $R_u = 20$ Ω and the ammeter resistance is 9 Ω, what is the current through the ammeter when the bridge is unbalanced with the standard resistor $R = 10$ Ω? Use Thévenin's theorem.

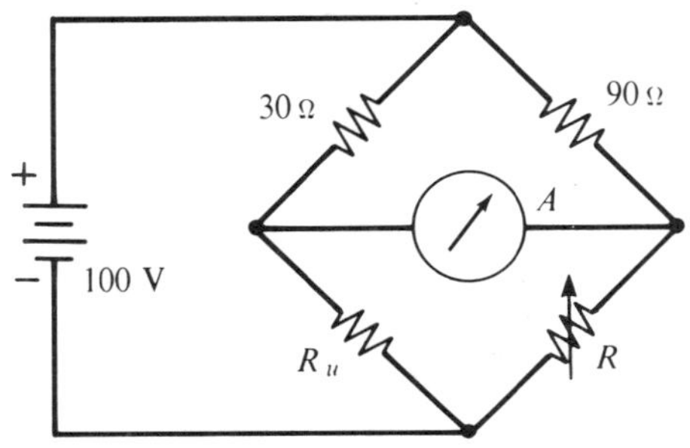

Fig. P–12

13. Measurements of resistance change of strain gauges require high sensitivity and are often made using a variation of the Wheatstone bridge shown in Fig. P–13. Resistors R_1 and R_2 provide a multiplier as usual, and resistor R_3 may be only 1–10 Ω for high sensitivity. Resistor R_u is the strain gauge. (a) Consider that ΔR_3 is set to zero and the bridge is balanced with no strain on the gauge, so that $R_u = R_S$. Now consider the gauge to be strained so that $R_u = R_S + \Delta R_S$. Show that when the bridge is rebalanced

$$\Delta R_S = \Delta R_3(R_2/R_1).$$

(b) The ratio between the relative resistance change and the relative strain change in the strain gauge is called the gauge factor F and is defined as

$$F = \frac{\Delta R_S/R_S}{\Delta l/l},$$

where l is the length of the wire and Δl is the change in length upon strain. A 120 Ω strain gauge with $F = 2$ is used in an equal-arm bridge ($R_1 = R_2$) that is initially balanced with no strain and rebalanced after straining. Compute the relative strain $\Delta l/l$ that will be present if a change of 0.1 Ω in resistor ΔR_3 is required to rebalance the bridge.

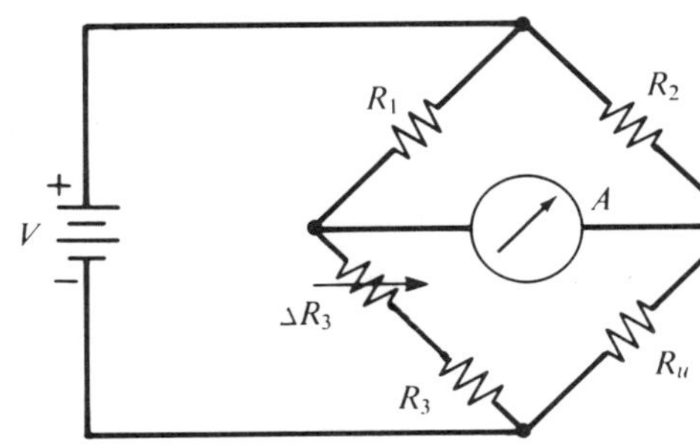

Fig. P–13

14. Consider a precision voltage divider designed to produce 1 V from a 5 V source. (a) If the load to be connected to the 1 V output of the divider has a resistance of 100 kΩ, what are the highest divider resistances that can be used so that the loading error is no more than 0.01%? Use Thévenin's theorem. (b) For a specific divider which satisfies the restrictions of part (a), draw a Norton equivalent circuit and compute I_N, the Norton equivalent current, and R_N, the equivalent shunt resistance.

15. For the Wheatstone bridge of problem 12, recalculate the current through the ammeter using a Norton equivalent circuit instead of a Thévenin equivalent circuit.

16. For many resistance measurement applications it is desirable to record resistance changes directly on a strip chart recorder or other device. Consider the Wheatstone bridge of Fig. 2–18. Consider that the null detector is replaced by a strip chart recorder. (a) Develop a general expression relating the output voltage v_o of the bridge (voltage between points x and y) to the resistance values. (b) Derive a relationship between the change in output voltage and the change in unknown resistance. [*Hint:* use calculus and find dv_o/dR_u.] (c) Under what conditions of bridge resistance will v_o vary linearly with R_u? (d) Under what conditions of bridge resistance will the measurement sensitivity be greatest? [*Hint:* Use calculus and set $d^2v_o/dR_u^2 = 0$.] (e) Is the output voltage linear with R_u under maximum sensitivity conditions?

Experiment 2–1 *Light-Sensitive and Temperature-Sensitive Resistive Devices*

Equipment: VOM
VRS
Fenwal UUA41J1 thermistor
NSL-346 C photoconductive cell
DRB
breadboard card

A photoconductive cell is used to illustrate how the resistance of a device can change as a function of the illumination, and how the measurement of current through the resistive cell can be related to the illuminance. A thermistor is then used to show the relationship between resistance and temperature for a very useful temperature transducer, again using a current meter to obtain resistance.

E2–1.1 Light Intensity Measurements

The VOM on one of its current scales can be used to measure the current through a photoconductive cell when a constant voltage is applied. A 10 V signal from the VRS is used to supply the constant potential.

Obtain a photoconductive cell. Mount the cell on the breadboard card. Connect the circuit shown in Fig. E2–1.

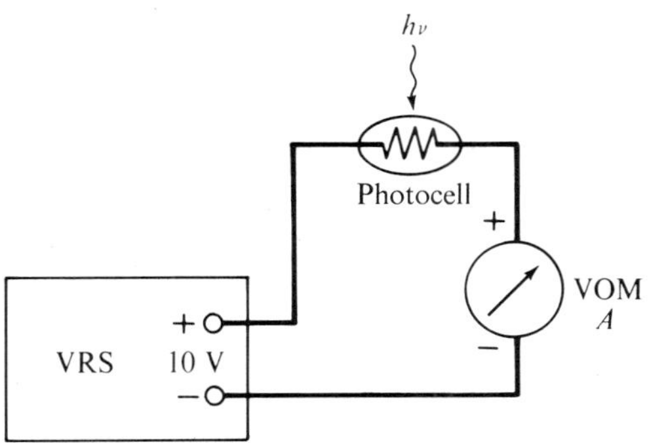

Fig. E2–1 Photocell measurement of light intensity.

For the following measurements, use the VOM on the most sensitive current scale that can be used without overdriving the meter. Allow room light to strike the photoconductive cell. *Measure the current through the cell. Calculate the resistance from Ohm's Law.* From a calibration chart obtain the "illuminance" striking the cell in foot-candles. Repeat the measurements while covering the photoconductive cell to eliminate as much light as possible. *Report the results in the table below.*

	Current through cell	Resistance of cell	Illuminance, fc
Room light			
Dark			

With room light striking the photocell, move your hand fairly close to the cell and *note how sensitive the cell is to slight changes in light intensity.*

E2–1.2 Temperature Measurements

A similar circuit to that used in the previous experiment can be used to measure the resistance of a thermistor. Since the VOM has a full-scale sensitivity of 50 μA and currents somewhat larger than this are expected, it will be necessary to convert the VOM to a 150 μA meter using a shunt.

Given that the VOM has a resistance of 5 kΩ on the 50 μA scale, *calculate the resistance necessary to convert the meter to read* 150 μA *full scale. Show the calculation here.*

Locate the UUA41J1 thermistor. This thermistor has a resistance of 10 kΩ at 25°C. Also locate the calibration

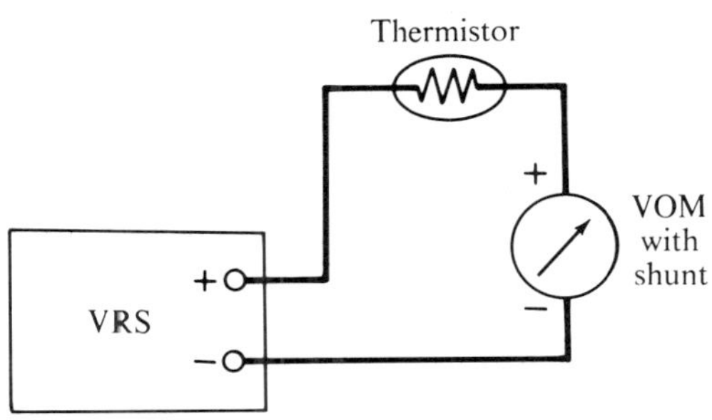

Fig. E2–2 Current measurement of temperature.

chart provided with the thermistor. Set up the circuit shown in Fig. E2–2. Use the DRB to provide the shunt resistance calculated above for the VOM. For ease in connecting the following circuits, mount the thermistor on a double banana jack plug. Set the VRS to 1.0 V, and *measure the current passing through the thermistor with the 150 μA meter. Measure the current for the thermistor in air. Calculate the resistance and use the calibration chart to calculate temperature. Now measure the temperature of your skin by holding the probe between the finger and thumb.* Obtain a beaker of ice water and immerse the probe in the water. *Again measure the resistance and calculate the temperature from the calibration chart.*

	Current through thermistor	Resistance of thermistor
Air		
Skin		
Ice water		

	Calculated temperature
Air	
Skin	
Ice water	

Estimate below the precision in °C of the temperature measurements.

Experiment 2–2 *Measurement of Thermistor Resistance with a Voltmeter*

Equipment: VRS
thermistor
DRB
DVM
recorder

In this experiment, a voltmeter is used to measure the voltage drop across a thermistor through which a constant current is applied. The temperature is varied and the voltage readings are related to resistance values from which temperature values are calculated.

The VRS can be converted into a constant current source by connecting a large resistance in series with it. Note that the series resistor should be much larger than the resistance of the thermistor for the current to be constant and independent of thermistor resistance. In this experiment a 10 MΩ series resistance is used, which is 1000 times the resistance of the thermistor at 25°C.

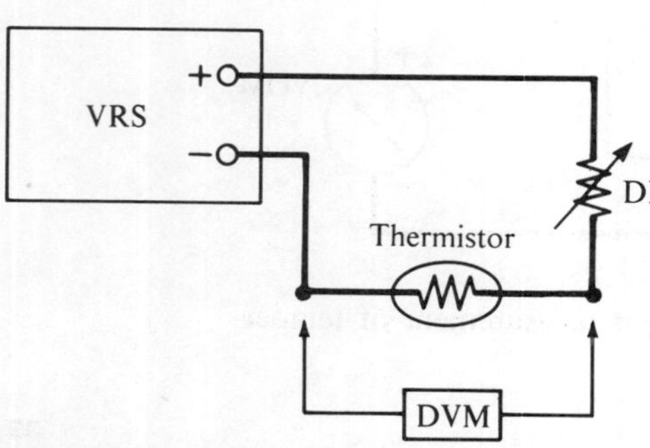

Fig. E2–3 Voltage measurement to obtain temperature.

Set up the circuit shown in Fig. E2–3.

Set the VRS at 10.0 V and the DRB at 10 MΩ (9.9999 MΩ). This applies a current of 1 μA (10^{-6} A) to the thermistor. *Use the DVM on the 1 V range to measure the voltage for the thermistor in air, between the fingers, and in ice water.*

	Voltage across thermistor	Resistance of thermistor
Air		
Skin		
Ice water		

	Temperature calculated
Air	
Skin	
Ice water	

Remove the DVM and attach the servo recorder across the thermistor. Use the 25 mV full scale range. Be certain the ground strip on the VRS is connected to the negative or COM VRS terminal. *Record the temperature of the thermistor in air for a few minutes.*

Estimate the response time of the thermistor by heating it between your fingers and recording the voltage as the resistance returns to its room temperature value (time to go from 10% to 90% of final value).

Response time = ______________________

Experiment 2–3 *Measurement of Thermistor Resistance with an Ohmmeter*

Equipment: thermistor
VOM

Temperature information was obtained in Experiments 2–1 and 2–2 by measuring current and voltage respectively, which were then related to resistance values from which temperature was calculated. It is shown in this experiment that an ohmmeter can be used directly to provide the resistance measurements that are related to temperature.

The VOM is used to measure resistance directly. The VOM is an ohmmeter of the voltmeter type with an internal battery to supply current. To measure resistance, the VOM is switched to one of the R scales. With the leads not touching, the VOM zero is adjusted (∞ resistance). Then, with the leads shorted, the ohms adjustment is used to set the ohms reading.

With the thermistor in the air, *measure the resistance with the VOM, and calculate the temperature from the calibration chart.* Be certain to adjust the 0 and ∞ positions prior to the measurement, and reset the zero each time the range is changed on the VOM. *Report the temperature of the thermistor held between your finger and thumb and immersed in ice water. Report the results below.*

	Resistance measured	Temperature calculated
Air		
Skin		
Ice water		

Estimate the precision of the above measurements.

Experiment 2–4 *Operational Amplifier Measurement of Thermistor Resistance*

Equipment: ADD
OA card
UUA41J1 thermistor
VRS
DVM

Temperature information is obtained in this experiment from a thermistor whose resistance is calculated from measurements of the output voltage of an operational amplifier circuit containing the thermistor. A constant current is applied to the thermistor by placing it in the feedback loop of an OA inverting amplifier to which a known input current is applied. Output voltage measurements are made and related to the thermistor resistance and thus to temperature. Alternatively, the thermistor is used as the input resistor for the OA and a constant input voltage is applied. The thermistor resistance determines the input current, which is then measured by a current comparison method and related back to the temperature of the thermistor.

Set up the circuit shown in Fig. E2–4, with the thermistor as the feedback resistor R_f and the VRS as the input voltage source. Use a precision input resistor R_{in} to apply an input current $v_{\text{in}}/R_{\text{in}}$ so that the output voltage will be $+1$ V for the thermistor at room temperature. *Derive the expected relationship between the thermistor resistance R_T and the measured output voltage v_o and report the chosen component values.*

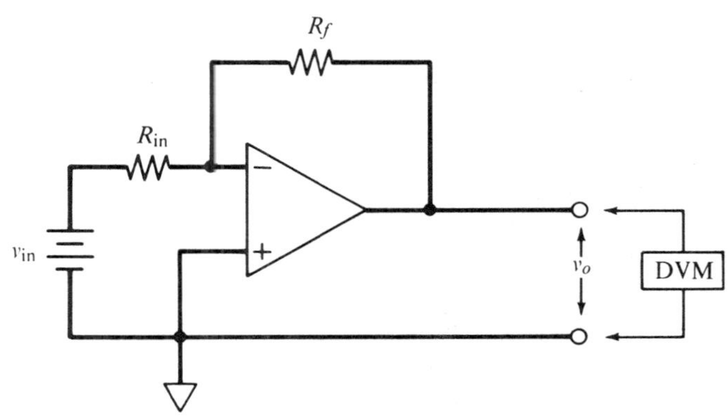

Fig. E2–4 OA circuit for measurement of resistance.

Measure the resistance of the thermistor R_T with the circuit of Fig. E2–4 for the thermistor in air, in ice water, and held between your thumb and finger. *Report the resistance measured and the temperature.*

	v_o	R_T	Temperature calculated
Air			
Skin			
Ice water			

Now change the circuit so that the thermistor is the input resistor R_{in} and the precision resistor is the feedback resistor. *Derive the expected relationship between R_T and the output voltage.*

Choose component values so that $v_o = +1$ V for the thermistor at room temperature, *and report the values picked.*

Again measure the output voltage and *report R_T and the temperature calculated for the thermistor in air, in ice water, and between your fingers.*

	v_o	R_T	Temperature calculated
Air			
Skin			
Ice water			

Experiment 2–5 ***Accurate Measurement of Thermistor Resistance with a Wheatstone Bridge***

Equipment: VRS
VOM
thermistor
DRB
breadboard card
two 100 kΩ, 1% resistors

Unknown resistances are most accurately measured by a comparison measurement, which is conveniently accomplished with a Wheatstone bridge. The resistance bridge is constructed and then used to measure the resistance of the thermistor. Precision resistors and a decade resistance box are used in making the bridge, and a multimeter is used on its most sensitive current scale as the null detector. The sensitivity, precision, and accuracy of the resistance (and temperature) measurements made with the bridge are compared with those of the methods used in the first three experiments.

Set up the circuit shown in Fig. E2–5 using 1% precision resistors for R_A and R_B. Use the decade resistance box for the variable standard resistor R_s.

Use the VOM as the null detector on its most sensitive current scale. Set the VRS on 10.0 V. Note that since $R_A = R_B$, $R_u = R_s$. Start with $R_s = 10$ kΩ, and vary the DRB reading until the null point (zero current) is reached.

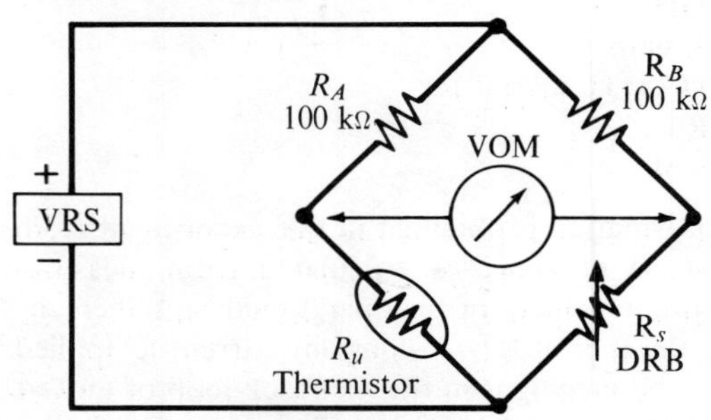

Fig. E2–5 Wheatstone bridge to measure temperature.

Recheck the null point by depressing the VRS ZERO button and making sure there is no deflection. *Measure the resistance of the thermistor in the air, in your fingers, and in ice water. Report the results below.*

	Unknown resistance, $R_u = R_s$	Calculated temperature
Air		
Skin		
Ice water		

Change R_A to a 10 kΩ, 1% resistor. Now $R_u = 10R_s$. *Repeat the above measurements.* Note that when R_A and R_B have different values the ratio becomes a multiplier factor. *Report your results below.*

	Resistance dialed	Unknown resistance, $10R_s$	Calculated temperature
Air			
Skin			
Ice water			

In the space below compare all the methods which were used to measure resistance. Consider sensitivity, accuracy, and precision in your discussion.

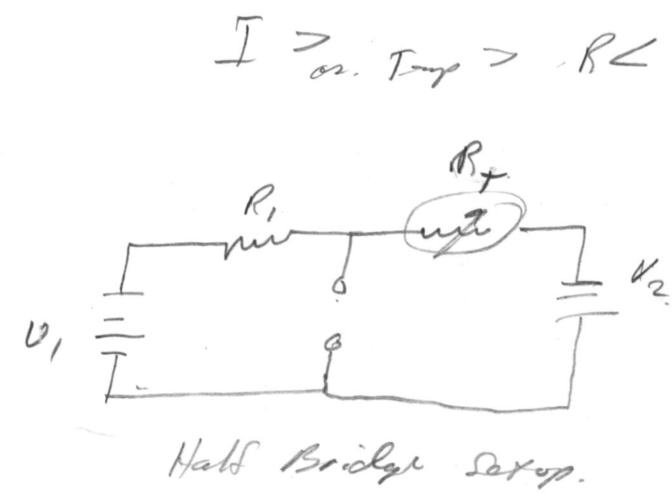

Experiment 2–6 *Linearization of a Transducer Transfer Function*

Equipment: VRS (two)
thermistor
DRB

The relationship between certain desired information in the physical/chemical domain and the output from a specific transducer is often a nonlinear function. In order to make it convenient to display this information on linear current or voltage measurement devices (DVM's, strip chart recorders), it is necessary to introduce linearization circuits. In this experiment, a network is constructed that provides a linear voltage output directly related to temperature, even though the resistance values from the thermistor transducer are nonlinear functions of temperature.

The resistance-temperature relationship of a thermistor can be approximated over a small temperature range by $R_T = Ae^{B/T}$, where R_T is the thermistor resistance, T is the absolute temperature, and B and A are constants which depend on the thermistor material. The constant B can be evaluated by plotting log R_T vs. $1/T$. The slope of this plot should be $B/2.303$. To make the linearization most accurate near 25°C, obtain values of R_T from the calibration chart at 15°C, 25°C, and 35°C. *In the space below tabulate R_T, log R_T, and $1/T$.*

R_T	log R_T	T, °K	$1/T$
		288	
		298	
		308	

For each 10 degree temperature difference calculate Δ log $R_T/\Delta(1/T)$ and average the two values. The constant B is the slope × 2.303. Report the values below.

Average slope = ____________________

Average B = ____________________

The circuit of Fig. E2–6 is often used to linearize the thermistor response. The value of R_1 is chosen so that the second order term in the series expansion of the exponential resistance-temperature relationship is zero, and thus resistance becomes linear in relation to temperature. If R_T is expanded in a series expansion and only the first two terms

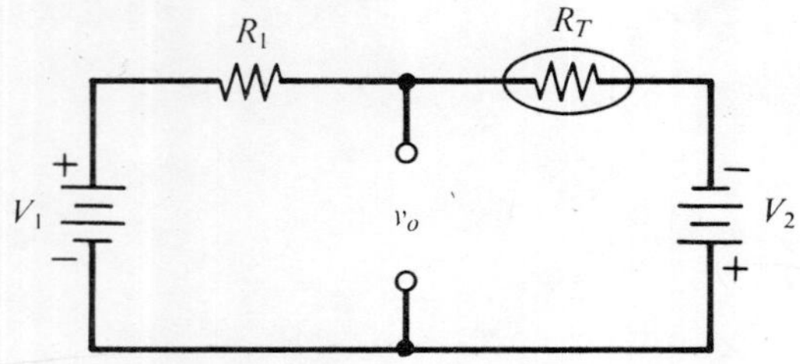

Fig. E2-6 Linearized thermistor network.

are kept, R_T is approximately given by $R_T = A[1 + B/T + B^2/(2T^2)]$. By choosing R_1 such that

$$R_1 = \left(\frac{B - 2T_0}{B + 2T_0}\right) R_T(T_0),$$

where $R_T(T_0)$ is the thermistor resistance at the temperature T_0 for which the output of Fig. E2-6, v_o, is zero volts, the second order term is approximately zero.

From the value of B found previously, choose the appropriate value of R_1 to linearize the thermistor near room temperature ($T_0 = 298°K$).

$R_1 =$ ____________________________

To make $v_o = 0$ V at the desired temperature, the voltages V_1 and V_2 are chosen from the relationships

$$V_1 = V\left(\frac{1}{2} - \frac{T_0}{B}\right) \quad \text{and} \quad V_2 = V\left(\frac{1}{2} + \frac{T_0}{B}\right),$$

where V is the total voltage $V_1 + V_2$. *Given that an appropriate value for V is 1 V, calculate what V_1 and V_2 should be.*

$V_1 =$ ____________________________

$V_2 =$ ____________________________

Since both voltages are now known, R_1 is known, and R_T can be found from the calibration chart at any temperature, the expected output voltage, v_o, can be calculated as a function of temperature. *For the temperatures given below, calculate v_o. For each 5 degree temperature change, calculate the slope of the v_o vs. T plot and see if it remains reasonably constant.*

Show a sample calculation of v_o.

Temperature, °C	v_o, V	$\Delta v_o/5°C$

Instead of two voltages in the linearization circuit, one source can be used and the resistors placed in a bridge where R_1 and R_T are one arm of the bridge. *Design below a bridge which would give a circuit equivalent to that of Fig. E2-6 and the same values of v_o as calculated above.*

Capacitive and Inductive Control Devices

Section 2–2

Capacitors and inductors play a major role in many instruments for the control of electric charge, current, and voltage. Circuits which are simple combinations of capacitors or inductors and other circuit elements provide a variety of important functions. The characteristics of practical capacitors and inductors and their applications in basic *RC* and *RL* circuits are described in this section. Capacitors are used with the versatile operational amplifiers in the widely applied integrator and differentiator circuits. These circuits and several applications in scientific measurements are described.

2–2.1 CAPACITORS

Capacitance is a measure of the quantity of charge required to create a given potential difference between two conductors. Since a finite capacitance exists between every pair of conductors, capacitance, like conductance, is a common property of all electronic devices. The property of capacitance in electric circuits affects the rate at which the circuit voltages can be changed or the duration over which voltage changes can be maintained. Devices which are made specially for their property of capacitance are called **capacitors.**

A capacitor consists basically of two pieces of conducting material separated by an insulator. This construction is represented pictorially in Fig. 2–29 as two metal parallel "plates" of area *a* separated by a distance *d*. The capacitor is "charged" by taking charged particles (such as electrons) from one plate and putting them on the other. The result of this separation of charge is a potential difference and an electric field between the two plates. The voltage developed by the charge separation is proportional to the amount of charge separated. The number of coulombs of charge per

Fig. 2–29 Pictorial representation of a capacitor.

volt of potential difference is the capacitance, C:

$$C = q/V. \tag{2–27}$$

The units of C are coulombs per volt, or **farads** (F). The farad is a measure of the capacity of a capacitor for storing charge at a given voltage.

The separation of charge necessary for the charging of a capacitor requires energy. A charged capacitor (one with a finite potential difference between the plates) stores electric potential energy. This stored energy can be converted to some other form of energy by "discharging" the capacitor, i.e., connecting a conductor or **load** which can produce heat or light or do mechanical work. However, the current through the load occurs at the expense of the charge on the capacitor. As the charge decreases, the capacitor voltage decreases as in Eq. (2–27) so that eventually the charge separation, the voltage, and the current are all zero. When a load is first connected to a capacitor charged to a voltage V, the first electron charge through the load loses an energy of $Q_e V$ joules, where Q_e is the electron charge. The voltage across the capacitor when the last electron charge is passed is essentially zero, so that the energy lost for this charge is also essentially zero. Since the capacitor voltage is proportional to the charge at any time, the average energy lost per electron is the same as the energy lost by the electron passed at half charge. Thus the energy or work obtained in discharging a capacitor completely is

$$W_d = \frac{qV}{2} = \frac{q^2}{2C} = \frac{CV^2}{2}, \tag{2–28}$$

which is also the amount of potential energy (in joules) stored by the capacitor.

By considering the energy stored in the capacitor, one can deduce the geometric factors affecting the capacitance. If the capacitor of Fig. 2–29 is charged to v volts and then the plates are moved apart to a distance $2d$, the energy in the system will be doubled, since the charge has been separated by twice the distance. From Eq. (2–28) it is clear that if W_d is doubled and q is constant, the voltage must be doubled. If the voltage in Eq. (2–27) is doubled for the same charge, the capacitance must be halved. Thus we see that the capacitance is inversely proportional to the conductor separation ($C \propto 1/d$). Again consider the capacitor with separation d charged to v volts. Now we will double the plate area with uncharged pieces of metal. The same charge is now distributed over twice the area and thus produces only half the field. By the same arguments as before, if v is halved when q is constant, C is doubled. Therefore, the capacitance is proportional to the area ($C \propto a$). If the space between the plates of a charged capacitor is at first completely empty (a vacuum), and then insulating material with polarizable molecules (a **dielectric**) is put in that space, the polarizable molecules in the dielectric tend to become aligned under the influence of the electric field. This process uses some of the energy stored in the field. Since q is not changed, v is reduced, which indicates that the capacitance has increased. The factor by which the capacitance is increased by the dielectric over that for a vacuum is called the **dielectric constant** of the material. Thus $C \propto K_d$. If these equations are combined into one,

$$C = \epsilon_0 \frac{K_d a}{d}, \tag{2–29}$$

where ϵ_0 is the combined proportionality constant and is called the permittivity of free space. The value of ϵ_0 is 8.854×10^{-12} farads per meter (Fm^{-1}).

If the dimensions and dielectric constant are known for a capacitor, its capacitance can be calculated from Eq. (2–29). For instance, a capacitor made of 2 squares of aluminum foil 10 cm on a side, separated by a cellophane ($K_d \approx 3.5$) sheet 0.02 cm thick, will have a capacitance of

$$C = \frac{(8.9 \times 10^{-12}\ \mathrm{Fm^{-1}})(3.5 \times 10^{-2}\ \mathrm{m^2})}{2 \times 10^{-4}\ \mathrm{m}} = 16 \times 10^{-10}\ \mathrm{F}.$$

It can be seen from this example that the farad is an extremely large unit. For this reason real capacitors are measured in microfarads ($1\ \mu\mathrm{F} = 10^{-6}$ F) and picofarads ($1\ \mathrm{pF} = 10^{-12}$ F). The capacitance in the above example would be given as 0.0016 μF or 1600 pF. Values for some common dielectric constants are given in Table 2–4.

Table 2–4 Common dielectric constants

Material	K_d
Air	1.0006
Glass	3.9–5.6
Plexiglass	2.8
Polyethylene	2.3
Polystyrene	2.6
Teflon	2.1
Mica	5.5
Paper	≈3.5
Water	78
Quartz	3.9
Oil	2.2
Paraffin	2.1

Practical Capacitors

The simplest capacitor consists of two sheets of metal foil separated by a dielectric. The dielectric can be paper, oil, mica, ceramic, or plastic film. The whole "sandwich" can be rolled, folded, or otherwise compacted. Then the unit is sealed from the effects of atmosphere. Some common capacitor types are shown in Fig. 2–30. The larger the area of the foil and the thinner the dielectric sheet, the higher the capacitance will be. However, if too high a voltage is applied across the capacitor, the dielectric will break down and a discharge will occur between the two foil sheets. This often results in permanent damage to the capacitor. The breakdown voltage rating may be increased by using a thick dielectric, but the capacitance will be correspondingly decreased. The capacitance of these units, which are the most frequently used capacitors, varies from 0.001 to 1 μF.

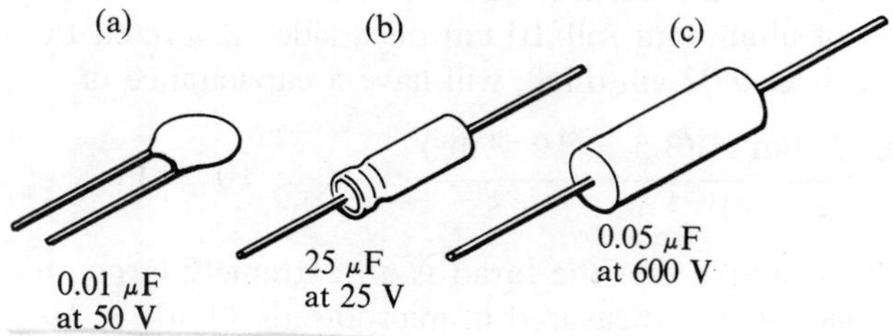

Fig. 2–30 Common capacitor types: (a) disk ceramic; (b) electrolytic; (c) paper or plastic film.

Capacitors can be obtained in various tolerance ratings, usually 20, 10, or 3%. Capacitors of higher accuracy can be obtained for special applications. Capacitors generally have a high temperature coefficient because of dimensional changes with temperature. In some very critical circuits it is necessary to consider the leakage current through the capacitor dielectric. A capacitor constructed for especially high leakage resistance may have a leakage resistance–capacitance product of 10^6 MΩ-μF.

Electrolytic capacitors are made of aluminum or tantalum foil with a surface that has an anodic formation of metal oxide film. The anodized metal foil is in an electrolytic solution. The oxide film is the dielectric between the metal and the solution. Because the dielectric is so thin, a high capacitance can be obtained in a small space. Most electrolytic capacitors must be used in a circuit where the polarity is always in one direction. If the polarity is reversed, the oxide will be reduced, thus destroying the dielectric. Gas will be evolved at the cathode and the capacitor may explode. Some tantalum electrolytic capacitors are made of two anodized metal foils connected by the electrolyte. Such capacitors are **bipolar** and can be used in ac signals or connected either way in the circuit. The oxide film is a relatively low resistance dielectric which results in significant leakage current.

The oxide in aluminum electrolytics tends to deteriorate with time. Electrolytic capacitors are generally used for filtering and transistor circuit applications where poor dielectric can be tolerated and where large capacitance in a small space is essential. They are generally rated from 1 μF to thousands of microfarads. Naturally, the physical size of the capacitor increases with increased capacitance and voltage rating. Tantalum capacitors are generally much smaller for a given capacitance than aluminum capacitors.

Variable Capacitors. Variable capacitors are usually small-valued capacitors not exceeding about 1000 pF. When larger variations are desired, a switch is often used to put various fixed-value capacitors into the circuit. There are two common types of variable capacitors, the mica trimmer capacitor and the variable air capacitor.

The mica trimmer capacitor is composed of two sheets of metal sandwiching a sheet of mica. The spacing between the metal sheets, and thus the capacitance, is adjusted by turning a machine screw. This kind of variable capacitor is used when adjustment is only occasionally necessary and when knowledge of the actual value of the capacitance is unimportant.

The variable air capacitors are adjusted from zero to full capacitance by turning a shaft 180°. If the movable plates are semicircular, the capacitance is nearly a linear function of the angular position of the shaft. End effects cause a small deviation from linearity at both limits of rotation.

There are many considerations involved in choosing a capacitor for a given circuit, such as size for a specific capacitance value, applied voltage effects, and stability in all environmental and circuit conditions. For example, the temperature can affect the capacitance by causing variations of the dielectric constant or changing the conductor area or spacing; or it can change the leakage resistance or affect the breakdown voltage. Moisture can make the dielectric more susceptible to breakdown and decrease the insulation resistance and capacitor life unless the capacitor is hermetically sealed.

There are some energy losses in a capacitor. The **capacitor power factor** is the ratio of power lost in the capacitor to the volt-ampere input, and it is usually expressed as a percentage. It is a complex function of frequency, temperature, applied voltage, and dielectric material. The power factor takes into account *all* losses, including leakage losses, dielectric absorption losses, and ohmic losses in the leads and conducting plates.

Many capacitors develop potential differences across their terminals shortly after complete discharge and thus produce unwanted dc voltages that can appear as offset or noise. This phenomenon is due to dielectric polarization effects and is known as **dielectric absorption,** sometimes called the "soaking effect." Capacitors with polystyrene or Teflon™ dielectrics are especially free from this effect and thus are well suited for integrators and other critical instrumentation applications.

Capacitive Transducers

Transducers in which the capacitance varies in response to the measured phenomenon are quite common. Because the capacitor is so sensitive to dimensional changes, it is often used where the parameter to be measured can cause a physical motion. The motion can cause a variation in the gap between the plates, a change in the area of the plates, or a change in the dielectric constant. Thus capacitive transducers exist for displacement, pressure, fluid level, and so on. Capacitive transducers have also been used to measure the dielectric constants of materials.

Combinations of Capacitors

When two or more capacitors are connected together in a circuit, the resulting equivalent circuit capacitance is related to the capacitance of the individual capacitors by simple relationships. Simple series and parallel combinations will be considered first. Then the analysis of a series-parallel network of capacitors will be illustrated.

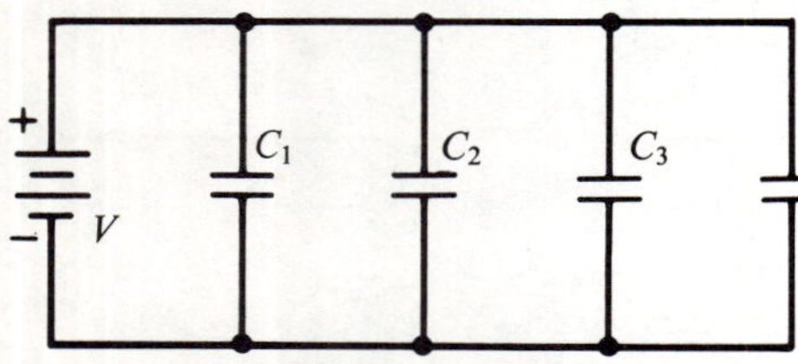

Fig. 2–31 Capacitors in parallel.

In the circuit of Fig. 2–31, several capacitors are connected in parallel to a voltage source V. Since the capacitors are in parallel, the voltage V appears across each one. It follows from the definition of capacitance, $C \equiv Q/V$, that the charge on C_1 is C_1V, the charge on C_2 is C_2V, and so on. Thus the total charge Q_T is

$$Q_T = V(C_1 + C_2 + C_3 + C_4). \tag{2–30}$$

The equivalent parallel capacitance C_p is Q_T/V, so that

$$C_p = C_1 + C_2 + C_3 + C_4. \tag{2–31}$$

Thus we see that the equivalent capacitance of capacitors in parallel is simply the sum of the capacitances. The combined capacitance of capacitors in parallel is always greater than the largest individual capacitance.

A series combination of capacitors connected to a voltage source V is shown in Fig. 2–32. It is assumed that each capacitor was completely discharged before being connected in the circuit. While charging, the current to each capacitor must be identical, since the capacitors are connected in series. When charged, the charge on each capacitor must be identical as indicated in the figure. The voltage v_1 across C_1 must be q/C_1; the voltage v_2 across C_2 must be q/C_2; and so on. The sum of the voltages across the capacitors must be equal to V (Kirchhoff's loop law) so that

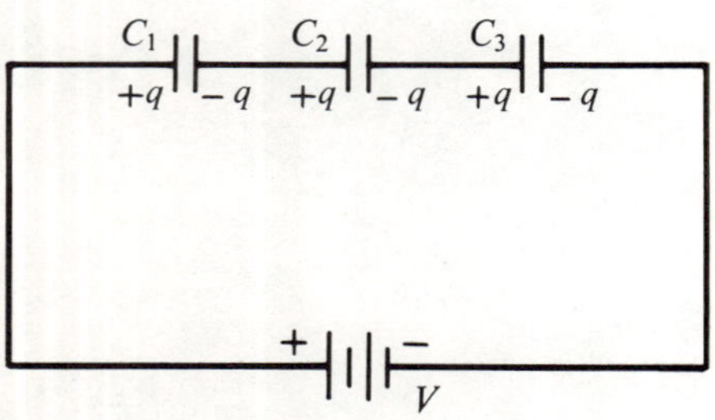

Fig. 2–32 Capacitors in series.

$$V = \frac{q}{C_1} + \frac{q}{C_2} + \frac{q}{C_3}. \tag{2–32}$$

The equivalent series capacitance $C_S = q/V$. If this expression is substituted in Eq. (2–32) for V and q is canceled, Eq. (2–33) results:

$$\frac{1}{C_S} = \frac{1}{C_1} + \frac{1}{C_2} + \frac{1}{C_3}. \tag{2–33}$$

The equivalent capacitance is the reciprocal of the sum of the reciprocals. For two capacitors, a simpler form of Eq. (2–34) exists:

$$C_S = \frac{C_1C_2}{C_1 + C_2}. \tag{2–34}$$

As Eq. (2–33) or Eq. (2–34) indicates, the capacitance of the series combination will always be *less* than the smallest individual capacitance.

The circuit of Fig. 2–33 is a series-parallel combination of capacitors connected to a voltage source of 16 V. This network can be solved by successive reduction as in the resistive network of Section 2–1.6. The equivalent capacitance of the parallel combination of C_2 and C_3 ($C_{2,3}$) is $C_2 + C_3$; $C_{2,3} = 3$ μF + 5 μF = 8 μF. This equivalent capacitance $C_{2,3}$ in series with C_1 gives an equivalent capacitance $C_{1,2,3} = 8 \times 8/(8 + 8) = 4$ μF. The series combination of C_4 and C_5 has an equivalent capacitance $C_{4,5} = 2 \times 6/(2 + 6) = 1.5$ μF. The parallel combination of $C_{4,5}$ and $C_{1,2,3}$ results in $C_T = 1.5 + 4 = 5.5$ μF. The total charge Q_T on the circuit is $C_TV = 5.5 \times 10^{-6}$ F × 16 V = 88 μC. This charge is split between $C_{1,2,3}$ and $C_{4,5}$: 1.5×10^{-6} F × 16 V = 24 μC on $C_{4,5}$ and 64 μC on $C_{1,2,3}$. Since C_4 and C_5 are in series, they each have a charge of 24 μC. This means the voltage on C_4 is 24 μC/2 μF = 12 V and the voltage on C_5 is 4 V. Now C_1 and $C_{2,3}$ are in series, with 64 μC of charge on each. Therefore, the voltage on C_1 is 64 μC/8 μF = 8 V. The voltage on C_2 and C_3 is 8 V. The charge on C_2 is 3 μF × 8 V = 24 μC; and on C_3, 40 μC. In this way the total equivalent capacitance is calculated and then the voltage on each capacitor and the charge across each capacitor are calculated.

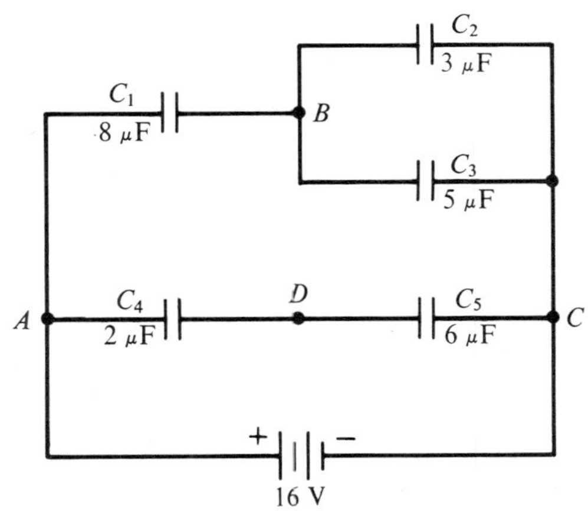

Fig. 2–33 Capacitor network.

Experiment 2–7 *The Capacitor as a Charge Storage Device*

Most measurement and control circuits are dependent in some way on the ability of a capacitor to store charge. Therefore, it is important to obtain laboratory experience on how capacitors perform when charged and when combined in various ways. In this experiment, the leakage characteristics of various types of capacitors are observed, and the distribution of charge in capacitor networks is measured. The voltage values obtained from the theoretical relationships are compared with the measured values in capacitor networks.

2–2.2 *RC* CIRCUITS

When a series combination of a resistor and a capacitor is connected to a voltage source, the values of the resistance and the capacitance affect the rate at which the capacitor can charge. Similarly, when a charged capacitor is discharged through a resistor, the discharge rate is affected by the capacitance and resistance values. Since resistance and capacitance are inherent parts of all circuits and devices, an understanding of the charging and discharging characteristics of *RC* circuits is important in determining the rate with which steady-state conditions will be reached after a change in the applied potential.

Series *RC* Circuit

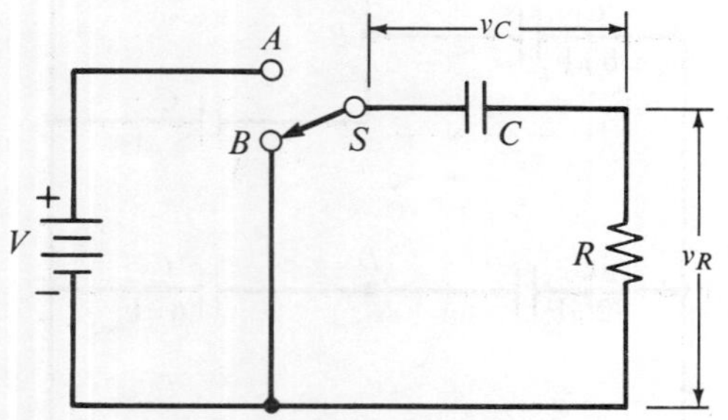

Fig. 2–34 A series *RC* circuit and step signal source.

There are several basic factors that govern the response of the simple *RC* circuit to a sudden change in voltage. In Fig. 2–34 a source of voltage V (of negligible internal resistance) can be suddenly impressed across the series *RC* circuit by turning switch S to A. From inspection of Fig. 2–34, assuming that the switch S is at A, the following statements can be deduced:

A. At every instant, the voltage across the capacitor is directly proportional to the stored charge q; that is, $v_C = q/C$.

B. The voltage across the resistance at any instant is $v_R = iR$.

C. The sum of the voltage drops around the circuit must equal the applied voltage at any instant (Kirchhoff's voltage law); that is,

$$V = v_R + v_C = iR + \frac{q}{C}. \tag{2–35}$$

D. The current at any instant is the same in all parts of a series circuit and is equal to v_R/R. If Eq. (2–35) is rearranged and solved for the current, the result is

$$i = (V - v_C)/R. \tag{2–36}$$

Suppose in the circuit of Fig. 2–34 that $V = 10$ V, $R = 100\ \Omega$, $C = 1\ \mu$F, and the switch has been at B for a sufficient time so that $q = 0$. Therefore, $i = 0$, $v_R = iR = 0$, and $v_C = q/C = 0$. Now the switch is turned to A. At the instant of closing, the capacitor has not had a chance to charge, and $v_C = 0$. Therefore, $i = (V - v_C)/R = V/R = 10/100 = 0.1$ A, and $v_R = 10$ V. Immediately the flow of electrons starts to charge the capacitor; so q and v_C increase and i and v_R decrease, each exponentially, as shown in Fig. 2–35. Note that the time scale is calibrated in units of *RC*. The product *RC* has the units of seconds,

$$RC = \left(\frac{V}{i}\right)\left(\frac{q}{V}\right) = \left(\frac{\text{volts}}{\text{coulombs/sec}}\right)\left(\frac{\text{coulombs}}{\text{volts}}\right) = \text{sec},$$

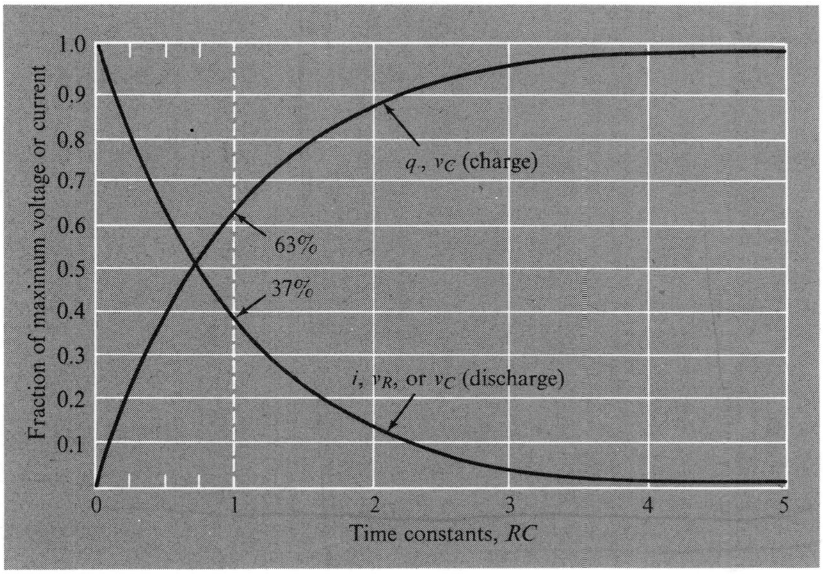

Fig. 2–35 Charge and discharge curves in the *RC* circuit.

and is called the **time constant.** The current i at any time t after turning the switch to A is given by

$$i = \frac{V}{R} e^{-t/RC}. \qquad (2\text{–}37)^3$$

After a time $t_1 = RC$, the current $i_t = (v/R)e^{-RC/RC} = (v/R)e^{-1} = (v/R) \times 0.368$. In other words, at time $t_1 = RC = 100 \times 1 \times 10^{-6} = 10^{-4}$ sec, the current is 36.8% of the current at the instant of impressing voltage V. This means that at time $t_1 = RC$ the voltage across the resistor is only 36.8% of its initial value and the capacitor is charged to 63.2% of the impressed voltage. The time $t_1 = RC$, which is referred to as the time constant of the *RC* circuit, is given the symbol τ. Table 2–5 tabulates values of v_C and v_R for different multiples of τ, both for the charging of the capacitor by impressing voltage V and for discharging a capacitor that is charged to a voltage V. Note that when $t = 4.6RC$ the capacitor is charged to 99.0% of the impressed voltage and v_R is only 1% of its initial value. For practical purposes the capacitor is often considered to be fully charged when $t \approx 5\tau$.

It is important to keep in mind that the voltage across a capacitor cannot change instantly; i.e., the capacitor voltage changes exponentially with time. From one point of view, the capacitor may be considered a short circuit at the first instant of impressing a sudden voltage change.

By watching the output of the series *RC* circuit, first across the resistor and then across the capacitor, various waveshapes similar to those in Fig.

Note 3. Derivation of Charging Equation.

Equation (2–37) can be derived by substituting dq/dt for i in Eq. (2–35) and solving the resulting differential equation. The result is

$$v_C = V(1 - e^{-t/RC}). \qquad (2\text{–}38)$$

To obtain the i vs. t relation, substitute Eq. (2–38) into Eq. (2–35),

$$v_R = V - v_C = Ve^{-t/RC}, \qquad (2\text{–}39)$$

and, since $v_R = iR$, it follows that

$$i = \frac{V}{R} e^{-t/RC}. \qquad (2\text{–}40)$$

Table 2–5 Output voltages across capacitor and resistor in series *RC* circuit

Time	Capacitor charging v_c % *V* applied	Capacitor charging v_R % *V* applied	Capacitor discharging v_c % *V* initial	Capacitor discharging v_R % *V* initial
RC	63.2	36.8	36.8	36.8
2*RC*	86.5	13.5	13.5	13.5
2.3*RC*	90.0	10.0	10.0	10.0
3*RC*	95.0	5.0	5.0	5.0
4*RC*	98.2	1.8	1.8	1.8
4.6*RC*	99.0	1.0	1.0	1.0

2–36 can be observed if the switch of Fig. 2–34 is turned on and off, or if a rectangular pulse source is used as the input. It is immediately apparent that the output waveform is greatly dependent on the relationship of the *RC* time constant τ to pulse width T_p. It is interesting to observe that the leading edge of the output across the resistor is always steep, so long as the input voltage has a steep leading edge. In contrast, the leading edge of the capacitor output is always changing exponentially. Note that the sum of voltages across the capacitor and resistor equals the input voltage at each instant for a given *RC* time constant. This can be observed by comparing the pairs of curves in Fig. 2–36c, d, and e.

It is worthwhile to observe that sharp positive and negative pulses can be obtained across the resistor when the *RC* time constant is much shorter than the pulse width. This type of response finds application in many circuits. In Fig. 2–36c the voltage across the capacitor is a rather linear sawtooth voltage. This suggests that the circuit can act as an integrator when the time constant is long in comparison with the pulse width. In fact, when the output is taken across the capacitor, the *RC* circuit is often referred to as an "integrator," because

$$v_o = v_c \simeq (1/RC) \int v_i \, dt. \tag{2–41}$$

Because *the output is the integral of the input only when the RC time constant is much longer than the pulse width,* such a "passive" *RC* circuit does not provide an exact integration. *RC* circuits in conjunction with operational amplifiers can provide very accurate integration of all waveforms, and are presented in Section 2–2.3.

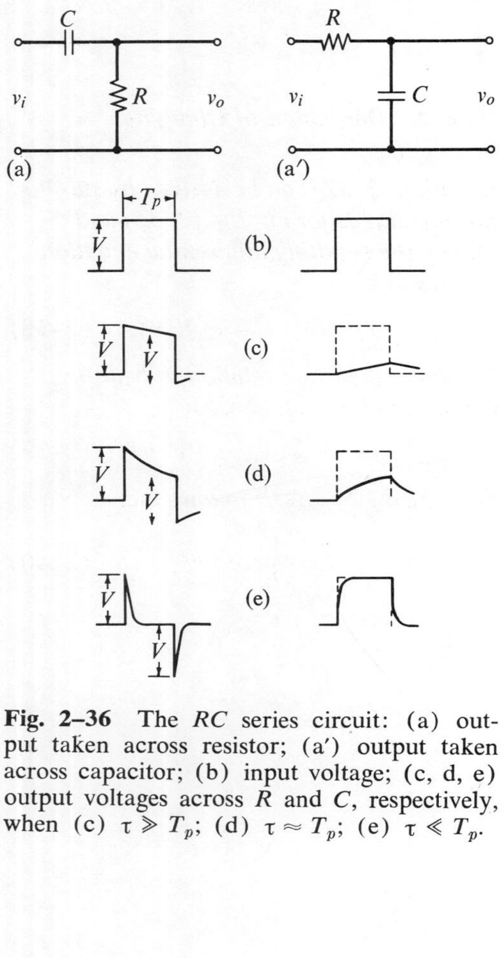

Fig. 2–36 The *RC* series circuit: (a) output taken across resistor; (a′) output taken across capacitor; (b) input voltage; (c, d, e) output voltages across *R* and *C*, respectively, when (c) $\tau \gg T_p$; (d) $\tau \approx T_p$; (e) $\tau \ll T_p$.

Parallel *RC* Circuit

To determine the shape and time constant of voltage changes in a circuit, it is only necessary to determine whether the output signal is across the capac-

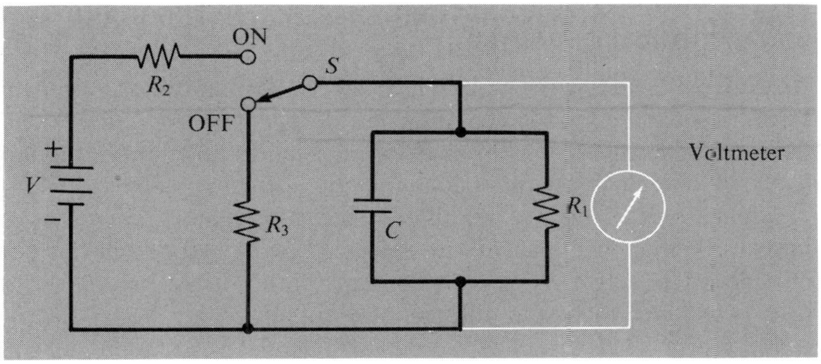

Fig. 2–37 Parallel *RC* switching circuit.

itance or the resistance, and what components determine the *RC* time constant in the case of each change. To take another example, consider the parallel *RC* circuit of Fig. 2–37. When switch *S* is thrown to the ON position, *C* will begin to charge toward the final voltage established by the voltage divider (R_1 and R_2), that is, $VR_1/(R_1 + R_2)$. The rate of charge is determined by *C* and by the output resistance of the charging circuit, which consists of the source *V* and the voltage divider. As was pointed out in Section 2–1.6, the equivalent resistance of a voltage divider is the parallel combination of the divider resistors, or $R_1R_2/(R_1 + R_2)$. Thus the charging time constant is $CR_1R_2/(R_1 + R_2)$. When switch *S* is turned OFF, capacitor *C* discharges toward zero through the parallel discharge paths R_1 and R_3. Thus the discharge time constant is $CR_1R_3/(R_1 + R_3)$. Since the voltage across the capacitor is being observed, the waveforms will resemble those in the right-hand column of Fig. 2–36.

Such parallel *RC* circuits are of great importance in switching and will be discussed in more detail in Section 2–3.

Experiment 2–8 *Control of Current and Voltage with the RC Network*

The charging and discharging of a capacitor through a resistor provides a current control function that is generally applicable in measurement and control systems. The *RC* network is used for controlling time delay circuits, for controlling the repetition rate of waveform generators, and for integration and differentiation of voltage signals. In this experiment the *RC* circuit is first tested with a step function input signal and then with a repetitive square wave. The types of output waveforms that can be produced by the *RC* circuit with a square wave input are observed and compared with theoretical curves.

2–2.3 ACTIVE *RC* INTEGRATORS, DIFFERENTIATORS, AND LOGARITHMIC CONVERTERS

An active *RC* circuit is one in which the voltage across or current through a section of the circuit is actively controlled. Active *RC* circuits can easily be made by using combinations of resistors and capacitors in the input and feedback circuits of operational amplifiers. Such circuits have a wide variety of applications as integrators, differentiators, oscillators, compensators, and filters. These applications will be explained and illustrated in this and other modules. The active *RC* circuit concept is introduced here with the often used active integrator and differentiator circuits.

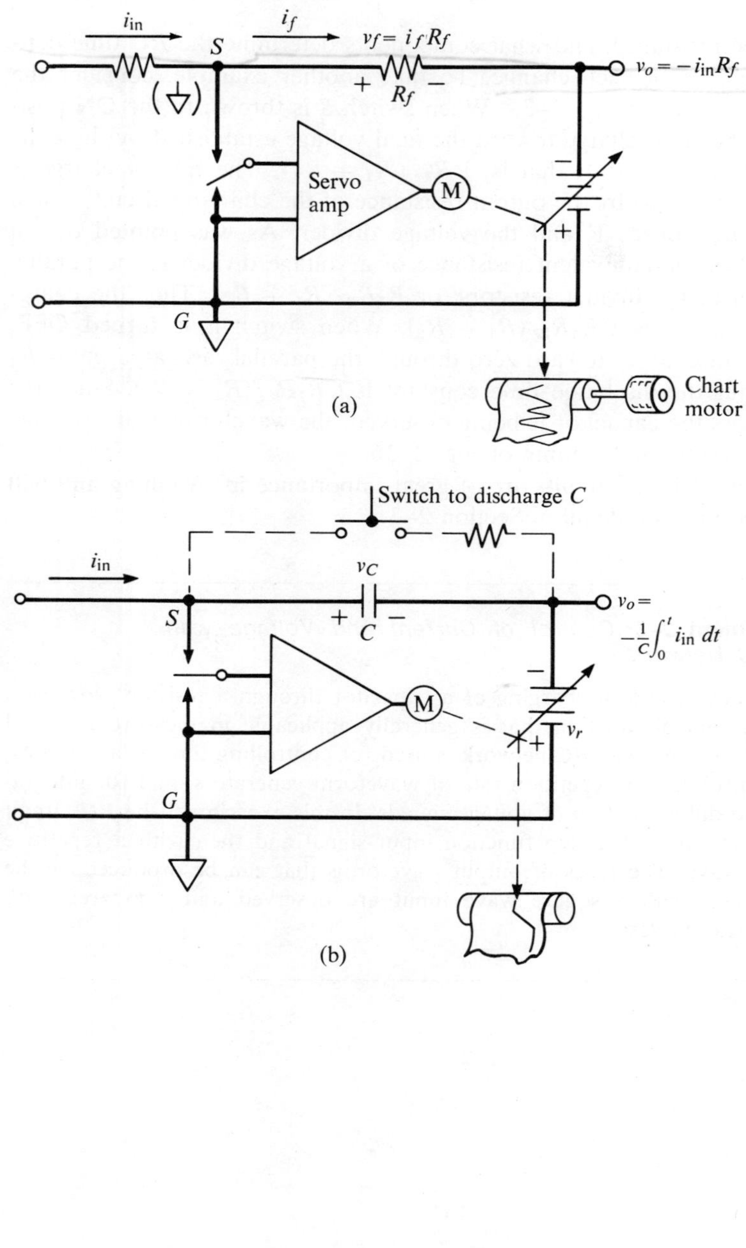

Fig. 2–38 (a) Servomechanical current follower. (b) Servomechanical current integrator.

Integrator

In many measurement and control applications it is desirable to obtain an accurate electronic integration of an input current or voltage. For example, the quantitative information from many transducers, such as those used in gas or liquid chromatography, is contained in the area of a peak-shaped signal. The use of an electronic integrator allows the quantitative information to be displayed directly without time-consuming manual area measurements. In other applications integration may be used to average the effects of random noise or to produce a linear time-base signal. The basic current follower systems described in Module 1 can perform the integration operation if a capacitor is substituted for the resistor as the feedback element.

The servomechanical current follower with resistive feedback is repeated in Fig. 2–38a. The slidewire output voltage v_o is automatically adjusted to maintain the reference current i_f (normally called the feedback current) equal to the input current i_{in} and thus to maintain the potential at the summing point S equal to that at point G. At balance, $v_o = -i_{\text{in}}R_f$, and the pen position, which is proportional to v_o, is thus related to the input current.

In Fig. 2–38b the feedback resistor is replaced with a capacitor C. As before, the servo system adjusts the slidewire output voltage to maintain a current null by keeping the potential at point S very nearly equal to that at point G. The reference or feedback current to point S comes from changing the charge on (and thus the voltage across) the capacitor C. For the servo system to balance, the slidewire output voltage v_o must equal the voltage across the feedback capacitor v_C and must be of opposite polarity. Since $v_C = q/C$, where q is the charge stored on the capacitor,

$$v_o = -v_C = -q/C. \qquad (2\text{–}42)$$

The charge stored on the capacitor is equal to the integral of the current during the charging period t, and since $i_{\text{in}} = i_f$,

$$v_o = -\frac{1}{C}\int_0^t i_{\text{in}}\,dt. \qquad (2\text{–}43)$$

Input voltages as well as currents can be integrated with the servo system by adding an input resistance between the source and the summing points. Just as was the case with the current follower, the servomechanical system can be replaced by an all-electronic operational amplifier for performing current and voltage integration. The OA voltage integrator illustrated in Fig. 2–39 operates on the same principle as the servo current integrator. The input current $i_{\text{in}} = i_f = v_{\text{in}}/R$. Substituting for i_{in} in Eq.

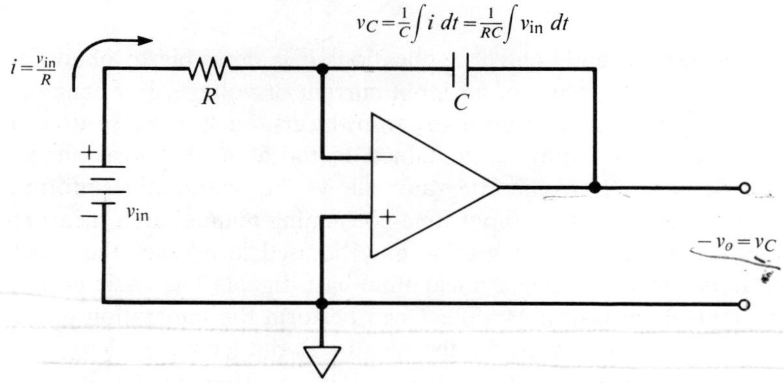

Fig. 2–39 Operational amplifier voltage integrator.

(2–43) gives

$$-v_o = v_C = \frac{1}{C}\int_0^t \frac{v_{in}}{R}\,dt = \frac{1}{RC}\int_0^t v_{in}\,dt. \tag{2–44}$$

Thus either currents or voltages can be accurately integrated by the servomechanical system or by the all-electronic operational amplifier. For accurate and precise integration, an OA with very low input bias current is highly desirable. Input bias currents can cause large errors because they are also integrated and contribute to v_o.

The integrator circuit can also be used as a capacitance-to-voltage converter. If the integrator circuit is used to place a known amount of charge on the capacitor, the OA output voltage will be inversely proportional to the capacitance as given by Eq. (2–42). The known amount of charge is usually generated by applying a constant voltage V through a known resistance R for a specific time. Thus $q = Vt/R$ and

$$C = -\frac{Vt}{v_o R}. \tag{2–45}$$

Experiment 2–9 *An Operational Amplifier Integrator*

When a capacitor is connected as the feedback device in an operational amplifier, the resulting integrating circuit can be used to measure the area under a current or voltage signal, to produce a linear sweep voltage (ramp), and to provide feedback control signals for controlling some other device. In this experiment the integrator is constructed and used as a linear sweep generator and for the integration of various waveforms.

Experiment 2–10 *Determination of Capacitance Using a Voltage Measurement Method*

If a known constant current i is used to charge a capacitor of capacitance C for a known time t, then the charge $q = it$, and the voltage v_C across the capacitor is $v_C = q/C$. Therefore, the capacitance can be determined by measuring the voltage after establishing a known charge on the capacitor. An accurate charge is placed on the capacitor by connecting it as the feedback element in an OA circuit. Since the output voltage v_o of the OA is equal to the voltage v_C across the capacitor, it is simple to measure v_C without disturbing the charge on the capacitor. The precision of the method is determined for various capacitors and the limitations of the method are noted.

Charge-Coupled Amplifier

It is clear from Eq. (2–42) that the integrator circuit is very useful for the measurement of charge, or more precisely, for the conversion of data from the charge domain to the voltage domain. Thus it is particularly useful with transducers having an output in, or easily converted to, the charge domain. Several charge-output transducers were described in Section 1–2.5 of Module 1. Since the charge on a capacitor is a function of the voltage across the capacitor and the capacitance value, both voltage and capacitance variations are easily converted into charge variations.

A common charge generating and measuring circuit is shown in Fig. 2–40. Since one terminal of C_{in} is at voltage v_{in} and the other terminal is held at the common potential by the operational amplifier, the charge on the capacitor is $q = C_{in}v_{in}$. If v_{in} varies by an amount Δv_{in}, the charge on C_{in} will vary by

$$\Delta q = C_{in}\,\Delta v_{in}. \qquad (2\text{–}46)$$

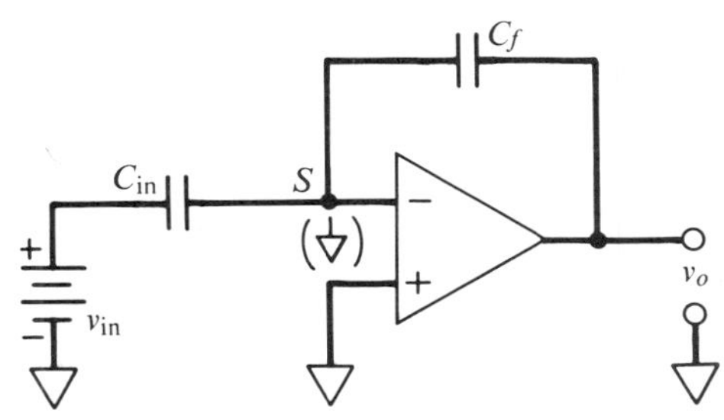

Fig. 2–40 Charge-coupled amplifier.

This same change in charge will appear across C_f, so that

$$\Delta v_o = -\Delta q/C_f = -\Delta v_{in}C_{in}/C_f. \qquad (2\text{–}47)$$

Thus the input voltage variations appear at the output multiplied by the factor C_{in}/C_f. This circuit is often used with biological transducers which produce voltage variations with a limited charge supply capability. It is necessary to short out C_f occasionally or bypass it with a large resistor to keep the output voltage from drifting off as a result of the integration of leakage currents.

The output voltage of the circuit of Fig. 2–40 will also vary with changes in the value of either C_{in} or C_f. Equation (2–46) can be rewritten to show the change in q due to a change in C_{in}:

$$\Delta q = v_{in}\, \Delta C_{in}. \tag{2–48}$$

In this case the output voltage will be

$$\Delta v_o = -\Delta q / C_f = -V_{in}\, \Delta C_{in}/C_f, \tag{2–49}$$

proving the circuit to be a linear capacitance-to-voltage converter.

Differentiator

The generation of an output voltage v_o that is proportional to the time rate of change (derivative) of the input voltage, dv_{in}/dt, is often an important operation in electronic measurement and control circuits. The derivative signal is readily obtained with an OA differentiator, as shown in Fig. 2–41. The input voltage is applied to the capacitor C, which is connected to the summing point of the OA. The charging or discharging current for C in the OA input circuit is $i_{in} = dq/dt$. Since $q = Cv$, the input current $i_{in} = C\,(dv_{in}/dt)$. The feedback current i_f through R is equal to i_{in} (that is, $i_f = i_{in}$), and the voltage drop across the feedback element, $i_f R$, is equal to the output voltage v_o, but of opposite polarity. Therefore,

$$v_o = -i_f R = -RC\frac{dv_{in}}{dt}. \tag{2–50}$$

The second and higher derivative signals can be obtained by connecting one or more differentiator circuits in sequence with the circuit shown in Fig. 2–41.

Some applications of the differentiator circuit can be seen by referring to Figs. 2–42a, b, and c. In Fig. 2–42a the derivative output signal is pro-

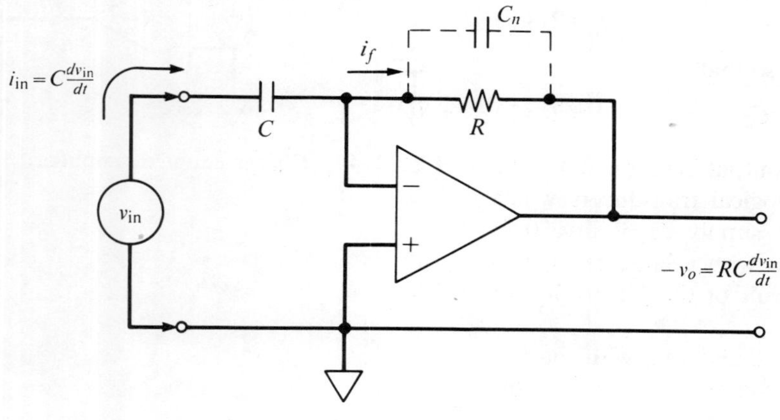

Fig. 2–41 Operational amplifier differentiator.

portional to the positive or negative slope of the input ramp voltage. The sudden transition from a positive to a negative voltage output when the slope reverses from positive to negative can be a very useful control signal. In Fig. 2–42b an information peak is superimposed on a variable background signal. If the second derivative of this signal is taken, there is negligible change in output voltage except when the information signal is present. The second derivative signal is also very useful in measurement and control circuits, and is used in several spectrometric applications as well as in analog computers.

The signals in Fig. 2–42c show how the first and second derivative signals of a sigmoid-type curve are independent of the dc level of the signal. There are several applications where it is necessary to locate the inflection point (maximum rate of change) of the sigmoid curve, and in this case either the first or the second derivative curve can be a valuable control signal. The second derivative output voltage is especially useful as a control signal because the output is zero until just before the inflection point, then rises to a peak in one direction and suddenly reverses and rapidly crosses the zero axis at the inflection point. Therefore, the second derivative signal can easily be used to control a switching circuit that will turn some device ON or OFF exactly at the time of the inflection point of the input signal.

Unfortunately, noise pulses that have a larger dv/dt than the input signal waveform will provide output noise that greatly degrades the useful output information. To minimize the problems of noise, the upper frequency response of the differentiator is generally limited by putting a small capacitor C_n across the feedback resistor, as shown in Fig. 2–41. However, there are many applications where the input signal is too noisy to differentiate in this way without greatly degrading the differentiator response, and it becomes necessary to use a much quieter integration technique to obtain the rate information. A circuit which allows rate information to be acquired in the presence of large amounts of noise is described next under integrating-type differentiators.

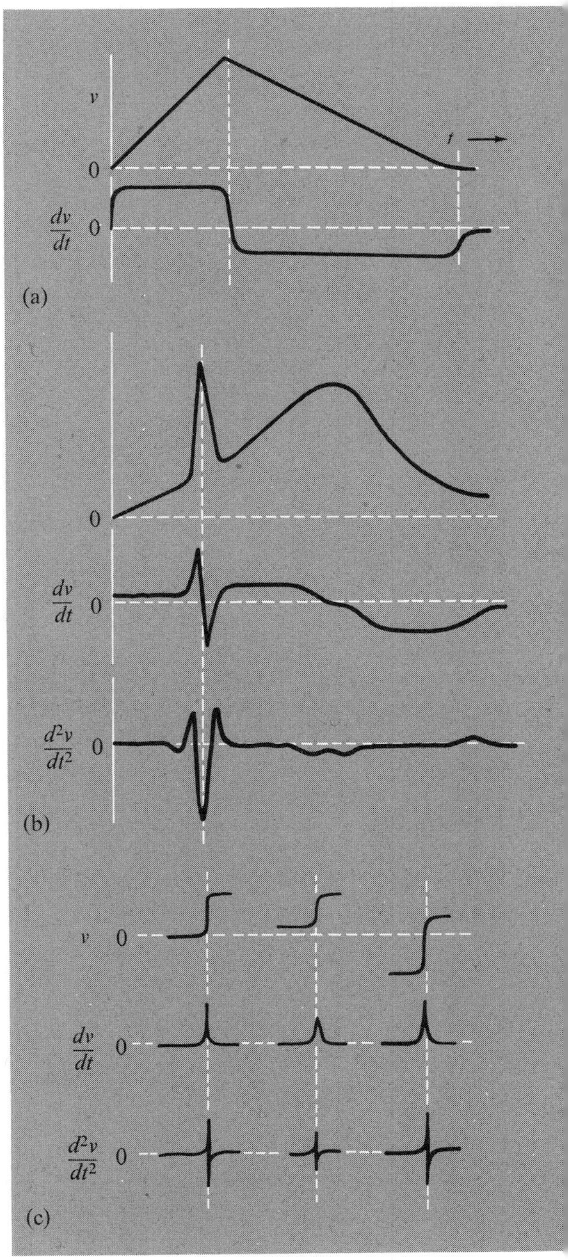

Fig. 2–42 Application of the differentiator. (a) First derivative output signal of a sawtooth input signal. (b) First and second derivative signals for a signal peak superimposed on a variable background. (c) Derivative signals for sigmoid input signals with various dc levels.

Experiment 2–11 *An Operational Amplifier Differentiator*

Information about the rate of change of a quantity with respect to time is frequently required for measurement or control purposes. In this experiment the rate of change of a voltage is measured by using an OA differentiator which produces an output voltage v_o that is directly proportional to the input rate of change, dv_{in}/dt. The differentiator circuit is connected, and also the test circuit is easily assembled by connecting an OA integrator which provides a known ramp voltage output as the test signal. The problems of using the conventional differentiator with noisy signals are noted, and corrective measures are tested to improve noise immunity.

Integrating-Type Differentiators

One technique for obtaining rate data utilizes a *comparison method.* An operational amplifier integrator is used to generate a variable reference curve whose output is maintained equal to the unknown signal. When the unknown and reference curves are equal and remain equal as determined by a null detector, the input voltage to the integrator must be equal to the rate of change of the unknown signal. That is, since the output v_o of the integrator is proportional to the integral of the input voltage v_{in}, then dv_o/dt is proportional to v_{in}.

A very good technique for generating reliable rate-of-change signals is based on the integration of two successive segments of the input signal and the subtraction of the resultant areas. At first, this might seem like a cumbersome method but it is relatively easy to implement and does exceptionally well in discriminating against noise present with the input signal. The principle of the method is illustrated in Fig. 2–43a, where the integration is made over two equal time increments Δt which are sequential, and the difference Δa between the resultant areas a_1 and a_2 is the parallelogram $ABCD$. The area of the parallelogram is given by

$$\Delta a = (\Delta t)l. \tag{2–51}$$

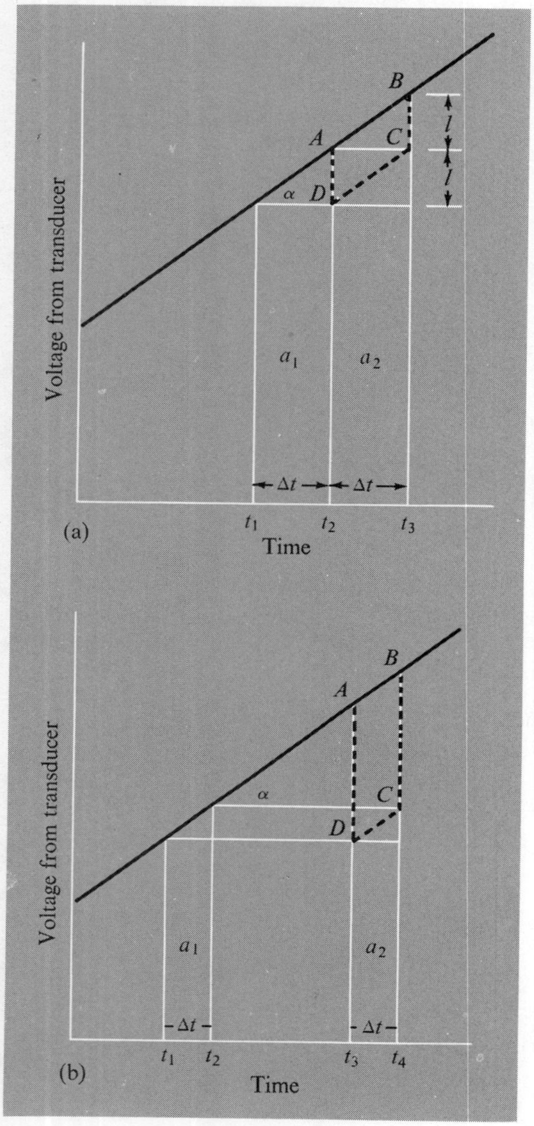

Fig. 2–43 Rate-of-change measurements using an integration method. (a) Typical rate curve showing measurement and subtraction of two sequential areas A_1 and A_2. (b) Measurement and subtraction of two areas a_1 and a_2 separated by time interval $t_3 - t_2$.

The slope S is defined as

$$S = \tan \alpha = \frac{l}{\Delta t}. \tag{2–52}$$

Substitution of Eq. (2–51) into Eq. (2–52) provides an expression relating the slope to the difference in area:

$$S = \frac{\Delta a}{(\Delta t)^2}. \tag{2–53}$$

Since Δt is a constant, the slope is directly proportional to the measured difference Δa. In Fig. 2–43b the two time increments are not consecutive but are separated by a time interval $t_3 - t_2$. In this case Eq. (2–53) can be rewritten as

$$S = \frac{\Delta a}{(t_3 - t_2)\Delta t + (\Delta t)^2}. \tag{2–54}$$

Since $(t_3 - t_2)$ is also kept constant, the slope remains directly proportional to the measured area. The total time for the measurement is $2\Delta t + (t_3 - t_2)$, and can vary from microseconds to seconds or more, depending on the application, specific circuits used, and type of noise problems.

Implementation with an Analog Integration System. The integration and subtraction operations illustrated in Fig. 2–43b can be implemented with the relatively simple analog circuit shown in Fig. 2–44a. This circuit uses two operational amplifiers, which are connected to one another and to a signal modifier by a switch network. During the first integration period, the input signal is directed through switches 1 and 2 to the integrator (OA2). If the signal modifier output is as illustrated in Fig. 2–44b, then line BC represents the signal applied to the integrator, which charges capacitor C, causing the output of OA2 to rise to a voltage v_1 as shown in Fig. 2–44c. During the second integration period the signal from the modifier is inverted by the gain-of-one inverter (OA1) and connected to the integrator by switch 3. The voltage represented by line DE in Fig. 2–44b is ap-

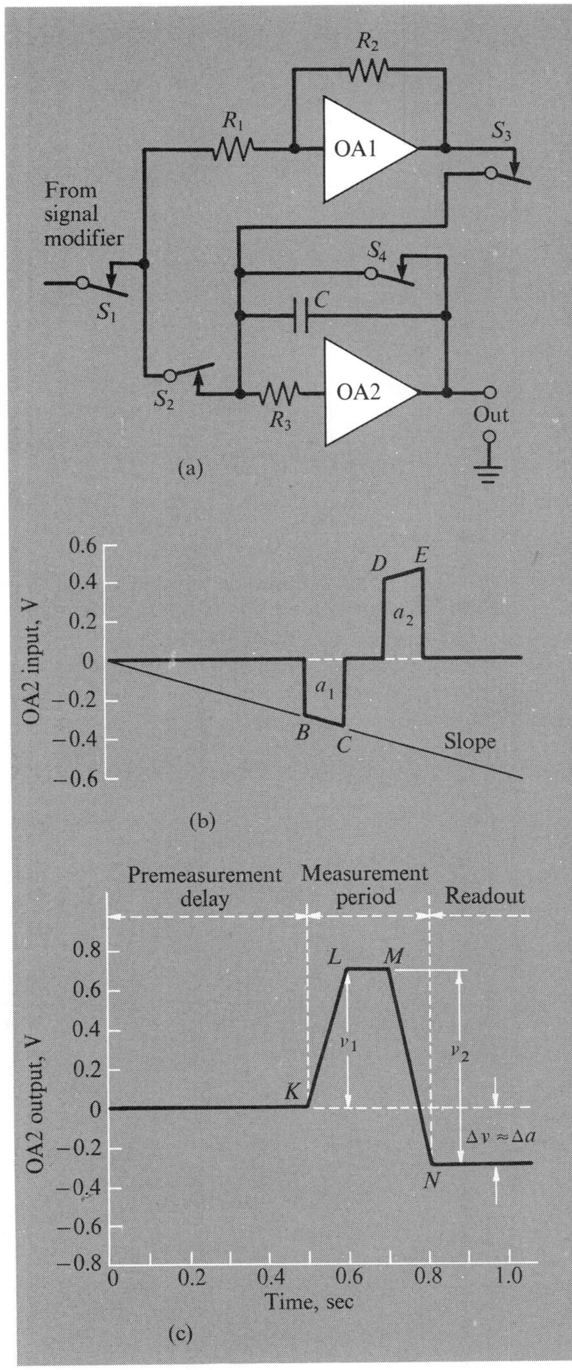

Fig. 2–44 Implementation of integrating rate meter. (a) Integration and subtraction circuit. (b) Rate signal and OA2 input. (c) Output signal of OA2.

plied to the integrator input, which discharges capacitor C, and the integrator output decreases in magnitude by the amount v_2 shown in Fig. 2–44c. The voltage difference Δv, which is read out at the end of the measurement period, is proportional to the difference in area, Δa, and is therefore proportional to the initial slope according to Eq. (2–54).

The switches S_1–S_4 can be activated with pulses generated by a logic circuit. Thus, upon receipt of a suitable trigger pulse, the logic circuit controls the start of the sequence of delay time, the length of the integration periods, the subtraction, and the readout of the result. The integration period is selected to be as long as feasible so as to average the noise most effectively. The shortest practical integration period is governed by the switching and OA response times and at present is about 0.01 to 0.1 msec, as will be discussed under high speed analog switches in Section 2–3.

Logarithmic Converters Using a Capacitor Discharge Technique

There are many measurement and control applications where it is important that the output signal be proportional to the logarithm of an input signal or proportional to the log ratio of two input signals. In spectrometric applications the logarithm of the intensity of a reference light beam P_r relative to an unknown sample light beam P_s [that is, $\log(P_r/P_s)$] is directly proportional to the desired information (concentration, photographic exposure, etc.).

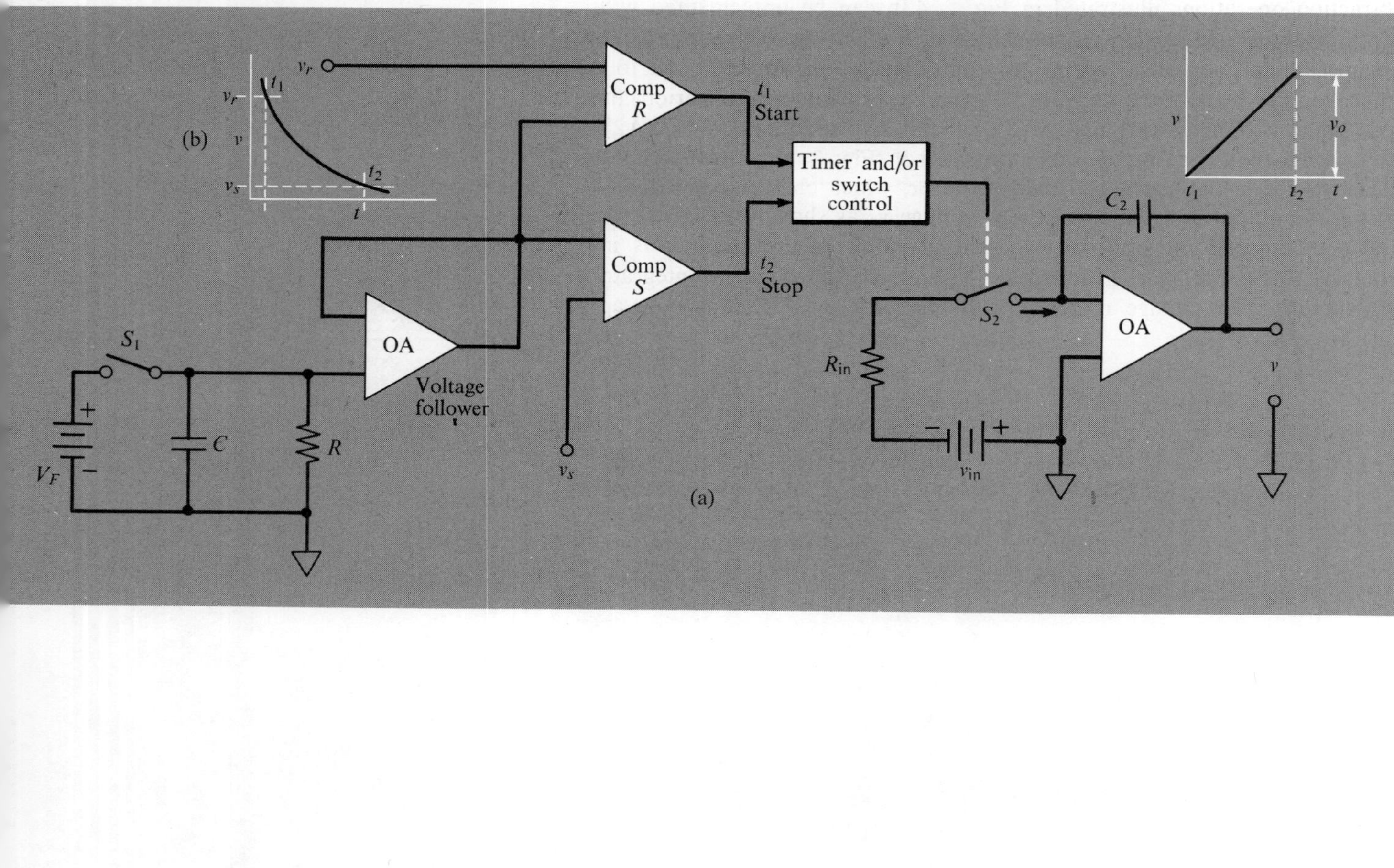

Fig. 2–45 Logarithmic converter using capacitor discharge method. (a) Block diagram of conversion system. (b) Discharge curve showing times at which inputs from log generator are equal to reference and sample voltages.

One method that has been used to generate an output signal that is proportional to the log ratio of two input signals is illustrated in Fig. 2–45. It encompasses several of the principles previously described in this section. The technique is as follows. A capacitor C in the log generator circuit is charged, by closing switch S_1 in Fig. 2–45, to a fixed stable voltage V_F that is always greater than the input signal voltages that contain the measurement information. The output of the log generator is fed to one input of both the reference comparator R and the sample comparator S. The input reference voltage, v_r, is fed to the other input of comparator R, and the input sample voltage, v_s, is fed to the other input of comparator S.

When switch S_1 is opened, the capacitor C discharges through R to provide a logarithmic discharge curve. The voltage follower in the log generator unloads the RC network and provides a low output resistance. When the voltage of the log generator decreases to a value equal to v_r at time t_1 (as shown in Fig. 2–45b), the comparator R rapidly changes its output level, starting a readout timer and/or closing switch S_2 at the input of a readout integrator. The capacitor in the log generator continues to discharge until the output is equal to v_s at time t_2. The rapid change in the output level of comparator S stops the timer and/or opens switch S_2. The readout can be from the timer, the integrator output, or both. If the integrator readout is used, switch S_2 closes at time t_1, and the capacitor C_2 is charged by the integrator input current $i_{\text{in}} = v_{\text{in}}/R_{\text{in}}$. At time t_2, switch S_2 opens and the output voltage of the integrator remains constant, so that

$$v_o = \frac{q}{C_2} = \frac{i_{\text{in}}(t_2 - t_1)}{C_2} = k_1(t_2 - t_1). \tag{2–55}$$

We can readily see how this output voltage v_o or the time interval $t_2 - t_1$ is proportional to log (v_r/v_s) by referring to Fig. 2–45b and considering the following basic equations. The voltage v_C across the capacitor C at any time t is

$$v_C = V_F\, e^{-t/RC}. \tag{2–56}$$

At time t_1 the voltage $v_C = v_r$, so that

$$v_r = V_F\, e^{-t_1/RC}. \tag{2–57}$$

At time t_2, the voltage $v_C = v_s$, so that

$$v_s = V_F\, e^{-t_2/RC}. \tag{2–58}$$

Therefore,

$$t_1 = RC \ln\, (V_F/v_r) \tag{2–59}$$

and

$$t_2 = RC \ln (V_F/v_s). \tag{2–60}$$

By subtracting Eq. (2–59) from Eq. (2–60) we obtain

$$t_2 - t_1 = RC \ln (v_r/v_s) = k_2 \log (v_r/v_s), \tag{2–61}$$

and substituting Eq. (2–61) into Eq. (2–55) gives the voltage output

$$v_o = k \log (v_r/v_s). \tag{2–62}$$

More recent logarithmic converters utilize the exponential current-voltage transfer functions of certain devices in conjunction with operational amplifiers to provide direct log readout. These converters are described in Module 3, *Digital and Analog Data Conversions.*

Experiment 2–12 *A Logarithmic Converter Using a Capacitor Discharge Method*

It is frequently necessary for a readout device to display linearly the logarithm or log ratio of voltage or current signals. In this experiment the log ratio of two voltage signals is determined by using a capacitor discharge method. The time difference involved in discharging a capacitor from a fixed reference voltage V_F to the two sample signals is measured and shown to be proportional to the log ratio. The precision and limitations of the method are determined.

2–2.4 ELECTRICAL INDUCTANCE

Inductance is that property of a device which resists a change in current through the device. All devices have electrical inductance associated with them to some extent and tend to impede or react against changes in current. Because of this impedance to changes in current, inductance is often an undesirable property in circuits, and it is important to understand the basic principles of inductance in order to minimize its effects in circuits where inductance is unwanted. On the other hand, some devices called **inductors** are purposely introduced in circuits to resist changes in current and thus to serve important control functions. The physical principles underlying the property of inductance are considered in this section, and several basic relationships are introduced. Practical inductors and their characteristics are described, and the effects of inductance in circuits in which a step change in voltage is applied are discussed. The important effects of inductance in alternating current circuits are considered in detail in Module 3.

Principles of Inductance

The basic principle of electromagnetic induction was introduced in conjunction with the generation of voltage in Section 1–3 of Module 1, *Electronic Analog Measurements and Transducers.* Faraday's law of electromagnetic induction is one of the basic laws of electromagnetism and will be reviewed briefly here in order to introduce the concept of **self-inductance** in devices.

In Fig. 2–46a, a magnet is shown moving toward a loop of wire which is connected to a current meter. Ordinarily no current would be expected because the electrical circuit (loop and current meter) contains no voltage source. However, when the magnet is moving, the meter deflects and a current exists in the circuit. When the magnet is at rest with respect to the coil, no current exists in the coil, and the meter shows no deflection. If the poles of the magnet are reversed so that the south pole is moved toward the coil, the meter deflects in the opposite direction. The current that appears in the circuit of Fig. 2–46a is called an **induced current** and is set up by an **induced electromotive force** (emf). In practice it makes no difference whether the magnet is moved toward the coil or vice-versa. Only the relative motion of the magnet and the coil is important.

The presence of an induced emf in a coil of wire can be demonstrated without the presence of a permanent magnet, as in Fig. 2–46b. Here two coils are placed near each other but at rest with respect to each other. If switch S is closed in the right-hand circuit containing the coil, the meter in the left-hand circuit momentarily deflects. When the current in the right-hand coil is steady, no current is measured by the meter. Again when the switch at the right is opened, an opposite, momentary deflection of the meter is observed. These experiments show that the emf in the circuit at the left is induced whenever the current in the other circuit changes. The magnitude of the induced emf depends on the rate of change of current and not on the magnitude of the steady current.

The important parameter that is common to the coils shown in Figs. 2–46a and b is the magnetic flux ϕ of the coil in which the induced emf appears. It makes no difference whether the flux is established by a permanent magnet as in Fig. 2–46a, or by a current loop as in Fig. 2–46b. The magnitude of the induced emf is given by Faraday's law of induction, as shown in Eq. (2–63):

$$v = -\frac{d\phi}{dt}. \tag{2–63}$$

Here the rate of change of flux is the parameter which determines the induced emf. The negative sign indicates the sign of the induced emf. If the rate of flux change is expressed in webers/sec, the emf v is in volts. For a

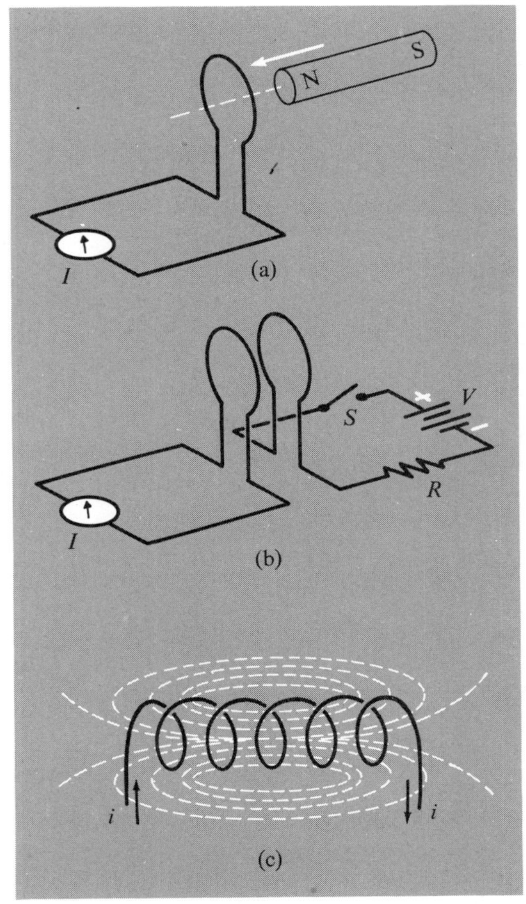

Fig. 2–46 Electromagnetic induction: (a) by a moving magnet; (b) by changing current in a nearby wire; (c) by nearby turns of the same coil.

coil of N turns, the induced emf is proportional to the number of turns, as shown in Eq. (2–64):

$$v = -\frac{N\,d\phi}{dt} = -\frac{d(N\phi)}{dt}. \tag{2–64}$$

Here the quantity $N\phi$ is the number of **flux linkages** of the coil.

The presence of two separate coils is not necessary for the inductive effect to appear. If the current through a single coil is changed, an induced emf appears in that same coil. This effect is called **self-inductance,** and the self-induced emf follows Faraday's law (Eqs. 2–63 and 2–64). A practical inductor may be simply a multiple-turn coil as shown in Fig. 2–46c. When current passes through the coil, the magnetic flux established in each turn is approximately the same, and the induced emf is given by Eq. (2–64), which shows the importance of the number of flux linkages, $N\phi$, in determining the induced emf.

In the absence of magnetic materials, such as iron, the number of flux linkages in a given coil is directly proportional to the current, as shown in Eq. (2–65):

$$N\phi = Li, \tag{2–65}$$

where the proportionality constant L is called the **inductance** of the coil. The relationship between induced emf and the current in the coil can readily be found from substitution for $N\phi$ in Faraday's law:

$$v = -\frac{N\,d\phi}{dt} = -L\frac{di}{dt}. \tag{2–66}$$

The self-induced emf is often called a **counter emf** because it is of opposite polarity to those changes in voltage applied to the coil which cause the changing current di/dt, as the minus sign in Eq. (2–66) indicates. As Eq. (2–66) also indicates, only a changing current causes such a self-induced emf. The counter emf always acts to oppose the current change.

Inductance L has the unit volt-second/ampere, which is given a special name, the **henry** (H). An inductance of 1 henry will induce a counter emf of 1 V when the current is changing at the rate of 1 A/sec. The units millihenry (mH) and microhenry (μH) are also used for many practical inductors.

The inductor reacts against a change in current through itself by storing electrical energy in the magnetic field. Work is required to increase the magnetic field; therefore, *the current in an inductive circuit cannot change instantaneously*. Equation (2–66) shows that an attempt to change the current abruptly will be opposed by an infinite emf induced in the coil. If the applied potential in an inductive circuit decreases, the energy stored in the magnetic field returns to the circuit in the form of an emf which tends to maintain the previous current level. In a perfect inductor, there is no opposition to a steady current (dc) because $di/dt = 0$.

Practical inductors and their characteristics are described below, and the effects of inductive circuits on various signals are discussed in Section 2–2.5 and in Module 3.

Practical Inductors

Inductors are never pure inductances because there is always some resistance in the coil windings and some capacitance between the coil windings. When choosing an inductor (occasionally called a choke) for a specific application, it is necessary to consider the value of the inductance, the dc resistance of the coil, the current-carrying capacity of the coil windings, the breakdown voltage between the coil and the frame, and the frequency range in which the coil is designed to operate.

Inductors are available with values ranging from several hundred henrys down to a few microhenrys. To obtain a very high inductance it is necessary to have a coil of many turns. The inductance can be further increased by winding the coil on a closed-loop iron core or ferrite[4] core. To obtain as pure an inductance as possible, it is desirable to reduce the dc resistance of the windings to a minimum. This can be done by increasing the wire size, which, of course, will increase the size of the choke. The size of the wire also determines the current-handling capacity of the choke. As was pointed out earlier, the work done in forcing a current through a resistance is converted to heat in the resistance. Magnetic losses in an iron core also account for some heating, and this heating restricts any choke to a certain safe operating current. The windings of the coil must be insulated from the frame as well as from each other. Heavier insulation, which necessarily makes the choke more bulky, is used in applications where a high voltage will be present between the frame and the winding. The losses sustained in the iron core increase as the frequency increases. They become so large at about 15 kHz that the iron core must be abandoned. This results in coils of reduced coupling efficiency, but fortunately very large inductance values are not used at high frequencies. The iron-core chokes are restricted to low-frequency applications, but ferrite-core chokes are good at high frequencies. Several practical inductors are illustrated in Fig. 2–47.

The use of precision inductors is rather uncommon. However, they are available and restrictions with respect to frequency, temperature, etc., must be taken into account.

Variable Inductors. In a variable inductor the magnetic coupling between the windings of the coil or the effective number of turns is varied in order to change the total inductance value. The magnetic coupling can be changed

Note 4. Ferrite.

Ferrite is a powdered, compressed, and sintered magnetic material consisting mainly of ferric oxide combined with one or more other metal oxides. It has high resistivity, which makes the eddy-current losses very low at high frequencies. It is used in magnetic core memories because it can be magnetized and demagnetized very rapidly.

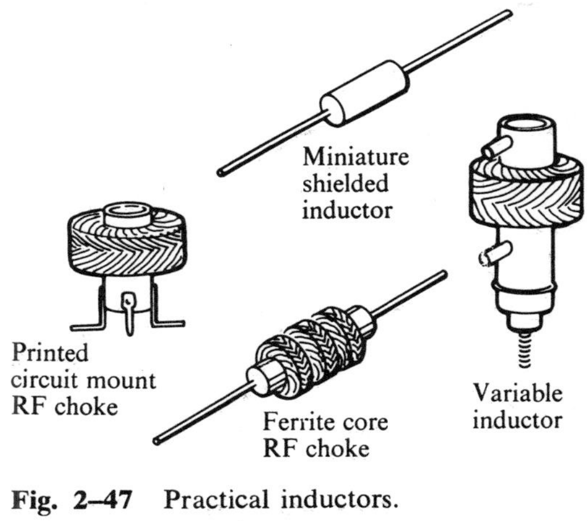

Fig. 2–47 Practical inductors.

by varying the orientation of part of the coil winding, or the effective number of turns can be varied by positioning a silver-plated brass core.

Another type of variable choke is called the swinging choke. The inductance of this choke increases with decreasing current. This type of choke is used to advantage in certain power supply circuits.

Large inductors, rated in henries (H), are used principally in power applications. The frequency in these circuits is relatively low, generally 60 Hz or low multiples thereof. In high frequency circuits, such as those found in FM radios and television sets, very small inductors are frequently used. These frequencies are generally in the 10 to 1000 MHz region and the chokes are of the order of microhenries (μH). In radio-frequency applications (broadcast band), chokes are frequently rated in millihenries (mH).

Inductive Transducers

Various sources of chemical, thermal, mechanical-displacement, or velocity input information can cause changes in inductive devices which can provide readily measured electrical signals. Inductive transducers are based on the relationship from Eq. (2–66),

$$L = N(d\phi/di), \qquad (2\text{–}67)$$

where L is the inductance in henries, N is the number of turns cut by the flux linkages, and $(d\phi/di)$ is the rate of change of flux with current.

The rate of change of flux with current can be changed by varying the length of flux path, the cross-sectional area, or the permeability. A physical movement of the core, a change in the distance between two or more coils, a change in the chemical composition of a portion of the flux path, or a stress on ferromagnetic material in the path, all can cause changes of inductance that can be measured.

2–2.5 *LR* Circuits

It was shown in the introduction to this section that an inductor reacts against—tends to impede—a change in current, and that it is impossible to change the current value instantaneously in a circuit that has inductance. In this respect inductive circuits are analogous to capacitive circuits, in which the voltage cannot be changed instantly. In each case, there will be a time lag between the attempt to change the voltage or current level in the circuit and the attainment of that change. The time of response of *LR* circuits can be analyzed in a way very similar to the method used for *RC* circuit response in Section 2–2.2. This is illustrated for the step signal response of a series *RL* combination below.

Series *LR* Response to Step Signals

A series *LR* circuit with a test step potential source is shown in Fig. 2–48a. When the switch is closed, $v_R = iR$ and $v_L = L\, di/dt$. From Kirchhoff's Law,

$$V = iR + L\frac{di}{dt}. \tag{2–68}$$

Solving this differential equation yields the following:

$$v_L = Ve^{-tR/L}, \tag{2–69}$$

$$v_R = V(1 - e^{-tR/L}). \tag{2–70}$$

The quantity L/R in these equations is the time constant for *LR* circuits. The curves for v_L and v_R as a function of L/R units are shown in Fig. 2–48b. The voltage v_L approaches zero and v_R approaches the steady-state value V. The curves of Fig. 2–48 verify that the *LR* circuit is an approximate differentiator when the output is taken across L and an integrator when the output signal is v_R.

The storage of energy in the magnetic field in an inductor can readily be seen from Eq. (2–68) for the *LR* circuit. If both sides of the equation are multiplied by i, the rate of doing work, Vi, is obtained:

$$Vi = i^2 R + Li\frac{di}{dt}. \tag{2–71}$$

The rate at which energy is delivered to the circuit, Vi, is seen to be equal to the sum of two terms. The i^2R term is the power dissipated in the resistor or the rate at which energy appears as heat in the resistor. Energy that does not appear as heat is stored in the magnetic field, and the rate of energy storage is given by the last term in Eq. (2–71). If the energy in the field is called U_B, the rate at which energy is stored may be written

$$\frac{dU_B}{dt} = Li\frac{di}{dt}. \tag{2–72}$$

If Eq. (2–72) is integrated, the total stored magnetic energy in an inductor of inductance L, which carries a current i, is given by

$$U_B = \tfrac{1}{2}Li^2. \tag{2–73}$$

This equation can be directly compared to the energy U_E stored in a capacitor which has a charge q, as given earlier in Eq. (2–28) as the work W_d available on discharge:

$$U_E = W_d = \frac{1}{2}\frac{q^2}{C}.$$

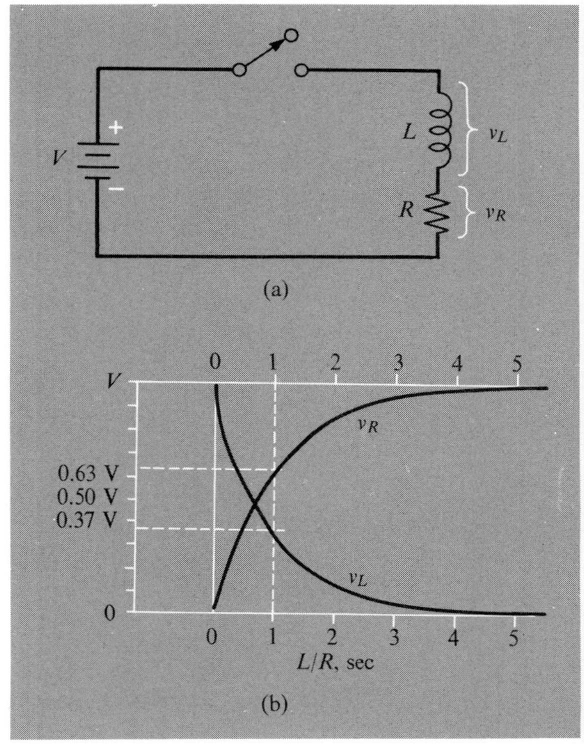

Fig. 2–48 *LR* circuits: (a) with step potential source; (b) voltage-time curves after switch closure.

With a capacitor the energy is stored in the electric field, whereas with an inductor the energy is stored in the magnetic field.

Experiment 2–13 *Response Characteristics of an LR Network*

Several different input waveform signals are applied to a series *LR* circuit and the resulting output waveforms are observed. These output signals are then compared with those from an *RC* network.

PROBLEMS

1. A 0.001 μF capacitor was charged to a voltage of 8.32 V. Calculate the work done in completely discharging the capacitor.
2. In the capacitive network of Fig. 2–31, $V = 5$ V, $C_1 = 0.1$ μF, $C_2 = 0.05$ μF, and $C_3 = 0.02$ μF. Assume that capacitor C_4 has been disconnected. Calculate the charge on each capacitor, the total charge q_T, and the equivalent parallel capacitance C_p.
3. The same capacitors and voltage source as in problem 2 are rewired in a series circuit like that of Fig. 2–32. Calculate the charge on each capacitor and the equivalent series capacitance, C_s.
4. The complex capacitive network shown in Fig. P-4 is wired. Calculate the equivalent capacitance of the network, the voltage across each capacitor, and the charge on each capacitor.

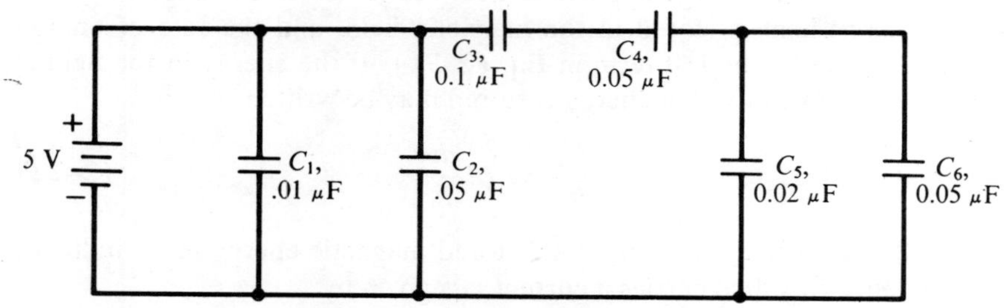

Fig. P–4

5. In the circuit of Fig. 2–34, $V = 1.0$ V, $R = 100$ kΩ, and $C = 0.1$ μF. Calculate the voltage across the capacitor, v_c, and the voltage across the resistor, v_R, at 0.5 msec, 1 msec, 5 msec, 10 msec, 50 msec, and 100 msec after switch S has been turned to position A.
6. In the circuit of Fig. 2–37, $V = 10.0$ V, $R_1 = 10$ kΩ, $R_2 = 40$ kΩ, and $C = 0.22$ μF. (a) How long after switch S is turned to the ON position does it take for capacitor C to become fully charged for practical pur-

poses? (b) If $R_3 = 10\ \Omega$, how long, for practical purposes, does it take for C to discharge?

7. (a) A 10 V rectangular pulse with a pulse width of 1.0 msec is applied to a series RC circuit at time t_0. If $R = 100\ \text{k}\Omega$ and $C = 1000$ pF, calculate the voltage across R and C, v_R and v_C respectively, at t_0, $t_0 + 0.1$ msec, $t_0 + 0.3$ msec, $t_0 + 0.5$ msec, $t_0 + 1$ msec, $t_0 + 1.3$ msec, and $t_0 + 1.5$ msec. (b) Repeat part (a) for $R = 1\ \text{M}\Omega$ and for times t_0, $t_0 + 0.5$ msec, $t_0 + 1.0$ msec, and $t_0 + 1.5$ msec.

8. A constant voltage of 5.0 V is applied to the OA integrator of Fig. 2–39. If $R = 5\ \text{M}\Omega$ and $C = 0.5\ \mu\text{F}$, how long will it take for the output voltage to decrease by 1.0 V?

9. Design an OA current integrator with components selected so that a 5.0 V output is equivalent to 1.000 microequivalents of charge (1 μeq = 96,487 μC).

10. The OA of Fig. 2–39 has an output voltage limit of ±12 V. The input resistor $R = 100\ \text{k}\Omega$, $C = 0.1\ \mu\text{F}$, and the inverting (−) input has an offset voltage of +50 μV with respect to the circuit common. The capacitor is initially shorted with a switch of 1 Ω resistance. (a) What is the initial output voltage when the switch is closed? (b) How long will it take the OA to reach its voltage limit after the switch is opened if $v_{in} = 0.00$ V? (c) What average input voltage is required to give an output integral accurate to 1%? (d) How long can the integration of the input voltage found in part (c) proceed before the OA reaches its output voltage limit? (e) What is the maximum tolerable offset at the inverting input for a 1000 sec integration accurate to 1% if the output voltage at 1000 sec is to be 10 V?

11. (a) Design an OA differentiator which will give an output of −5.0 V for an input rate of charge of +8 V/sec. (b) Design a differentiator to give an output of +1.0 V for an input rate of charge of +15 V/sec.

12. In the logarithmic converter of Fig. 2–45, $V_F = 1.0$ V, $R = 1\ \text{M}\Omega$, and $C = 0.5\ \mu\text{F}$. The readout integrator has $v_{in} = 500$ mV, $R_{in} = 100\ \text{k}\Omega$, and $C_2 = 1.0\ \mu\text{F}$. What will the readout voltages of the integrator be for the following values of input signals v_r and v_s? (a) $v_r = 0.5$ V, $v_s = 0.05$ V. (b) $v_r = 0.5$ V, $v_s = 0.1$ V. (c) $v_r = 0.5$ V, $v_s = 0.25$ V.

13. An unknown inductor produces an induced voltage of 2 V when the current through it is changing at 15 A/sec. What is the inductance?

14. What is the induced voltage produced by a 300 mH choke if the current through it is changed at a rate of 5 A/sec?

15. In the series LR circuit of Fig. 2–48a, $L = 200$ mH, $R = 100\ \Omega$, and $V = 1.5$ V. Calculate the voltage across the inductor, v_L, and that across the resistor, v_R, at 0.5 msec, 1 msec, 2 msec, 10 msec, and 100 msec after the switch has been closed.

Experiment 2–7 *The Capacitor as a Charge Storage Device*

Equipment: VRS
ADD
FET OA card
DVM or VOM
10 kΩ resistor
0.22 μF, 0.50 μF, and 1.0 μF (non-electrolytic) capacitors
multiple connector card

Most measurement and control circuits are dependent in some way on the ability of a capacitor to store charge. Therefore, it is important to obtain laboratory experience on how capacitors perform when charged and combined in various ways. In this experiment, the leakage characteristics of various types of capacitors are observed, and the distribution of charge in capacitor networks is measured. The voltage values obtained from the theoretical relationships are compared with the measured values in capacitor networks.

E2–7.1 Charge Storage by the Capacitor

In this experiment a dc voltage is applied to a resistor and to a capacitor. The capacitor is observed to store charge. A voltage follower is inserted to isolate the resistor or capacitor from the voltage measurement device.

Set up the circuit shown in Fig. E2–7. Use a multiple connector card on the ADD to connect the resistor or capacitor. Arrange a touch contact for the + lead of the VRS. When the OA card is plugged into the ADD the common connection to the OA that is shown in Fig. E2–7 occurs through the ADD power supply common. Therefore connect the VRS (−) and DVM (−) terminals to the ADD common.

First use the 10 kΩ resistor. *Measure the voltage with the VRS connected. Remove the touch contact (+ VRS lead). Observe the DVM (or VOM).*

Does the resistor hold charge?

Replace the resistor with a 1 μF capacitor. *Measure the voltage with the touch contact connected. Remove the touch contact.*

Does the capacitor hold charge?

Wait for a few minutes and observe the voltage as it slowly decays. The capacitor is discharging through the very high input resistance of the follower. Such *RC* circuits will be studied in later experiments.

E2–7.2 Capacitor Combinations

In this experiment various series and parallel combinations of capacitors will be studied. Relationships for the voltage across capacitors in series and in parallel will be derived and compared to measured values.

Parallel Charged Capacitors in Series. Connect the capacitor charging circuit shown in Fig. E2–8. Use the multiple connector card and patch wires to connect the capacitors together. Use the follower to isolate the capacitors from the DVM or VOM.

Measure the voltage across the VRS terminals and across each capacitor.

V_{VRS} ______ $v_{1\,\mu F}$ ______ $v_{0.5\,\mu F}$ ______

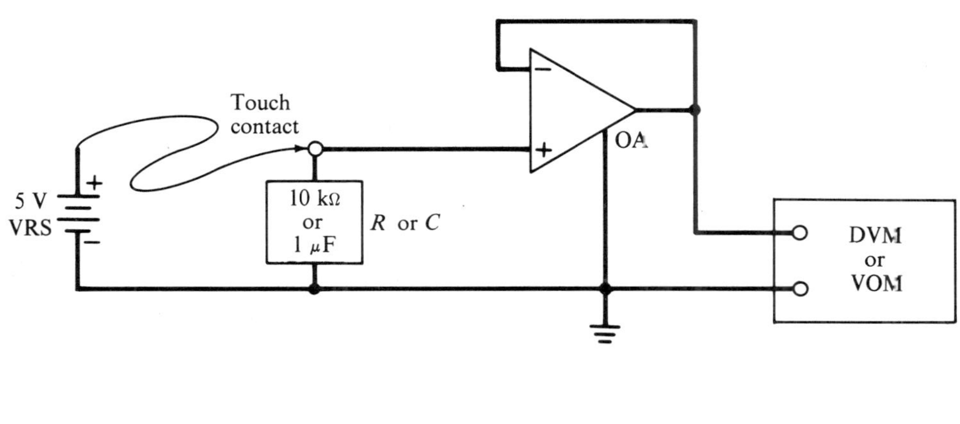

Fig. E2–7 Series resistor or capacitor circuit.

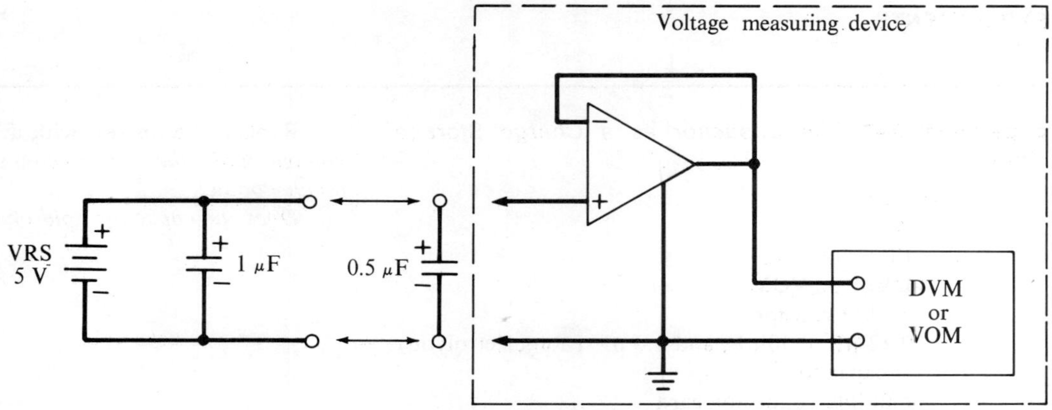

Fig. E2–8 Capacitor charging circuit.

Remove the voltage measurement circuit. With the capacitors charged, disconnect the VRS. Remove the wires connecting the two capacitors together. Connect the + terminal of the 1 μF capacitor to the − terminal of the 0.5 μF capacitor as shown in Fig. E2–9. Be careful not to discharge either capacitor when removing the patch wires.

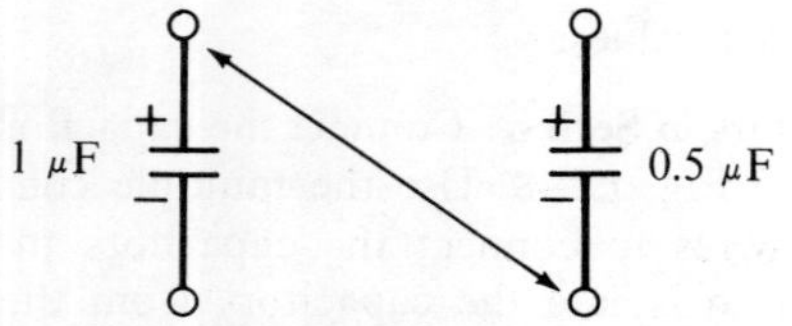

Fig. E2–9 Series capacitors.

Now measure the voltage across each individual capacitor with the follower-DVM (VOM). Then measure the total voltage from the − end of the 1 μF capacitor to the + end of the 0.5 μF capacitor (v_s). Discharge the capacitors by shorting each with a wire. Recharge the capacitors as in Fig. E2–8 and rewire them in series. *Obtain triplicate results.*

$v_{1\,\mu F}$ ______ $v_{0.5\,\mu F}$ ______ v_s ______

Explain your results below.

This circuit is known as a voltage doubler and is often used in power supplies when high voltages are desired.

Again charge the two capacitors by the circuit of Fig. E2–8. With the capacitors charged, removed the VRS. Carefully connect the +terminal of the 1 μF capacitor to the +terminal of the 0.5 μF capacitor. Measure the voltage across each capacitor with the follower-DVM (VOM). Then measure the total potential from the − end of the 1 μF capacitor to the − end of the 0.5 μF capacitor (v_s). *Obtain duplicate results.*

$v_{1\,\mu F}$ ______ $v_{0.5\,\mu F}$ ______ v_s ______

Explain the results below.

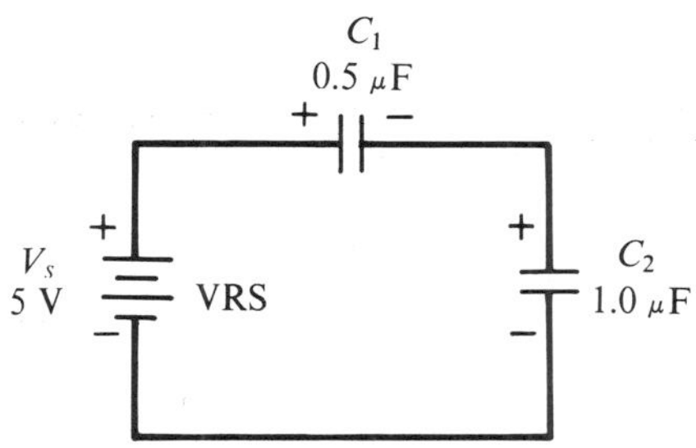

Fig. E2–10 Series charging circuit.

Series Charged Capacitors Placed in Parallel. Connect the series capacitor charging circuit illustrated in Fig. E2–10 on the multiple connector card. This circuit is a capacitive voltage divider, and the total source voltage, V, will divide across C_1 and C_2 according to the relationships

$$v_{C_1} = V\left(\frac{C_2}{C_1 + C_2}\right) \quad \text{and} \quad v_{C_2} = V\left(\frac{C_1}{C_1 + C_2}\right).$$

In the space below derive the voltage divider relationship for v_{C_1} and v_{C_2}.

Use the follower-DVM or follower-VOM (Fig. E2–8) to measure the source voltage and the voltage across each capacitor. *Compare below the experimental values of v_{C_1} and v_{C_2} with those calculated from the voltage divider relationships.* Discharge the capacitors and *repeat the above measurements using a* 0.22 μF *capacitor in place of the* 0.5 μF *capacitor.*

C_1	$(V)_{meas}$	$(v_{C_1})_{meas}$	$(v_{C_1})_{calc}$	$(v_{C_2})_{meas}$	$(v_{C_2})_{calc}$
0.5 μF					
0.22 μF					

Explain any deviations.

With the 0.22 μF capacitor in series with the 1.0 μF capacitor, charge the two capacitors. Remove the voltage measuring device from the circuit of Fig. E2–10. Remove the VRS leads from the charged capacitors. Without discharging the capacitors, remove C_1 and place it in parallel with C_2 so that the positively charged plate of C_1 is connected to the positively charged plate of C_2 and the two negatively charged plates are connected together. Measure the voltage across the parallel combination (v_p). Discharge the capacitors and repeat the procedure of charging them in series and then placing them in parallel until triplicate measurements have been taken. Replace the 0.22 μF capacitor with the 0.5 μF capacitor and repeat the charging and measurement of the voltage across the parallel combination until triplicate measurements have been made. *Report the results below. Average the triplicate measurements.*

C_1, μF	C_2, μF	v_p	$(v_p)_{avg}$
0.22	1.0		
0.22	1.0		
0.22	1.0		
0.5	1.0		
0.5	1.0		
0.5	1.0		

Knowing that the charge on each capacitor before the capacitors are connected in parallel is $q = v_C C$ and that capacitors in parallel form an equivalent capacitance which is the sum of the individual capacitance values, *derive a relationship for the expected values of v_p.*

Compare below the calculated and the average measured values of v_p.

C_1	$(v_p)_{meas}$	$(v_p)_{calc}$	Percent difference
0.22 μF			
0.50 μF			

Charged Capacitor in Parallel with Discharged Capacitor. Use the VRS set at 1 V to charge a 1 μF capacitor. Disconnect the VRS. Attach the follower-DVM(VOM), and measure the voltage $v_{1\,\mu F}$. Place a discharged 0.5 μF capacitor in parallel with the charged 1 μF capacitor and measure the voltage of the parallel combination, v_p. Remove the 0.5 μF capacitor and discharge it. Recharge the 1 μF capacitor, remove the VRS, and again place the uncharged 0.5 μF capacitor in parallel. *Repeat until triplicate measurements have been made.*

Trial	$v_{1\,\mu F}$	v_p	$(v_p)_{avg}$
1			
2			
3			

Explain the results on the basis of the equations for a charged capacitor.

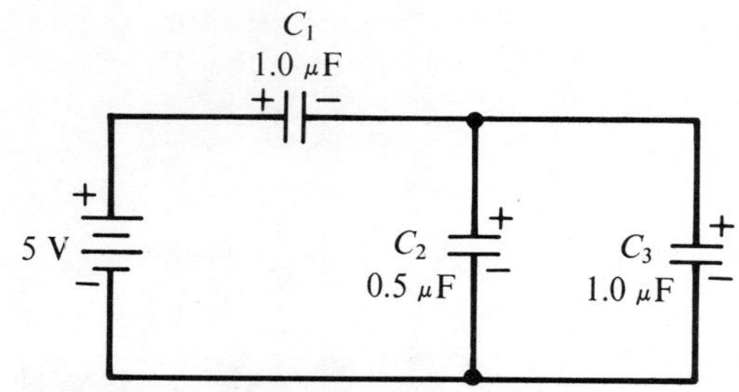

Fig. E2–11 Series-parallel capacitors.

Series-Parallel Combination. Connect the circuit shown in Fig. E2–11. Measure the source voltage V and the voltage across each capacitor. Compare the values obtained to expected values.

$(V)_{meas}$	$(v_{C_1})_{meas}$	$(v_{C_2})_{meas} = (v_{C_3})_{meas}$

Show your calculations here.

$(v_{C_1})_{calc}$	$(v_{C_2})_{calc}$

Remove the voltmeter and the source. Carefully place C_1 in parallel with the other two capacitors so that the + plate of C_1 is connected to the + plates of C_2 and C_3 and the − plates are connected together. *Calculate below the expected voltage across the parallel combination.*

Measure the voltage in triplicate. Be sure to discharge all capacitors and recharge them using the circuit of Fig. E2–11. *Compare the measured and expected values of* v_p.

Trial	$(v_p)_{meas}$	$(v_p)_{avg}$	$(v_p)_{calc}$	Percent difference
1				
2				
3				

Experiment 2–8 *Control of Current and Voltage with the RC Network*

Equipment: ADD
relay card with multiple connector card
VRS
recorder
10 MΩ resistor
two 0.1 μF and 1 μF capacitors
DRB
sine-square or function generator
oscilloscope

The charging and discharging of a capacitor through a resistor provides a current control function that is generally applicable in measurement and control systems. The *RC* network is used for controlling time delay circuits, for controlling the repetition rate of waveform generators, and for integration and differentiation of voltage signals. In this experiment the *RC* circuit is first tested with a step function input signal and then with a repetitive square wave. The types of output waveforms that can be produced with a square wave input are observed and compared with theoretical curves.

Before beginning this experiment read Equipment Appendixes 1 and 2, which describe the binary information module section of the ADD and the relay card. *Answer the questions.*

There are 8 logic level two-position switches on the binary information module (BIM). When switch *A* is in the "0" position, *what voltage appears at the A output on the top row of the BIM?* ______

What voltage appears when A is in the "1" position? ______

______

To what connector on the relay card should a signal be applied to control the movable contact? ______

______

E2–8.1 *RC* Response to a Step Function

Connect the circuit shown in Fig. E2–12. Use the relay card for the switch and the multiple connector card attached to it for *R* and *C*, as shown pictorially in Fig. E2–13.

The relay is activated by pushbutton switch *A* on the binary information module. When *A* is in the "0" state, the *RC* circuit is connected to ground. When *A* is in the "1" position, the capacitor can be charged through resistor *R*. For the initial experiment, use $R = 1$ MΩ, $C = 1$ μF. With switch *A* in position "0", adjust the recorder zero to the far right side of the chart. If a multispeed recorder is used, select a chart speed of 2 in./min. With the chart moving, push *A* to the "1" position so that the capacitor charges. *Record the charging curve until the voltage across C remains*

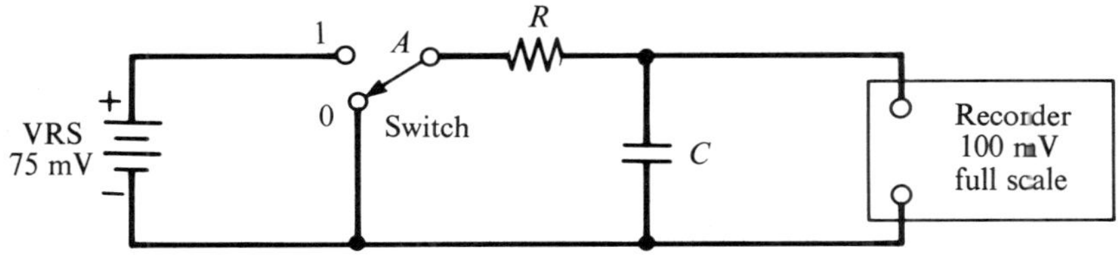

Fig. E2–12 Series *RC* circuit.

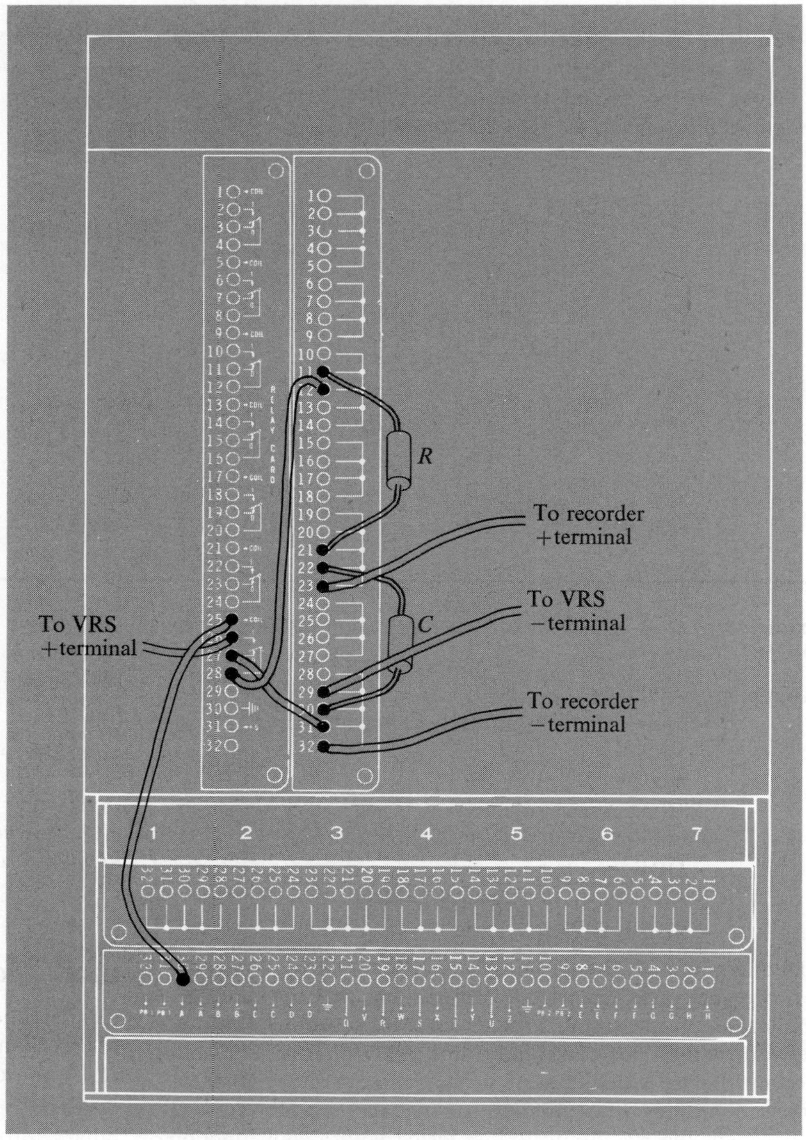

Fig. E2–13 Pictorial diagram of *RC* circuit of Fig. E2–12.

steady with time. Now push switch *A* to the "0" position and record the capacitor discharge curve. Since the *RC* time constant is 1 sec, accurate measurements of the charging and discharging times cannot be made on the recorder. Change *R* to 10 MΩ. *Repeat the above steps and measure the time required for the capacitor to charge to 63% of its final value* V_{C_f} *and compare this to the RC time constant* τ. *From the discharge curve, measure the time for the capacitor to discharge to 37% of its initial value before discharge,* V_{C_i}, *and compare this to the time constant.*

Time constant τ	Time to reach $0.63V_{C_f}$	Time to reach $0.37V_{C_i}$

How many time constants does it take for the capacitor to charge for practical purposes?

____

Place a second 1 μF capacitor in parallel with the first and *repeat the above measurements.*

Time constant τ	Time to reach $0.63V_{C_f}$	Time to reach $0.37V_{C_i}$
________	________	________

Why is this circuit an approximate integrator?

Reverse the positions of *R* and *C* in Fig. E2–13. Use $R = 1$ MΩ, $C = 1$ μF. Change switch *A* to the "1" position and *record the voltage-time curve.* Change switch *A* to the "0" position. Repeat for $R = 10$ MΩ, $C = 1$ μF; and for $R = 10$ MΩ, $C =$ parallel combination of two 1 μF capacitors. *For the same RC values, compare the voltage-time curves obtained across C with those obtained across R. Explain the voltage-time curves below.*

E2–8.2 *RC* Circuit Response to Square Waves

A square wave can be considered as a series of repetitive step functions. In this experiment the response of a series *RC* circuit to square waves will be studied. The differentiation and integration behavior of an *RC* circuit will be observed.

Before beginning this experiment read Equipment Appendix 3, which discusses the sine-square generator and the function generator. *Answer the questions below.*

What range of frequencies is available from the sine-square generator?

What different kinds of waveforms can be produced by the function generator?

Set up the circuit shown in Fig. E2–14 on the breadboard card, using the DRB for *R* and a 0.1 μF capacitor for *C*. Do not use the oscilloscope probe for this experiment. Instead use banana jack cables and an adapter.

***RC* Differentiator.** Set the square wave generator for an output of about 1 kHz and a 1 V peak-to-peak amplitude. With $C = 0.1$ μF, vary *R* in steps from about 100 Ω to 100 kΩ. *Observe the waveforms on the scope and sketch in the space*

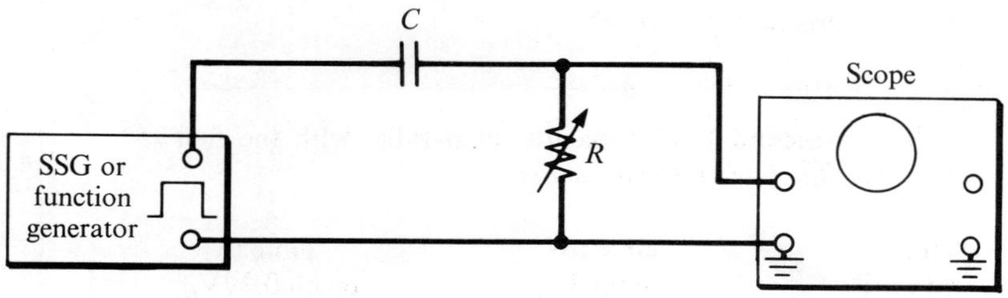

Fig. E2–14 Circuit for measuring response of *RC* circuit to square waves.

below three cases observed where: (a) the output across the resistor is a series of spikes; (b) the output is an undistorted square wave; and (c) the output is a square wave with a slight tilt on the top. *Note the value of R and C beside each sketch.*

For the case in which the output is a series of spikes, *measure the time it takes for the voltage to decay to* $1/e$ *times its maximum value, and compare that time to the RC time constant of the circuit.*

Time to decay to $1/e \times V_{\text{max}} =$ ________________

RC time constant = ________________

Explain below why this circuit is called a differentiator.

***RC* Integrator.** Reverse *R* and *C* in Fig. E2–14 so that the scope is connected directly across the capacitor. Use the same values of *R* and *C* as in the previous experiments and *sketch the curves obtained below.*

Compare the waveforms obtained for the differentiator with those obtained for the integrator by drawing them side by side for the same values of R and C.

Experiment 2–9 *An Operational Amplifier Integrator*

Equipment: VRS
ADD
FET OA card
VOM or DVM
wristwatch or stop watch
oscilloscope

When a capacitor is connected as the feedback device in an operational amplifier, the resulting integrating circuit can be used to measure the area under a current or voltage signal, to produce a linear sweep voltage (ramp), and to provide feedback control signals for controlling some other device. In this experiment the integrator is constructed and used as a linear sweep generator and for the integration of various waveforms.

Before beginning this experiment read Equipment Appendix 4, which describes the digital timing module (DTM) section of the ADD. *Answer the following question.*

What frequency timing signals are available from the digital timing module?

E2–9.1 Linear Sweep Generator

Wire the circuit shown in Fig. E2–15. Balance the OA by pressing the VRS zero button and adjusting the balance potentiometer until $v_o = 0$ (keep the shorting wire shorted).

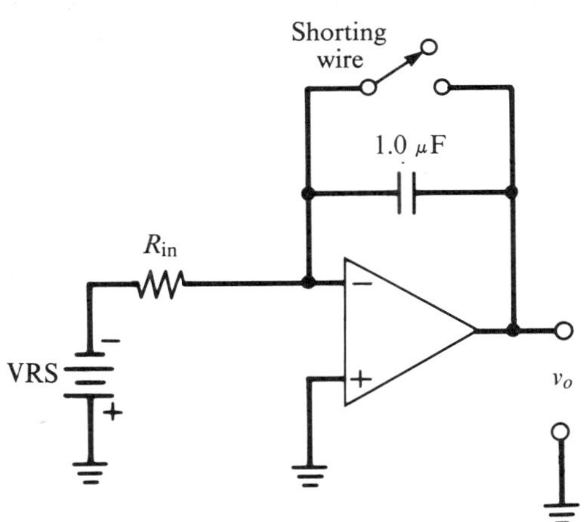

Fig. E2–15 Linear sweep generator.

With $R_{in} = 1$ MΩ and the VRS set at 100 mV, measure the time required for the output to rise from 0 to 1, 0 to 3, 0 to 5, and 0 to 8 volts. Use a wristwatch or stop watch to time the linear sweep. *Observe the output on the VOM or DVM. Compare the experimental times to those calculated from the theoretical slope.* The output should be

$$v_o = -\frac{1}{RC}\int_0^t v_{in}\,dt.$$

Since v_{in} is constant,

$$v_o = -\frac{v_{in}\,t}{RC}, \qquad \frac{\Delta v_o}{\Delta t} = -\frac{v_{in}}{RC}.$$

Record your data below.

Δv_o	$(\Delta t)_{expt}$	$(\Delta t)_{theor}$
1 V		
3 V		
5 V		
8 V		

Is the sweep generator accurate within component tolerances?

Change R_{in} to 10 MΩ and connect the output of the sweep generator to a 100 mV recorder. Record the linear sweep. Discharge the capacitor when the recorder gets to full scale by shorting it with the shorting wire. Repeat the integration, and stop the integration at exactly 50 mV by pressing the zero button on the VRS.

Why does the output remain constant or drift only slowly?

E2–9.2 Integration of Waveforms

Set up the circuit shown in Fig. E2–16. The 1.0 μF capacitor on the output of the DTM is to remove the dc component of the unidirectional square wave. Set $R_{in} = 100$ kΩ and $C_f = 0.1$ μF.

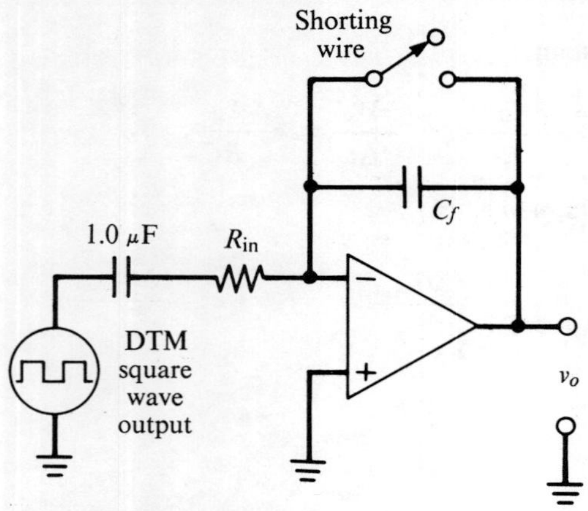

Fig. E2–16 OA integrator.

Observe first the output of the DTM on the scope. Draw the waveform below. Label the axes properly.

Next observe the integrator output, and draw the waveform observed in the space below.

Change R_{in} to 10 kΩ *and observe the waveform at the output of the integrator. Draw the observed waveform below.*

Explain how the observed waveform arises, assuming the OA integrator to be a perfect integrator.

Experiment 2–10 *Determination of Capacitance Using a Voltage Measurement Method*

Equipment: VRS with current probe
ADD
OA card
1 μF capacitor
watch

If a known constant current i is used to charge a capacitor of capacitance C for a known time t, then the charge $q = it$, and the voltage v_c across the capacitor is $v_c = q/C$. Therefore, the capacitance can be determined by measuring the voltage after establishing a known charge on the capacitor. An accurate charge is placed on the capacitor by connecting it as the feedback element in an OA circuit. Since the output voltage v_o of the OA is equal to the voltage v_c across the capacitor, it is simple to measure v_c without disturbing the charge on the capacitor. The precision of the method is determined for various capacitors, and limitations are noted.

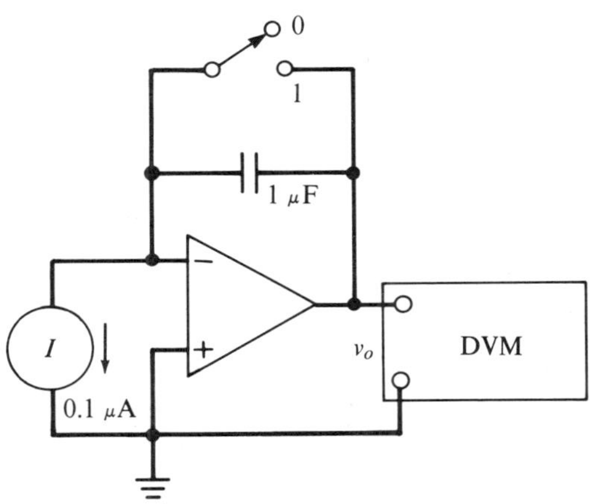

Fig. E2–17 Circuit for measuring capacitance.

Construct the circuit shown in Fig. E2–17. Use the VRS on 100 mV, $-S$ polarity, with the voltage-to-current probe on 1 MΩ to provide the 0.1 μA constant current source. Use the relay controlled by PB1 to short the capacitor.

Make certain the amplifier is balanced before beginning this experiment. Apply the constant current. Short the feedback capacitor by depressing PB1. Unshort the capacitor and begin timing the application of current on a watch. At the end of 10 sec, depress the VRS zero button and measure the voltage. *Repeat for triplicate measurements.* From the known amount of charge added, $q = it$, and the measured voltage, v_o, *calculate the capacitance C. Repeat for a* 0.5 μF *capacitor. Average the triplicate results for each capacitor.*

Trial	Nominal C	v_o	C_{calc}	C_{avg}
1	1 μF			
2	1 μF			
3	1 μF			
1	0.5 μF			
2	0.5 μF			
3	0.5 μF			

From the above measurements, estimate the precision of measuring capacitance by this method. Discuss the limitations of this technique.

Experiment 2–11 *An Operational Amplifier Differentiator*

Equipment: VRS
ADD
two OA cards or dual OA card
relay card
VOM

Information about the rate of change of a quantity with respect to time is frequently required for measurement or control purposes. In this experiment the rate of change of voltage is measured with an OA differentiator which produces an output voltage v_o that is related to the input rate of change dv_{in}/dt. The differentiator circuit is connected, and the test circuit is an OA integrator which provides a known ramp output voltage.

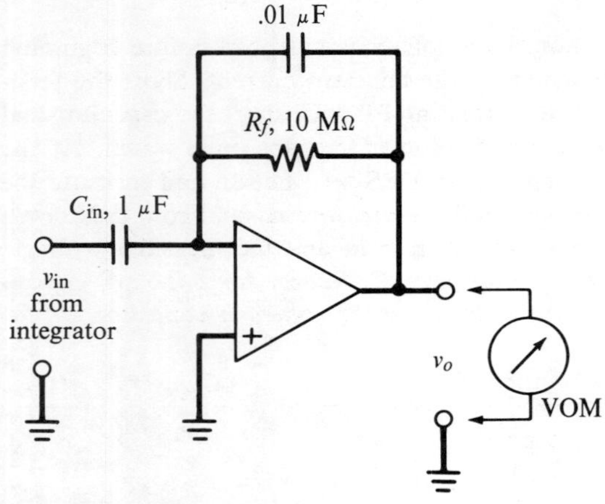

Fig. E2–18 OA differentiator.

Construct the linear sweep generator of Fig. E2–15. Use the relay card as the shorting wire. Arrange the circuit so that switch A on the binary information module can be used to short the feedback capacitor C. Change the VRS polarity so that a positive voltage is applied at the input. Set $R_{in} = 1$ MΩ and the VRS on 0–500 mV. Balance the OA as before.

With a second OA, wire the differentiator shown in Fig. E2–18. Connect the ramp generator output to the differentiator input. Note that the 0.01 μF capacitor around the differentiator feedback resistor is intended to minimize high-frequency noise.

Set the input voltage to the integrator at 0 V and unshort the integration capacitor. *Observe the differentiator output.* Any voltage is due to the integrator OA bias current. Change the VRS voltage to 100 mV, and *measure the differentiator output. Repeat for VRS voltages of* 200 *and* 300 mV.

Compare the differentiator output with that expected from $v_o = -RC(dv_{in}/dt)$. Results may be slightly different from those predicted from theory because of the input bias current of the integrator.

VRS voltage	Theoretical integrator slope	Measured differentiator output	Theoretical differentiator output
0			
100 mV			
200 mV			
300 mV			

Experiment 2–12 *A Logarithmic Converter Using a Capacitor Discharge Method*

Equipment: VRS
relay card
DVM or VOM
two OA cards or dual OA card

It is frequently necessary for a readout device to display linearly the logarithm or log ratio of voltage or current signals. In this experiment the log ratio of two voltage signals is determined by using a capacitor discharge method. The time difference involved in discharging a capacitor from a fixed reference voltage V_F to the two sample signals is measured and shown to be proportional to the log ratio. The precision and limitations of the method are determined.

E2–12.1 Manual Logarithmic Converter

Set up the circuit shown in Fig. E2–19. Use the VRS set at 5 V for V_F. Use the DVM or VOM to measure the output of the log generator circuit. Use the relay card controlled by BIM switches as switches S_1, S_2, and S_3. Assume that $v_r = 1$ V and $v_s = 0.1$ V. When the output of the follower reaches v_r, close switch S_2 and open S_3 simultaneously to allow the readout integration capacitor to charge. When the follower output reaches v_s, open switch S_2 and measure the output voltage v_o of the integrator. *Report the value of* v_o *and compare it to the expected relationship. Obtain triplicate results.* Choose a few other values of v_r and v_s and obtain triplicate results at each value. *Record all the results in the table below.*

v_r	v_s	$(v_o)_{meas}$	$(v_o)_{calc}$

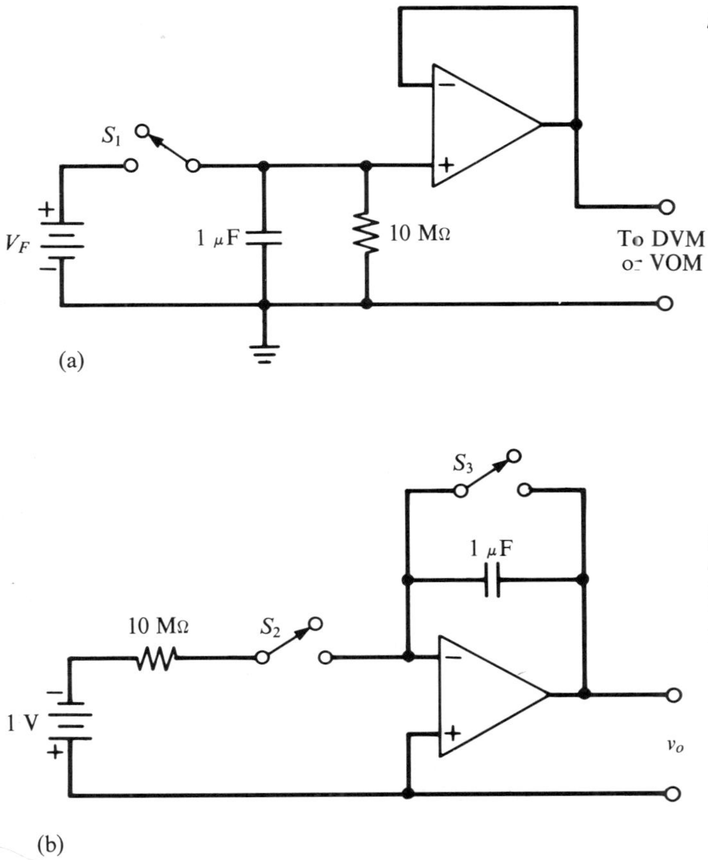

Fig. E2–19 Manual logarithmic converter: (a) log generator; (b) integration readout.

Discuss the observed measurement precision and possible limitations of this method.

E2–12.2 Automated Logarithmic Converter (Optional)

The logarithmic converter can be automated by assembling the circuit of Fig. 2–45. Test the circuit for various values of v_r and v_s and measure its logarithmic response.

Experiment 2–13 *Response Characteristics of an LR Network*

Equipment: ADD
oscilloscope
VRS
multiple connector card
20–200 mH choke

Several different input waveform signals are applied to a series *LR* circuit and the resulting output waveforms are observed. These output signals are then compared with those from an *RC* network.

Locate the inductor. Wire the circuit shown in Fig. E2–20 on a multiple connector card placed in the analog digital designer.

Use the dual trace mode on the oscilloscope by placing the VERT MODE switch in the ALT position. Place the TRIGGER MODE switch on CH.1 and the triggering controls on INT/DC/+/NORM. Set the TIME BASE control on 2 msec/cm and the CH.1 and CH.2 sensitivities at 0.1 V/cm. Apply dc voltages from the VRS of 100 mV and 200 mV.

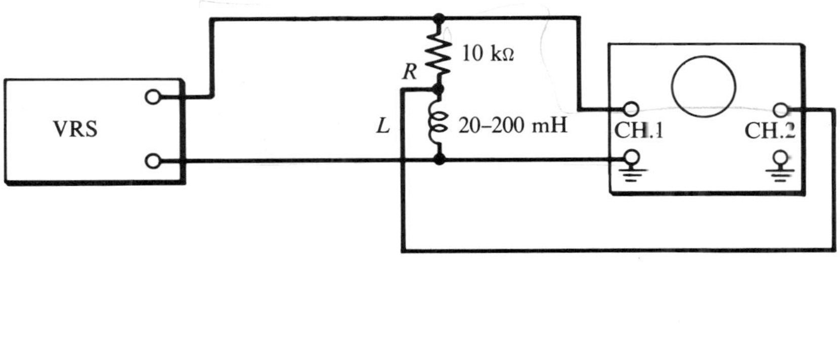

Fig. E2–20 Series *RL* circuit.

Does any potential appear across the inductor? Why?

Now press the VRS chop *S* button IN, and *observe both the waveform from the VRS and that across L. Draw the observed waveforms below and explain how the waveform across the inductor arises.*

Remove the VRS from the above circuit, and instead apply a 100 Hz square wave from the digital timing module. Lower the scope CH.1 and CH.2 sensitivities to 2 V/cm. Be sure to connect the ADD common to the end of *L* that is attached to the scope common.

Draw the waveforms observed below.

Replace the inductor with a 0.1 μF capacitor and *observe the waveforms. Draw the waveforms and explain why they are so different from those obtained with the inductor.*

Remove the capacitor and replace it with the choke as before. Change the input signal from the DTM square wave output to the DTM sawtooth wave output. Leave the frequency at 100 Hz. *Observe and record the waveforms.* Change the inductor to a 0.1 μF capacitor, and *record the waveforms beside those observed from the inductor. Discuss reasons for any differences noted.*

Electronic Switching

Section 2–3

One of the most familiar devices used for the control of electrical signals or power is the switch. A switch is a device with two states, conducting and nonconducting, which is used to ensure that a particular circuit is either complete or open. Since the control of electrical quantities is prerequisite to their application, significant developments in control devices have opened up possibilities for new tools and opportunities for the scientist and instrument designer alike. The recent (and continuing) development of inexpensive semiconductor switches capable of changing states of conductance in the submicrosecond time scale is a major factor in the current computer and instrument revolution.

A switch is composed of two conductors which are joined when the circuit is complete and separated by an insulator when the circuit is open. Because the junction of conductors is basic to all switches, considerable insight into the operation and limitations of practical switches can be gained by first considering the nature of the junctions between conductors, their characteristics, and their mechanism of conduction. After this introduction, the large variety of modern switching devices can then be categorized and discussed according to the type, arrangement, and number of junctions employed. The later portions of this section are devoted to an introduction to the applications of switches in the control of information transmission, in waveform generation, and in the control of ac power.

2–3.1 MECHANISMS OF CONDUCTION AT JUNCTIONS

A conductor has been shown to be a material that has a significant concentration of charge carriers. When two dissimilar conducting materials are joined, the junction between them is not necessarily a conductor. A nonconducting junction occurs when the charge carriers of the two materials interact in such a way as to deplete the concentration of charge carriers in the junction region, or when there is no mechanism for the transport of charge carriers from either material across the junction. On the other hand, when the junction region is populated with charge carriers that can readily move from one material to the other, the junction is conducting. The characteristics of junctions that always conduct are important, since such junctions provide all the electrical connections between components, devices, and instruments. Junctions that conduct under some conditions but not under others provide a useful function as switches which change conducting state in response to a change in the condition(s) controlling the conduction process. The following discussion begins with the metal-metal junction and then proceeds through the pn semiconductor junction, the metal-semiconductor junction, and the metal-gas/vacuum junction.

Metal-Metal Junctions

The energy of conduction electrons in a metal can be considered to be the combination of two types of energy, chemical and electrical. The chemical portion of the electron energy depends on several properties of the material, including the temperature, and is called the **chemical potential energy,** μ_e. The electrical portion of the electron energy results from the electrical potential of the material or electric fields within it and is equal to $-Q_eV$, where Q_e is the electron charge and V is the electrical potential of the material with respect to some arbitrary reference. The combination of chemical potential energy and electrical energy is called the **electrochemical potential energy,** $\bar{\mu}_e$, so that

$$\bar{\mu}_e = \mu_e - Q_eV. \tag{2–74}$$

The minus sign appears in Eq. (2–74) because electrons are negatively charged. For positively charged carriers the sign in Eq. (2–74) is +.

Contact Potential. When two metals A and B are joined as in Fig. 2–49, conduction electrons pass from one to the other until the electrochemical potential energy of the electrons in both is equal, that is, $\bar{\mu}_e^A = \bar{\mu}_e^B$. Expressing the electrochemical potential energy of the electrons in each material as the sum of the chemical and electrical contributions, we obtain

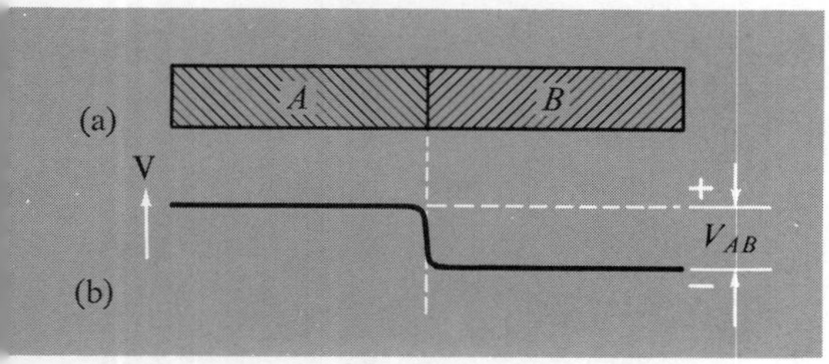

Fig. 2–49 (a) Junction of dissimilar metals. (b) Potential profile for $\mu_e^A > \mu_e^B$.

$$\mu_e^A - Q_eV_A = \mu_e^B - Q_eV_B,$$

from which the difference in potential between A and B can be obtained:

$$V_A - V_B = V_{AB} = \frac{1}{Q_e}(\mu_e^A - \mu_e^B). \tag{2–75}$$

From Eq. (2–75) we can see that if metals A and B have the same composition and temperature ($\mu_e^A = \mu_e^B$), the potential difference between them is zero. On the other hand, if A and B are dissimilar, electrons will cross the junction until a potential difference is developed which depends on (and offsets) the chemical potential difference between the metals. This potential difference, *which occurs at all junctions between dissimilar conductors,* is called the **contact potential.** The contact potential is characteristic of the two materials at any given junction temperature.

An attempt to measure the contact potential is illustrated in Fig. 2–50. The voltmeter leads of material C are connected to A and B respectively. Assume that the entire system is at the same temperature, so that the voltage measured will be simply the sum of the contact potentials for the three dissimilar metal junctions in the circuit:

$$V = V_{CA} + V_{AB} + V_{BC}. \tag{2–76}$$

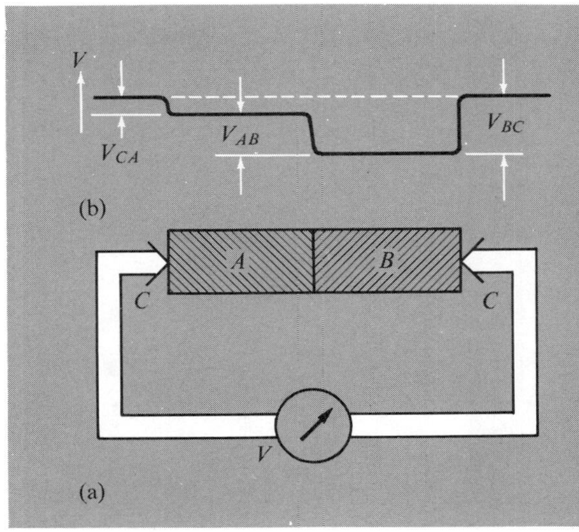

Fig. 2–50 (a) Attempt to measure contact potential between A and B. (b) Potential profile showing cancellation of contact potentials.

If we substitute into Eq. (2–78) the equivalent of Eq. (2–75) for each junction,

$$V = \frac{1}{Q_e}(\mu_e^C - \mu_e^A) + \frac{1}{Q_e}(\mu_e^A - \mu_e^B) + \frac{1}{Q_e}(\mu_e^B - \mu_e^C) = 0. \tag{2–77}$$

Thus the contact potentials are shown to cancel exactly. This is reasonable if one considers that the electron energy is constant throughout the system and hence an electron will experience no change in energy as it moves from C on one side of the voltmeter through A and B to C on the other side. Thus the contact potentials in an isothermal (constant temperature) system will always cancel regardless of the number of dissimilar metal junctions in the circuit. The potentials generated in nonisothermal circuits are discussed later in this section.

The contact potential difference between any two metals in electronic equilibrium can be estimated by considering that no energy is lost or gained by moving an electron from A to B across the junction, and so similarly there must be no energy gained or lost by taking an electron from A to a place far from either metal and then placing it in B. The energy required to remove an electron from A is called the **work function,** $-Q_e\Phi_A$, where Φ_A is called the **Richardson potential** for the material. The energy required to remove the electron from the electric field of A is $-Q_eV_A$. Adding the energies required for each step of the process and setting the sum equal to zero, we obtain

$$(-Q_e\Phi_A + Q_eV_A) - (-Q_e\Phi_B + Q_eV_B) = 0,$$

from which we get

$$V_A - V_B = (\Phi_A - \Phi_B). \tag{2–78}$$

Equation (2–78) indicates that the contact potential between two metals is simply equal to the difference in their Richardson potentials. Since work function measurements are possible, but greatly affected by the material surface, the tabulated values of work functions should be taken as approximate values in most cases. However, since values of the Richardson potential are from 1 to 5 V for most metals, the contact potential between dissimilar metals is often several volts! Yet, because of the precise cancellation of contact potentials in a complete circuit, such junctions are used routinely to transmit accurately signals only microvolts in amplitude.

Thermal Potentials with Zero Current. When two portions of a conductor are at different temperatures, there will be a flow of heat from the hotter portion to the colder. One of the mechanisms for the transmission of heat energy is the motion of the mobile charge carriers. Thus the transport of heat and the motion of charge are closely related for most conducting materials. The thermoelectric phenomena which are based on the relations between heat and charge flow are significant in several areas of scientific measurements. It was shown above that there is a net flow of electrons under conditions of nonuniform electrochemical energy of conduction electrons. Now we wish to consider also the flow of electrons due to a nonuniform temperature. A general relationship for the total flux of electrons, J_e (particles/sec-cm^2), is

$$J_e = -K_\mu \frac{d\bar{\mu}_e}{dx} - K_T \frac{dT}{dx}. \tag{2–79}$$

The first term to the right of the equality sign indicates that the flux of electrons due to a nonuniform $\bar{\mu}_e$ is proportional to the change in $\bar{\mu}_e$ per increment of distance along the wire. The flux of electrons due to the nonuniform temperature is similarly proportional to the temperature gradient, dT/dx, as given by the right-hand term. The proportionality constants are specific for a given material at a given temperature.

Consider a conducting wire A which is at temperature T_1 at one end and T_2 at the other, as shown in Fig. 2–51. Since there is no electrical connection to the wire, there is no net current; that is, J_e in Eq. (2–79) is zero. Since there is a continuous heat flow through A, heat must be added to the warmer end and removed from the cooler end to maintain the temperatures T_2 and T_1. Setting Eq. (2–79) equal to zero, we obtain

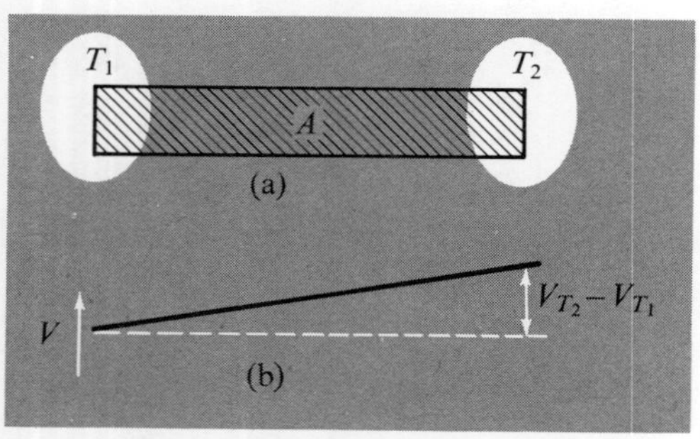

Fig. 2–51 (a) Conductor with a temperature gradient. (b) Potential profile for the conductor.

$$\frac{K_T}{K_\mu} = \frac{-d\bar{\mu}}{dx} \cdot \frac{dx}{dT} = \frac{-d\bar{\mu}}{dT}. \tag{2–80}$$

The ratio of proportionality constants K_T/K_μ has physical significance as the relative response of the electron to gradients of temperature and electrochemical energy, and has the units J C^{-1} deg^{-1}. This ratio is given the symbol S^*. Thus

$$S^* \equiv \frac{K_T}{K_\mu} \qquad (2\text{–}81)$$

and for $J_e = 0$ (the case of Eq. 2–80),

$$d\bar{\mu}_e = -S^*\, dT, \qquad (2\text{–}82)$$

which tells us that a temperature change will result in a change in $\bar{\mu}_e$ in proportion to the value of S^* for the material at that temperature. Equation (2–82) can be integrated to find the change in $\bar{\mu}_e$ for A between T_1 and T_2:

$$\bar{\mu}_{eT_2} - \bar{\mu}_{eT_1} = -\int_{T_1}^{T_2} S^*\, dT.$$

The $\bar{\mu}_e$ terms can be separated into their chemical and electrical parts from Eq. (2–74):

$$\mu_{eT_2} - Q_e V_{T_2} - (\mu_{eT_1} - Q_e V_{T_1}) = -\int_{T_1}^{T_2} S^*\, dT.$$

This equation can be solved for the potential difference $V_{T_2} - V_{T_1}$:

$$V_{T_2} - V_{T_1} = \frac{1}{Q_e}(\mu_{eT_2} - \mu_{eT_1}) + \frac{1}{Q_e}\int_{T_1}^{T_2} S^*\, dT. \qquad (2\text{–}83)$$

Thus we see that a potential is developed across a wire as a result of the temperature difference. This potential is due partly to the difference in the chemical potential of the electron at the two temperatures (the first term to the right of the equal sign) and due partly to the average value of S^* in the temperature range from T_1 to T_2 (right term of Eq. 2–83).

An arrangement such as that shown in Fig. 2–52 might be used to attempt to measure the thermally generated potential in A; B and b are the voltmeter leads of identical composition but having different temperature gradients. The voltmeter is at a temperature T_R, which may be equal to T_1 or T_2 in some situations. The total measured voltage will be the sum of the potentials generated in the conductors by the thermal gradients and the contact potentials between A and the voltmeter leads. The latter might not cancel in this case, since the junctions are not at the same temperature:

$$V = (V_{bT_R} - V_{bT_1}) + V_{bA} + (V_{AT_1} - V_{AT_2}) + V_{AB} + (V_{BT_2} - V_{BT_R}). \qquad (2\text{–}84)$$

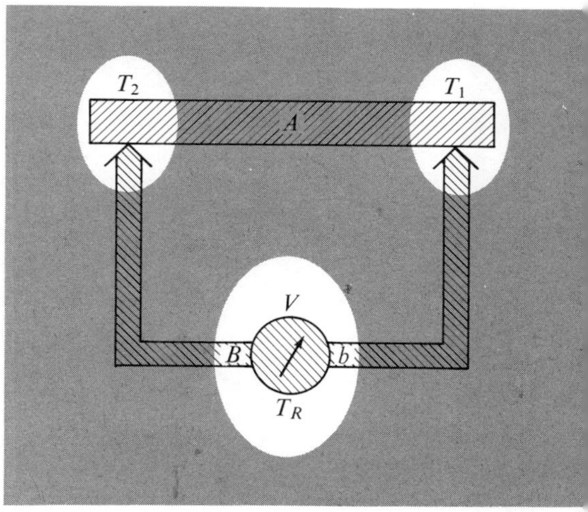

Fig. 2–52 Measurement of thermally generated potential.

If we substitute the equivalent of Eqs. (2–83) and (2–75) for each temperature gradient and junction into Eq. (2–84), we obtain

$$V = +\frac{1}{Q_e}(\mu^b_{eT_R} - \mu^b_{T_1}) + \frac{1}{Q_e}\int_{T_R}^{T_1} S^*_b\, dT + \frac{1}{Q_e}(\mu^b_{eT_1} - \mu^A_{eT_1})$$

$$+\frac{1}{Q_e}(\mu^A_{eT_1} - \mu^A_{eT_2}) + \frac{1}{Q_e}\int_{T_1}^{T_2} S^*_A\, dT + \frac{1}{Q_e}(\mu^A_{eT_2} - \mu^B_{eT_2})$$

$$+\frac{1}{Q_e}(\mu^B_{eT_2} - \mu^B_{eT_R}) + \frac{1}{Q_e}\int_{T_2}^{T_R} S^*_B\, dT,$$

in which we find that all the terms in chemical potential cancel ($\mu^b_{eT_R} = \mu^B_{eT_R}$ because it is the same material at the same temperature):

$$V = \frac{1}{Q_e}\int_{T_R}^{T_1} S^*_b\, dT + \frac{1}{Q_e}\int_{T_1}^{T_2} S^*_A\, dT + \frac{1}{Q_e}\int_{T_2}^{T_R} S^*_B\, dT$$

and $S^*_b = S^*_B$ at any given temperature, so

$$V = \frac{1}{Q_e}\int_{T_2}^{T_1} S^*_B\, dT + \frac{1}{Q_e}\int_{T_1}^{T_2} S^*_A\, dT = \frac{1}{Q_e}\int_{T_1}^{T_2} (S^*_A - S^*_B)\, dT. \tag{2–85}$$

We see that the portion of the thermally generated potential which is due to the change in chemical potential with temperature within each material exactly cancels the change in contact potential between materials with temperature. Thus the potential developed when dissimilar conductor junctions are at different potentials (the **Seebeck effect**) is entirely due to the difference in the value of S^* between the two materials.

Of particular interest is the amount of change in the thermally generated voltage for a given change in temperature. Equation (2–85) can easily be solved for dV/dT:

$$\frac{dV}{dT} = \frac{1}{Q_e}(S^*_A - S^*_B) = S_A - S_B = S_{AB}. \tag{2–86}$$

The quantity dV/dT is called the **thermopower** or **thermal sensitivity** of A relative to B and is given the symbol S_{AB}. Table 2–6 gives the thermopower of many materials relative to platinum.

The reader will recognize the thermally generated potential difference as being the principle of operation of the thermocouple used to measure temperature difference. It can also be the cause of spurious potentials which can seriously affect the accuracy of low-level voltage measurements. For measurement of temperature, one would choose the metals A and B to have as large a value for S_{AB} as possible. Also, the less S_{AB} varies with

Table 2–6 Thermopower $S = dV/dT$ of thermoelement made of materials listed against platinum (reference junction kept at a temperature of 0°C), μV/°C

Bismuth	−72	Silver	6.5
Constantan	−35	Copper	6.5
Nickel	−15	Gold	6.5
Potassium	− 9	Tungsten	7.5
Sodium	− 2	Cadmium	7.5
Platinum	0	Iron	18.5
Mercury	0.6	Nichrome	25
Carbon	3	Antimony	47
Aluminum	3.5	Germanium	300
Lead	4	Silicon	440
Tantalum	4.5	Tellurium	500
Rhodium	6	Selenium	900

From Kurt S. Lion, *Instrumentation in Scientific Research—Electrical Input Transducers,* McGraw-Hill, New York, N.Y., 1959.

temperature, the more linear the temperature-voltage transfer function will be. It is of interest that the potential developed by the temperature gradients in Fig. 2–52 is independent of the temperature T_R. This is because the potential caused by the temperature gradients $T_R - T_2$ and $T_1 - T_R$ cancel except for $T_1 - T_2$.

In the thermocouple circuit of Fig. 2–52 the voltmeter and its connecting wires are made of material B. It is frequently desirable to make the thermal junctions out of materials other than the material used for the voltmeter connections. This introduces additional dissimilar metal junctions. A typical arrangement is shown in Fig. 2–53. An analysis similar to that used with Fig. 2–52 shows that the potentials generated in the voltmeter leads C and c cancel (providing the wires are homogeneous) and the potential measured depends only on the relative thermopower of A and B and the temperatures T_1 and T_2. It is essential that the CA and Bc junctions be held at the same temperature, which becomes the reference temperature in the measurement of the AB junction temperature T_1.

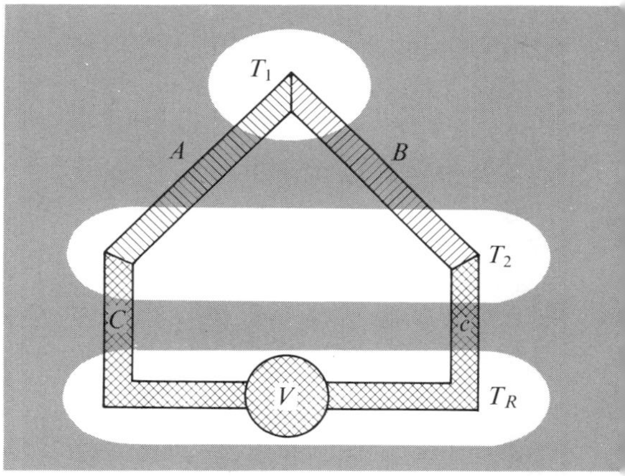

Fig. 2–53 Practical thermocouple measurement arrangement.

It should be clear from the above discussion that to avoid thermally generated potentials, all wires should be homogeneous or free of thermal gradients, junctions between dissimilar metals should be avoided where possible, and when dissimilar metal junctions are unavoidable, the junctions should be held at a common temperature. The uniformity of composition of a wire is often tested by measuring the potential between the two ends of the wire with the wire ends held at the same temperature (as in

Fig. 2–52 with the wire at A and $T_1 = T_2$). Now, by means of a flask of very cold liquid, a cold spot is moved along the length of the wire. If the wire is homogeneous, the effects of the opposing thermal gradients on either side of the cold spot will exactly cancel. Experimenters using thermocouples for exact temperature measurements will test the wires used in the thermocouple circuits this way, rejecting all wire sections giving a measurable potential difference as the cold spot was passed through them.

Thermoelectric Effects with Current. It has been demonstrated above that the motion of charge carriers can be due to a temperature gradient as well as an electrochemical energy gradient. It is also then necessarily true that heat flow can be due to a gradient of the electrochemical energy of charge carriers as well as a gradient of temperature. If there is a current in a conductor, there must be an electrochemical potential gradient. This gradient, in turn, will result in a flow of heat in the wire (and a temperature gradient along it). This is referred to as the **Thomson effect.** The rate at which heat must be added to a section of a current-carrying wire to keep it at a constant temperature is then equal to the heat removed by the Thomson effect minus the heat generated by the energy lost in passing the current through the resistance of the wire (Joule heating). Thus

$$\Delta W = \tau i \, \Delta T - i^2 R, \qquad (2\text{–}87)$$

where ΔW is the rate of heat addition required, τ is the Thomson coefficient, ΔT is the temperature change in the section of wire, and R is the resistance of the section of wire. The first term to the right of the equals sign represents the Thomson effect, the next the Joule heating. The Thomson coefficient is related to S^* for the material by the equation

$$\tau = \frac{T}{Q_e}\left(\frac{\partial S^*}{\partial T}\right) = T\left(\frac{\partial S}{\partial T}\right), \qquad (2\text{–}88)$$

where $S = S^*/Q_e$. Thus we see that the Thomson coefficient is proportional to the change in S with temperature for the material. The units of τ are volts/deg. The Thomson coefficient can have either sign. Thus the Thomson effect may tend to heat or cool the wire, depending on the sign of τ *and on the direction of the current* (sign of i). Note that the Joule heating term warms the wire regardless of the sign of i. Thus the Thomson effect is reversible while the Joule heating is not.

Another heating and cooling effect that results from the motion of charge occurs at the junctions of dissimilar materials. It is now convenient to define a term Q^*_{AB} as the heat transported per particle as it traverses the junction from A to B. Q^*_{AB} is related to S^*_{AB} by the equation

$$Q^*_{AB} = TS^*_{AB} + (\bar{\mu}^A_e - \bar{\mu}^B_e), \qquad (2\text{–}89)$$

where the difference in $\bar{\mu}_e$ between A and B is due to the iR voltage developed across the junction, or

$$Q^*_{AB} = TS^*_{AB} - Q_e iR_j.$$

Thus heat must be added to or taken from a junction between materials of different S^* for every electron passing through the junction if the junction is to remain at a constant temperature. This is called the **Peltier effect.** The rate of heat addition ΔW is obtained by multiplying Q^* by the rate of electrons crossing the junction, or i/Q_e. When this is done,

$$\Delta W = \frac{Q^*_{AB} i}{Q_e} = \frac{-iT}{Q_e} S^*_{AB} - i^2R_j,$$

in which $S^*_{AB}/Q_e = S_{AB}$ and $-i^2R_j$ is the irreversible joule heating in the junction. Now,

$$\Delta W = -iTS_{AB} - i^2R_j, \tag{2–90}$$

where $-TS_{AB}$ is called the **Peltier coefficient,** π_{AB}. In its final form, then,

$$\Delta W = i\pi_{AB} - i^2R_j. \tag{2–91}$$

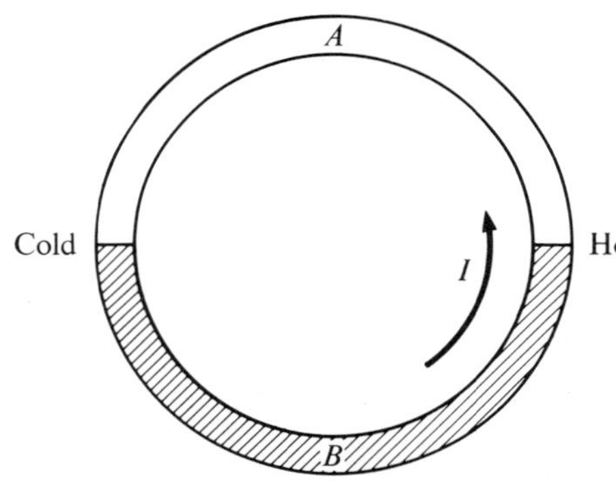

Fig. 2–54 Illustration of Peltier effect.

The Peltier effect is reversible and π may have either sign; thus heating or cooling may occur at a junction with current. A simple circuit with two dissimilar materials would have two junctions, one AB and one BA as shown in Fig. 2–54. If π_{AB} is positive, then according to Eq. (2–90) heat must be added to the left-hand junction to keep it at a constant temperature (assuming $i\pi_{AB} > i^2R_j$). Therefore, the left-hand junction will become cool. The current traverses the right-hand junction from B to A. Since $\pi_{BA} = -\pi_{AB}$, the right-hand junction will become warm. Thus whenever current is forced through a pair of dissimilar junctions, one becomes warmer and the other becomes cooler. This is the basis for **thermoelectric heating/cooling** devices. Since the Peltier effect is reversible, it also follows that if the cold junction were heated by an external source and the hot junction were cooled, a current in the direction shown would be generated. Stacks of such "thermocouple junctions with current" are used as relatively inefficient heat-to-electric-power converters called **thermoelectric generators.**

pn Semiconductor Junctions

A pn semiconductor junction is generally made by changing the dominant dopant from acceptor to donor type in a single crystal of semiconductor. The region where the majority charge carrier changes from electrons to holes is called the pn junction. The holes from the p region and the electrons from the n region are free to cross the junction. Since the chemical energy for electrons is higher in the n region than in the p region, some electrons will cross from the n to the p region. Similarly, some holes will cross to the n region as a result of the higher chemical energy for holes in the p region. The transport of + charges into the n region and — charges into the p region develops a potential difference between the n and p regions. This potential presents an energy barrier sufficient to offset the energy of the majority carriers on either side of the junction, so that at equilibrium there is no net flow of charge across the boundary. A contact potential is thus established with the n region positive with respect to the p region in a manner somewhat similar to the previously discussed metal-metal junction.

The drift of holes across the junction reduces the concentration of electrons in the junction region, since the product of the electron and hole concentrations must remain constant as given in Eq. (2–11). Similarly, the concentration of holes is lower at the junction than in the rest of the p region. The concentration profiles for holes and electrons are shown in Fig. 2–55b. In the example chosen, the dopant concentration is slightly less in the n region than in the p region. As shown by the concentration profiles, the total concentration of charge carriers is greatly depleted in the junction region, decreasing to the intrinsic value at its minimum. This region is called the **depletion region.** Thus the junction region is essentially nonconducting compared to the remainder of the n and p doped regions. The contact potential appears across the insulating junction region as shown in Fig. 2–55c. The magnitude of the contact potential is about 0.3 V for germanium pn junctions and about 0.6 V for silicon.

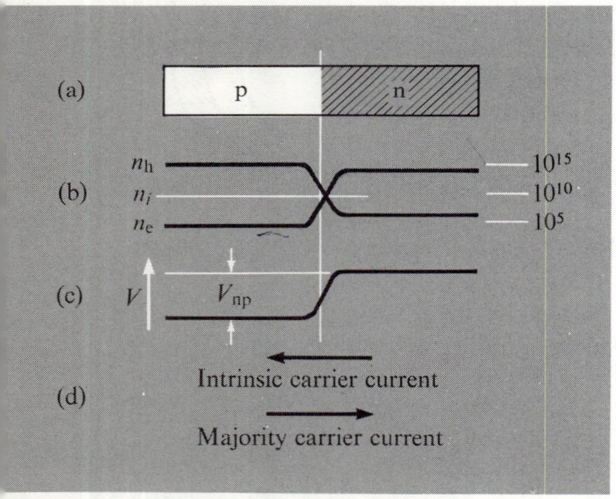

Fig. 2–55 (a) pn junction with (b) carrier concentration, (c) potential profiles, and (d) junction current directions.

As shown above, the concentration of charge carriers in the depletion region is essentially that of the intrinsic semiconductor. The charge carriers in intrinsic semiconductor material are electron-hole pairs generated by thermal excitation. However, there is a strong electric field in the depletion region causing the generated holes to move to the p region and the electrons to move to the n region. The current due to the carriers in the intrinsic region is thus from right to left, as shown in Fig. 2–55d. The magnitude of the intrinsic carrier current, I_i, depends on the rate of carrier generation in the depletion region, which is a function of temperature only. The net flow of + charge from right to left due to the intrinsic carrier current would tend to reduce the contact potential across the junction. However, as the potential barrier is decreased, the number of majority carriers on each side with enough energy to cross the junction increases. The current due to majority carriers, I_m, is from left to right and increases exponentially with increasing temperature and decreasing junction potential as given in Eq. (2–92):

$$I_m = Ke^{-Q_e V_{np}/kT}, \tag{2–92}$$

where k is the Boltzmann constant J/°K and T is the temperature in °K. Since the net current across the junction must be zero, the intrinsic and majority currents must cancel exactly. For any given temperature, there is only one value of the junction potential for which this will be true. One can qualitatively see how the higher intrinsic carrier generation rate of germanium over silicon results in the lower junction potential for germanium.

The Biased pn Junction. A **bias** is an external voltage applied to the material with the pn junction. In order to apply an external voltage source, it is necessary to make metal contacts to the semiconductor. The resulting device is shown pictorially in Fig. 2–56a. At both metal-semiconductor junctions, a contact potential is established. As discussed in the previous sections for a series of metal-metal junctions, the sum of the contact potentials must be zero in any circuit with the same material at both extremes. The potential profile through the contacts and junction is shown in Fig. 2–56b. We will assume in this discussion that the metal-semiconductor contacts are good conductors for current in both directions. Metal-semiconductor junctions are discussed in greater detail in the next section.

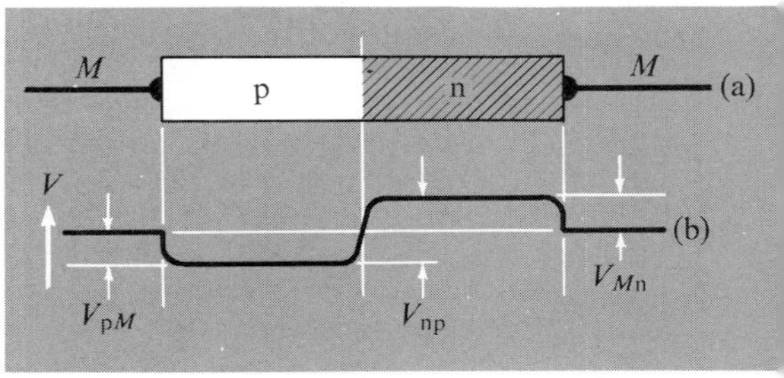

Fig. 2–56 pn junction with metal contacts and potential profile.

Since the connecting wires, the contacts, and the doped semiconductors are all very much better conductors than the depletion region, essentially all of the applied bias voltage will appear as a change in the potential difference across the pn junction. This is shown by the family of potential profiles in Fig. 2–57b. The change in V_{np} will have a great effect on the majority carrier current I_m as given by Eq. (2–92), but will have little effect on the intrinsic current I_i. At zero applied bias, $I_m = I_i$, so

$$I_i = I_m^0 = Ke^{-Q_e V_{np}^0/kT}, \tag{2–93}$$

where I_m^0 and V_{np}^0 are the values for the majority current and junction potential respectively at zero bias. At nonzero bias, the junction potential is $V_{np}^0 - V$ and a net forward current, I, can cross the junction. The net current is equal to the difference between I_m and I_i. Thus

$$I = I_m - I_i = Ke^{-Q_e(V_{np}^0 - V)/kT} - I_i. \tag{2–94}$$

If I_i from Eq. (2–93) is substituted in Eq. (2–94) and the equation is rearranged, we obtain the Shockley equation,

$$I = I_i(e^{Q_e V/kT} - 1). \tag{2–95}$$

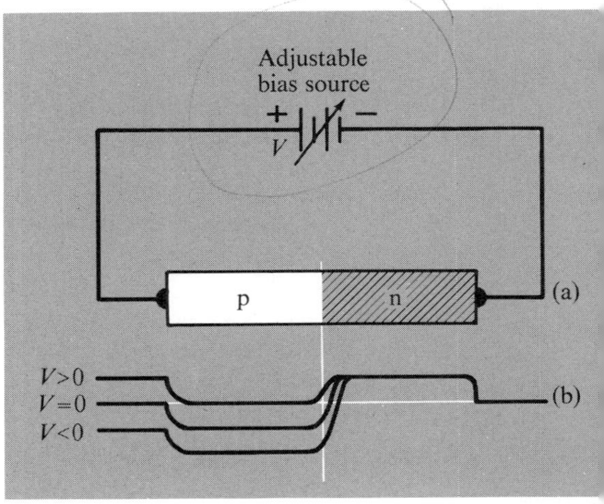

Fig. 2–57 (a) Biased pn junction. (b) Potential profiles showing the effect of bias voltage on the pn junction potential.

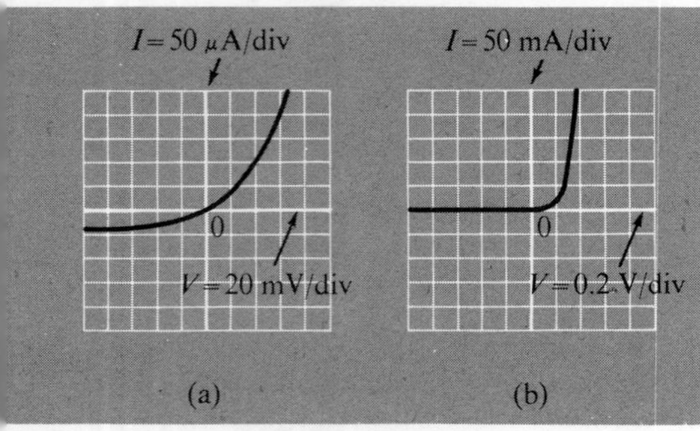

Fig. 2–58 Current-voltage characteristic of a germanium pn junction: (a) near zero bias; (b) complete forward characteristic.

A plot of the current I through the junction vs. the applied bias voltage V is shown in Fig. 2–58 for a germanium pn junction. The exponential rise in current for increasing positive values of bias voltage is clearly seen in Fig. 2–58a. For many germanium pn junctions the observed I-V curve fits the Shockley equation very well. However, to fit the observed behavior for silicon diodes, it is necessary to add a factor η in the denominator of the exponential. This has been ascribed to charge carrier recombination in the depletion region. In Fig. 2–58b, which shows a larger range of the I-V behavior, it is seen that the forward bias required for substantial forward current is 0.3 to 0.4 V. This corresponds to the bias required to offset the junction contact potential essentially completely. For a silicon pn junction, 0.6 V of forward bias is required for substantial forward currents. As the bias voltage is increased in the reverse direction, the magnitude of I_m decreases rapidly as a result of the increased junction potential. After −0.1 V bias, the current is simply I_i, approximately −40 μA for the germanium pn junction at room temperature shown in Fig. 2–58. A comparable silicon pn junction might have an I_i of −40 nA. Of course, I_i increases rapidly with increasing temperature since it is directly related to the conductivity of the intrinsic semiconductor. Thus the pn junction is a good conductor with a small forward bias applied, but a very poor conductor for even large values of reverse bias voltage. The pn junction is described later as a switch whose state (closed or open) is determined by the polarity of the bias voltage. Such a switch is called a **rectifier** and a junction with these characteristics is called a **rectifying junction.** A device consisting of a rectifying junction with two connectors is called a **diode.**

Reverse-Bias Breakdown. The reverse- or zero-biased pn junction can be thought of as two conductors separated by a thin layer of insulator (the depletion region). Thus the junction has the property of capacitance. As the reverse bias voltage is increased, the electric field strength in the depletion region increases as shown in Fig. 2–57b. This has the effect of increasing the thickness of the depletion layer somewhat. The resulting variation in capacitance is the basis of operation of voltage-controlled capacitors called **varactors.** The electric field strength in the depletion region can be increased to a value where the insulating capability of the junction region breaks down. Breakdown can occur by any of three mechanisms: **thermal instability, avalanche multiplication,** or **tunneling.** The heat generated in the junction region under reverse bias is IV watts. If V increases to a value which produces heat more rapidly than it is conducted away from the junction, the temperature will rise. This causes an increase in the reverse current, which causes a further increase in the heat generated. There is a maximum voltage and temperature beyond which this regenerative process proceeds to the thermal destruction of the device containing the junction. This is breakdown by thermal instability.

The thermally generated (intrinsic) charge carriers in the depletion region are accelerated toward the region where they are majority carriers by the electric field in the junction region. The maximum velocity attained by these charge carriers depends on the electric field strength and the distance traveled between collisions of the carriers with other particles. Thus the momentum of the charge carrier at the instant of collision increases with increasing electric field strength. When the momentum of the charge carriers is equal to the energy of ionization of the semiconductor atoms, electron-hole pairs will be produced upon collision. These charge carriers will also be accelerated to attain the momentum required for further ionizing collisions, and so on. The bias voltage at which the avalanche multiplication breakdown takes place depends primarily on the thickness of the depletion layer. Clearly, the thinner the depletion region, the lower the junction potential needs to be to establish a given electric potential gradient. In the construction of pn junctions, the depletion layer thickness is controlled by the dopant concentrations. The greater the concentration of majority charge carriers, the thinner the depletion region, because a smaller fraction of the charge carriers is required to create the depletion region. The avalanche breakdown voltage varies from several hundred volts to 10 V as the concentration varies from 10^{15} to 10^{17} atoms/cm^3. Over this same range of concentration, the depletion layer thickness at breakdown varies from about 100 to 0.1 μm. The avalanche breakdown voltage increases somewhat with increasing temperature. The higher the breakdown voltage, the wider the range of reverse voltages for which the device is useful. Some devices are made to be operated at their breakdown voltage (at a safe current). Breakdown diodes and their applications are discussed in a later section.

At an extremely high electric field strength of about 10^6 V/cm, the valence electrons in the p region have an energy comparable to the conduction electrons in the n region. When the p region valence electron energy exceeds the n region conduction electron energy, it may cross the pn junction directly by a process called **tunneling.** Similarly, an n region conduction electron having a higher energy than a valence band hole can move through the junction to fill the hole. Thus current can be in either direction across the junction by the tunneling mechanism. Sufficiently high fields for tunneling are obtained by heavy doping so that the depletion layer is extremely thin. A depletion layer of 10^{-6} cm (0.01 μm) would have a field strength of 10^6 V/cm with only 1 V of junction potential. This is achieved with a dopant level of the order of 10^{19} atoms/cm^3. Thus, as the dopant concentration increases, the breakdown voltage decreases and the breakdown mechanism changes from avalanche (about ~8 V) to tunneling (below ~4 V). This is shown by the *I-V* curves (a and b) in Fig. 2–59. Between 4 V and 8 V, a combination of both mechanisms

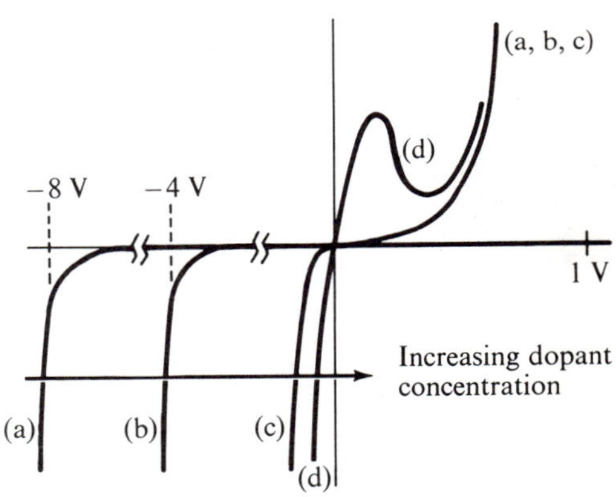

Fig. 2–59 *I-V* curves for several high dopant concentrations: (a) avalanche breakdown; (b) tunnel breakdown; (c) backward diode; (d) tunnel diode.

operates simultaneously. The tunnel breakdown voltage decreases with increasing temperature—the opposite of the avalanche temperature effect. Diodes with a breakdown voltage of about 6 V have nearly zero change in breakdown voltage vs. temperature, since the temperature effects of the two mechanisms cancel.

A further increase in dopant concentration could produce tunneling breakdown with only 0.6 V of junction potential. Since this is the zero bias junction potential of a silicon pn junction, the junction is in breakdown with no applied bias! Of course, the tunneling current across the junction is equal in both directions, so that no net current flows. However, only a small value of reverse bias is needed to produce a large reverse current. On the other hand, a forward bias decreases the junction potential and thereby eliminates the tunneling effect. Very little forward current passes until the forward bias offsets the contact potential. Such a device, which conducts in the reverse direction better than in the forward, is called a **backward diode.**

A still further increase in doping produces a junction that is still in tunneling breakdown for small values of forward bias. This produces good conduction in both directions near zero bias. For higher values of forward bias, the tunneling current decreases and the total forward current falls off until the normal majority carrier mechanism causes it to increase once more. A device with this type of junction is called a **tunnel diode.**

Metal-Semiconductor Junctions

The structure and characteristics of metal-semiconductor junctions follow directly from the discussion of metal-metal and pn semiconductor junctions. When a metal and a semiconductor are joined, a contact potential is established as electrons cross the junction to the material in which they have the lower energy. The equilibrium contact potential just offsets the difference in chemical energy of conduction electrons in each material.

The contact potential between a metal and a semiconductor is characteristic of the specific combination at a given temperature and may have either sign. Depending on the sign of the contact potential and the majority carrier of the semiconductor, the junction is either of the **depletion** or **enhancement** type. The four possible combinations of carrier type and contact potential polarity are shown in Fig. 2–60. The top case in the illustration is a junction to an n-type semiconductor with the metal positive. Because the charge carrier concentration in the semiconductor is generally many orders of magnitude less than that in the metal, virtually

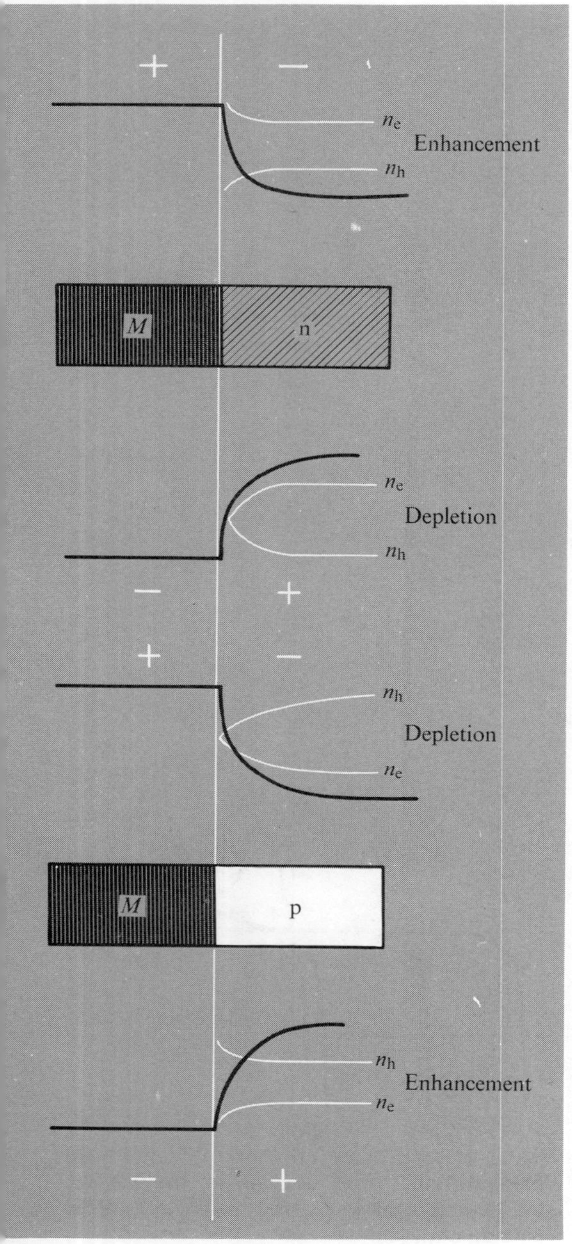

Fig. 2–60 Potential and carrier concentration profiles of contacts to n and p semiconductor for each contact voltage polarity.

all of the contact potential gradient occurs in the semiconductor. The junction potential requires an excess of electrons in the semiconductor at the junction. Thus the concentration of majority carriers at the junction is *enhanced* by the contact potential. Since there is a high concentration of charge carriers on both sides of the junction, conduction across the junction occurs readily in both directions. A junction which has a straight-line current-voltage curve passing through the origin is called **ohmic** because it obeys Ohm's Law for both positive and negative voltages.

The opposite effect occurs if the contact potential is of the opposite sign. As shown for the case of the n-type semiconductor which is more positive than the metal, the potential gradient requires a reduction in the electron concentration at the junction. This is a depletion junction similar to the pn semiconductor junction. At normal dopant concentrations, the depletion metal-semiconductor junction is rectifying, i.e., it conducts much better in one direction than the other. Rectifying diodes have been manufactured using depletion metal-semiconductor junctions; these include the copper-oxide, selenium, and point-contact germanium diodes. In most applications pn junction diodes have replaced these earlier types of diodes. A depletion junction can be made ohmic by using a very high level of dopant. This decreases the depletion region thickness to the point where tunneling can occur or, in the extreme case, where it is so thin that it is simply transparent to the electrons. As Fig. 2–59 shows, a tunneling junction is ohmic for small applied voltages. This is generally not a limitation, since the tunneling junction is such a good conductor that the junction bias is always small. An ohmic contact on n-type semiconductors is achieved in practice by evaporating a gold-antimony alloy onto the semiconductor. When this is heated, the antimony diffuses into the semiconductor, creating a region of very high dopant concentration. A contact wire is then soldered or welded to the gold spot.

The case of junctions between metals and p-type semiconductors is exactly analogous to the discussion of the n-type above. When the metal is positive, there is depletion; and when the metal is negative, there is enhancement. Ohmic contacts to p-type semiconductors are made by depositing a gold-arsenic alloy on the material to produce a highly doped junction.

Rectifying metal-semiconductor junctions known as Schottky diodes are used for extremely rapid switching. These diodes are discussed in Section 2–3.3.

Other Junctions

Several other junctions across which charge flows are of interest. One is the junction between a metal and a vacuum. Early in this discussion of junctions, the work required to move an electron from a metal M through a vacuum to a distant place was defined as the work function, $Q_e\Phi_M$. Any electron in M having this much energy can theoretically escape from the metal. As the temperature of the material increases, the number of electrons at the surface having sufficient energy to escape increases. Some materials which have a low work function and a high melting point can be heated to a temperature where electrons literally boil out of the metal. When this is done in a vacuum, the metal becomes positively charged and an increasing number of "free" electrons are attracted back to the metal. When a steady state is reached, the rate of evaporation of electrons from the metal is equal to their rate of return and a constant number of electrons exists in a cloud around the metal. This cloud of electrons is called a **space charge.** If another conductor **(electrode)** is placed in the vacuum and made more positive than the hot metal electrode, some of the electrons in the cloud will move in the direction of the positive electrode. Thus a current of electrons is established in the vacuum. If one electrode is hot and the other is cold, the electrons can only go from the hot electrode (**cathode**) to the cold electrode (**anode**) and not the other way. Such a device has rectifying characteristics and is called a **vacuum tube diode.** The **thermionic emission** of electrons and their acceleration by electric fields in a vacuum are the basis of operation of many important scientific devices, including the cathode ray tube.

The junction between the metal and the gas in a **neon lamp** or **gas discharge tube** is also of interest. A neon lamp consists of two cold metal electrodes in a glass envelope with a low pressure of neon gas. At low voltages between the two electrodes, there is virtually zero current because there are essentially no charged particles to act as charge carriers. As the applied voltage is increased, the electric field strength between the electrodes increases. Because of the shape of the electrodes, the electric field is not uniform and eventually, at some voltage V_s, the field strength is large enough to ionize a neon gas atom. The positive ion and electron are then accelerated toward the negative and positive electrodes respectively. By the time these accelerated particles collide with a neutral neon atom, they have acquired more than enough energy to ionize the atom. Thus an avalanche-type breakdown of the neon gas into a conducting plasma of neon ions, electrons, and neon atoms occurs.

Once conduction is initiated, some means of limiting the current to a safe value is required to avoid the destruction of the device by Joule heating. If the current is limited to a safe value, the voltage across the device will adjust itself to the minimum voltage necessary to maintain the applied current. In order to maintain current, it is necessary to maintain a sufficient concentration of ions and electrons in the gas and to transport charge across both metal-plasma junctions. Since the electrons are so much more mobile than the ions, they carry most of the current. The field strength in the gas region will be that required to produce ionizing collisions. At the positive electrode (anode), the current is composed of electrons crossing from the plasma to the metal. The potential across this junction is approximately the Richardson potential for the metal. At the negative electrode (cathode) electrons must cross the junction from the metal to the plasma. Since the metal is not hot, surface electrons must gain the work function energy from some other source. In this case it is from the energy of the positive ion bombardment of the cathode. Thus the electric field in the region of the cathode must be strong enough to accelerate the ions to the work function energy. These effects produce the potential energy profile shown in Fig. 2–61. The electrons emitted from the cathode are given a high energy as they are accelerated by the steep gradient in the cathode region. The collision of these electrons with neon atoms produces highly excited atoms which lose some of their energy by emitting light. Thus a sheath of light is produced about the cathode.

The presence of neon ions in the high field region about the cathode is due to collisions with the emitted electrons. Because of inefficiencies in the energy transfer processes required for ion production and ion bombardment emission, there is a minimum current density which will support this process. Thus the ion bombardment process occurs only on a portion of the electrode area which is small enough to maintain the minimum current density. If the total current is increased, the "active" portion of the electrode area increases as evidenced by an increase in the size of the illumination sheath. The voltage across the cathode-plasma junction is thus essentially independent of the applied current.

The voltage required to maintain conduction, V_m in Fig. 2–61a, is clearly less than that required to start it, V_s. The resulting current-voltage curve is shown in Fig. 2–61b. When the applied voltage reaches V_s, the conduction process begins and the voltage drops to V_m. As the current increases, the maintenance voltage remains almost constant since the junction potentials vary only slightly. The gas discharge tube is seen to switch from nonconducting to conducting with a constant voltage when the applied bias voltage exceeds V_s. A practical circuit includes a source of voltage V greater than V_s connected to a series combination of the discharge tube and a current limiting resistor R. The voltage across the tube will be V_m and the current will be $(V - V_m)/R$. When the bias voltage is reversed, the same processes occur but at opposite electrodes. Differences in electrode material and geometry of the two electrodes will result in different values of V_s and V_m for forward and reverse conduction.

Gas discharge tubes are used as switches, as constant voltage devices, and as indicator lamps. For neon indicator lamps, typical values of V_s and V_m are 87 V and 60 V respectively. Geometry, gas pressure, and gas purity affect these values considerably. Other gases (e.g., argon) are used

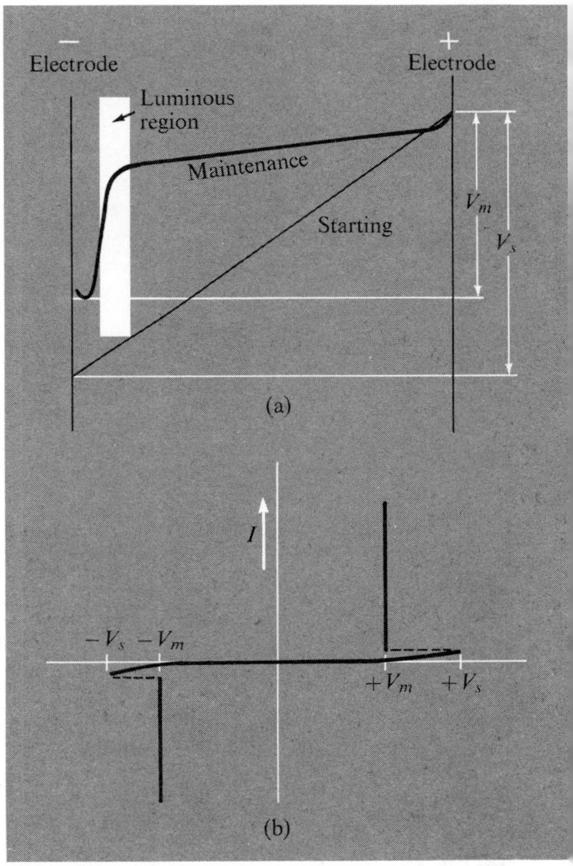

Fig. 2–61 Gas discharge tube: (a) potential profile; (b) current-voltage curve.

for other voltage ranges. At low gas pressures, a higher field in the plasma results in higher energy collisions and the entire plasma region is illuminated as in the all-too-common neon sign.

2–3.2 SWITCHING PRINCIPLES AND MECHANICAL SWITCHES

The switch has been described as a device which determines whether a particular circuit path is conducting or nonconducting. An ideal switch could be placed in any circuit and act as a solid connection when in its closed state and as a perfect insulator in its open state. Since no switch is ideal, there are many kinds of switches available with various combinations of features and compromises. Mechanical switches have a pair of metal contacts which touch when the switch is closed and are apart when the switch is open.

Mechanical switch contacts can be actuated in several ways: manually (toggle and rotary switches), mechanically (lever switches), electromechanically (relays and solenoids), and magnetically (magnetic reed relays). Purely electronic devices, including many types of transistors and semiconductor diodes, are also used as switching elements. These solid state switching devices are discussed in the next section (2–3.3). Regardless of the switch type, the basic requirements for a switching device are the same: the device must have two terminals between which an electrical current can be conducted; the device or the circuit must be designed so that there are only two stable states of conductivity in the circuit; and the device must be capable of being "switched" from one state to the other. The appropriate choice of switch depends on the desired method of actuating the switch, the voltage and current levels in the switched circuit, and the connection accuracy and switching speed required. The considerations of accuracy and speed are first discussed in relation to a simple switch operating in a generalized circuit. The descriptions of practical mechanical switches, including manual switches and relays, in this section are intended to aid in the practical application of these basic switch characteristics and limitations.

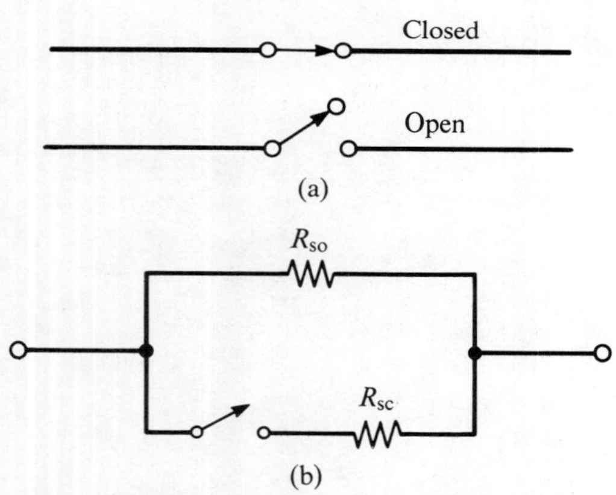

Fig. 2–62 (a) Switch symbol. (b) Schematic representation of nonideal switch.

Switch Resistance

An ordinary mechanical switch (e.g., light switch) has two electrical states: **open** and **closed** (or OFF and ON). The ideal switch has zero resistance between its two connections when closed and zero conductance (infinite resistance) when open. Neither ideal is completely achieved by practical switches.

The schematic symbols that are often used for open and closed switches are shown in Fig. 2–62a. However, in considering the action of a real switch in an electronic circuit it is frequently important to represent

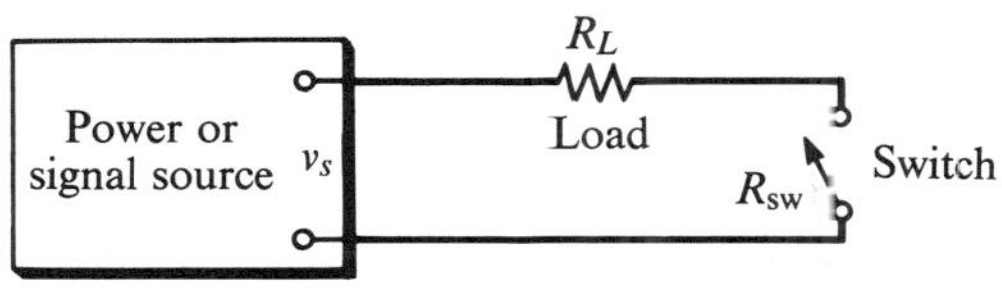

Fig. 2–63 Generalized switching circuit.

the nonideal open and closed switch resistances R_{so} and R_{sc}, as illustrated by the schematic for a nonideal switch in Fig. 2–62b.

In the generalized switching circuit shown in Fig. 2–63, the current in the circuit (through the load) is controlled by the switch. Another way to view the circuit, in consideration of the nonideal switch, is that the source voltage v_s is divided between the resistance of the load and the resistance of the switch. To transfer the maximum signal to the load the closed switch resistance, R_{sc}, should be very small compared to R_L. Similarly, if the open switch resistance is not very much greater than R_L the current will not be effectively turned off by the switch. For purely resistive switches, when R_{so}, R_{sc}, and R_L are known, the effectiveness of a given switch in a desired application can be evaluated using Kirchhoff's laws and the circuit of Fig. 2–63.

Current-Voltage Curves and the Load Line. As the earlier discussion on conducting junctions demonstrated, many junction types used in switches have nonohmic characteristics; i.e., the current through the switch is not a linear function of the voltage across the switch terminals. Therefore, the dc analysis of how the switch performs in a circuit is made most readily by a graphical solution which utilizes the **current-voltage curve** of the device and the **load line** of the circuit. The graphical solution is illustrated in this section for a simple ohmic switch, and in the next section for the pn junction diode and other solid state switching devices. The electrical characteristics of a resistive device for which the resistance is not constant over its operating range are often best shown by a current-voltage curve. A current-voltage (*I-V*) curve is a plot of the voltage *V* across a device vs. the current *I* through it. The *I-V* curve of a resistor is shown in Fig. 2–64. This line is straight as expected from Ohm's Law. A large resistance has a more horizontal *I-V* curve, whereas a lower resistance has a more vertical curve. The current-voltage curve of a perfect conductor would coincide with the vertical coordinate, and the line for an infinite resistance would coincide with the horizontal coordinate.

A switch has two *I-V* curves, one for the resistance of each state. The *I-V* curves for a perfect switch, which has zero ON resistance and infinite OFF resistance, will fall on the vertical and horizontal coordinates respectively. Exaggerated deviations from the ideal switch characteristics are shown in Fig. 2–65.

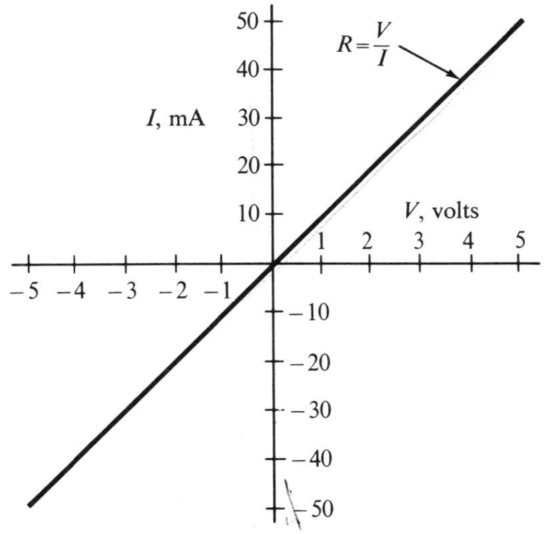

Fig. 2–64 Current-voltage (*I-V*) curve for a resistor.

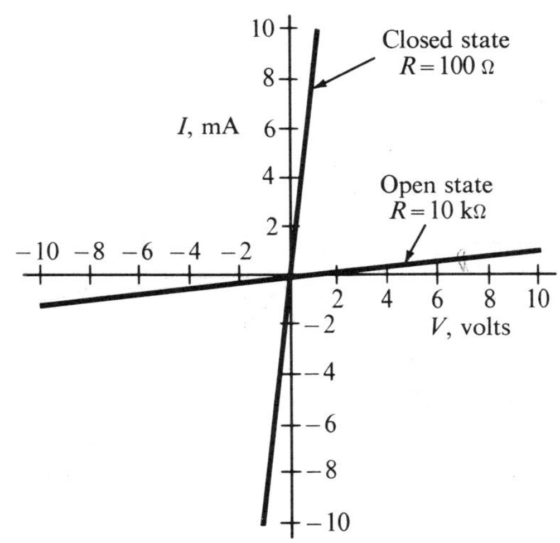

Fig. 2–65 Current-voltage curves for a less-than-ideal switch.

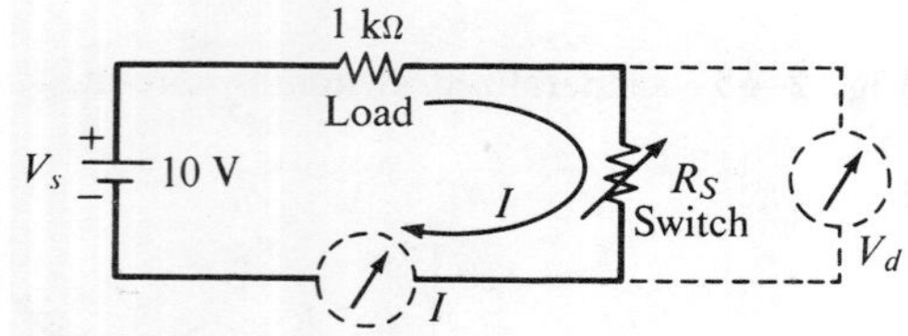

Fig. 2–66 Simple circuit with switch.

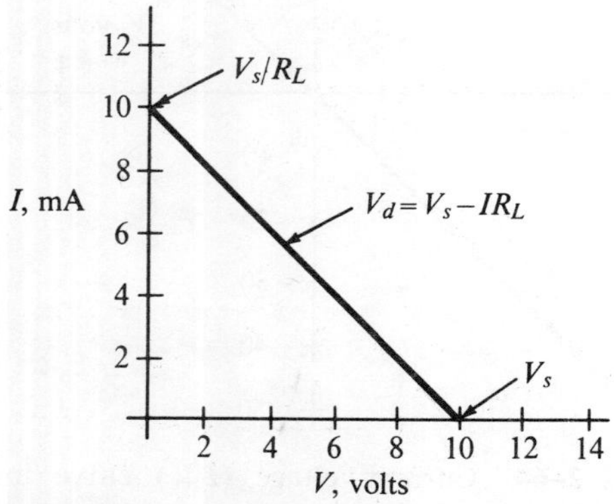

Fig. 2–67 Load line for the circuit of Fig. 2–66.

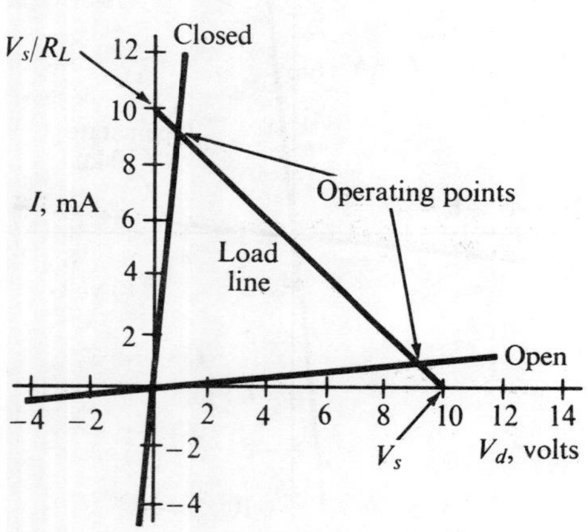

Fig. 2–68 Load line and characteristic curve combined to give operating points.

To determine the **operating points** for a switch, i.e., the voltage across and current through a switch in its ON and OFF states, it is only necessary to superimpose the *I-V* curves for the switch, such as those in Fig. 2–65, on the load line for the switch circuit. This is illustrated below for the circuit of Fig. 2–66, where a simple switch is used to control the current delivered to a resistive load.

Note that the switch is shown as a device with a variable resistance R_S. From Ohm's Law

$$I = V_s/(R_L + R_S). \tag{2–96}$$

Rearranging yields $IR_S = V_s - IR_L$, where IR_S is equal to V_d, the voltage across the device. Therefore,

$$V_d = V_s - IR_L. \tag{2–97}$$

This is a linear equation relating I and V_d, and a current-voltage line can be drawn as shown in Fig. 2–67. It is called "load line" because it is determined only by the source voltage V_s and the load resistance R_L. An easy way to obtain the load line is to locate the intercepts and join them by a straight line. From Eq. (2–96) it is seen that if $V_d = 0$, then $I = V_s/R_L$ and if $I = 0$, then $V_d = V_s$. For any value of R_S, the operating values of I and V_d must be located on the load line.

Figure 2–68 shows a superposition of the current-voltage curves of Fig. 2–65 on the load line of Fig. 2–67. Since the switch *I-V* curve and the circuit load line both define the possible current-voltage relationships for the system, the actual circuit current and switch voltage must be given by the intersection of the *I-V* curve and the load line. Each intersection is called an **operating point** of the switch *in that circuit*. Since the switch has an *I-V* curve for each state, there are two operating points as shown in Fig. 2–68. The action of each new switching device can be studied using this graphical method.

Experiment 2–14 *Characteristic Curves and Load Line for an Ideal Switch*

Experience is gained with a curve tracer for obtaining characteristic curves of an ideal switch. The load in a switch circuit is varied to show the variation of the load line and to gain experience with graphical techniques which are also applicable in studying diode and transistor characteristics in subsequent experiments.

Switch Capacitance

In the previous discussion of the nonideal switch we considered only the reduction in amplitude of the voltage across the switch caused by its nonideal ON and OFF resistances. There are also factors inherent in the switch device or circuit which affect the maximum speed of switching. One such factor is the capacitance, which is unavoidable in the construction of the device or circuit. This capacitance results in a certain "inertia" in the switch and affects the time required to go from the ON to the OFF state or vice versa.

To account for the switch capacitance the schematic of the nonideal switch given in Fig. 2–62 can be modified to have an equivalent capacitance C as shown in Fig. 2–69.

The determination of the maximum switching speed of the nonideal switch of Fig. 2–69 in the switch circuit of Fig. 2–66 requires an analysis of the charge and discharge of the capacitance C through the effective circuit resistances.

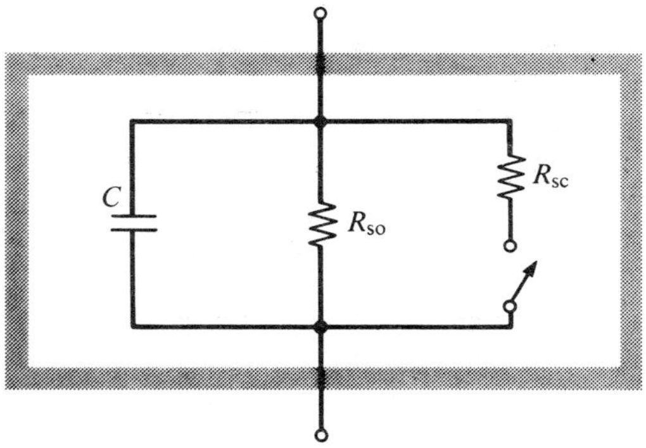

Fig. 2–69 Schematic representation of a nonideal practical switch.

Switch Inertia. The nonideal switch represented by the schematic diagram of Fig. 2–69 is shown connected in a series circuit in Fig. 2–70, and the effect of the equivalent switch capacitance will now be noted. The resistances R_{so} and R_{sc} are the open and closed values for switch S, and R_L is the load resistance of the circuit.

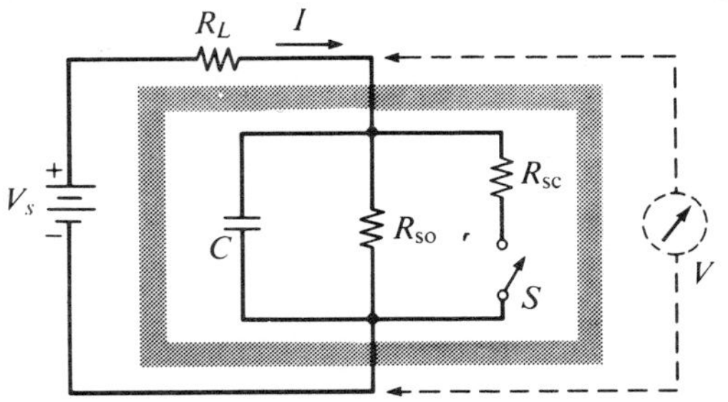

Fig. 2–70 Switch circuit with nonideal switch.

Assume that the switch S is initially closed, so that the voltage V_{sc} across the switch is

$$V_{sc} = V_s \frac{R_{sc}}{R_L + R_{sc}}$$

as determined by the voltage divider R_L and R_{sc}. Now if the switch is opened, the voltage across the open switch will rise exponentially as the equivalent capacitance C is charged to the maximum open switch voltage determined by the voltage divider R_L and R_{so},

$$V_{so} = V_s \frac{R_{so}}{R_L + R_{so}},$$

as illustrated in Fig. 2–71.

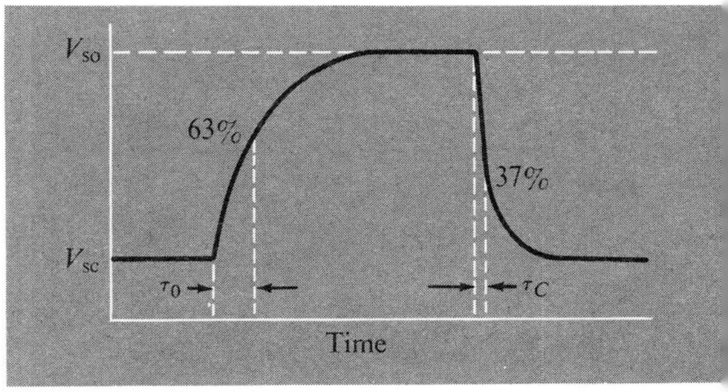

Fig. 2–71 Effect of capacitance on transition between voltage leads.

The charging time constant τ_0 will be determined by C and the output resistance of the voltage divider R_L and R_{so}. Thus

$$\tau_0 = CR_LR_{so}/(R_L + R_{so}). \qquad (2\text{–}98)$$

If $R_{so} \gg R_L$ as is usually the case, then $\tau_0 \approx CR_L$.

The discharge time constant τ_C is determined by C and the parallel discharge paths R_{so}, R_{sc}, and R_L. Since it is assumed that for an effective switch, $R_{so} \gg R_{sc}$,

$$\tau_C = CR_L R_{sc}/(R_L + R_{sc}). \tag{2–99}$$

In most cases $R_{sc} \ll R_L$ so that $\tau_C \approx CR_{sc}$.

In the above circuit, if $C = 100$ pF, $R_L = 10$ kΩ, $R_{so} = 1$ MΩ, and $R_{sc} = 100$ Ω, then $\tau_0 = CR_L = 100 \times 10^{-12} \times 10^4 = 10^{-6}$ sec $= 1$ μsec, and $\tau_C = CR_{sc} = 100 \times 10^{-12} \times 10^2 = 0.01$ μsec. Note that $\tau_C \ll \tau_0$, which is true for the typical switch. Therefore, a circuit with significant switch capacitance reaches a steady state more rapidly when the switch is closed than when it is opened. There are other factors inherent in certain devices such as diodes and transistors which determine switching speed, and these will be discussed in the next section.

Actuating Switch Contacts Electrically

When an electrical signal is used to actuate a switch, there are generally two electrical circuits involved: the actuating circuit and the switched circuit, as shown in Fig. 2–72. The degree of interaction between the actuating and switched circuits depends on the type of switch used and the circuit. Some switching devices have inherently one or two connections common between the switch and the actuating element. The three possible degrees of interconnection for simple switches are shown in Fig. 2–73.

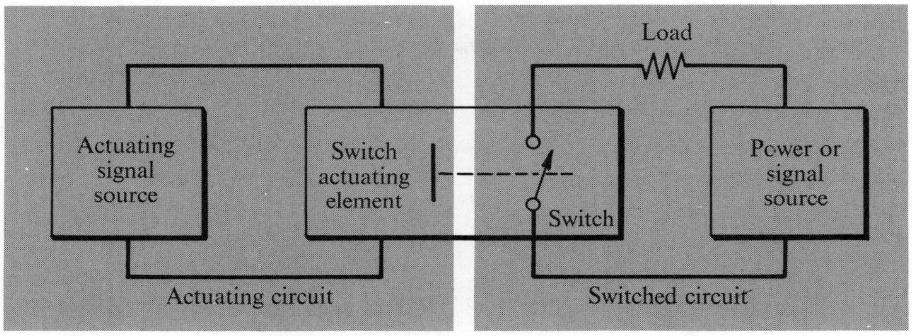

Fig. 2–72 Electrically actuated switching circuit.

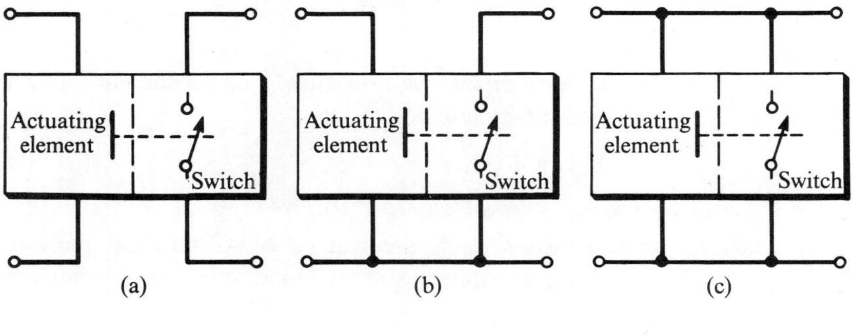

Fig. 2–73 Switching devices: (a) 4-terminal; (b) 3-terminal; (c) 2-terminal.

Examples of four-, three-, and two-terminal devices are the relay, transistor, and diode, respectively. As the descriptions of specific switching devices in this section illustrate, the degree of interaction between the actuating and switched circuits inherent in the switch is a major factor in defining its range of applicability.

Mechanical Switches

Mechanical switches are available in a wide variety of types with characteristics for many different applications. All mechanical switches require some form of pressure to close (make) and open (break) the switch contacts. Mechanical switch contacts can be actuated by the pressure of a human hand, in which case they are known as **manual switches.** Contacts can also be actuated by mechanical pressure, such as the pressure exerted by a cam rotated by a motor. These switches are known as **mechanically actuated switches.** Finally, switch contacts can be actuated electrically by electromagnets or permanent magnets. Such remotely controlled mechanical switches are known as **relays.** In this section a few of the most important types, characteristics, and applications of mechanical switches are presented and some of their advantages and limitations are discussed.

Manual Switch Types. Manual switches are used in applications where complete operator control over opening and closing switch contacts is desired and where switching speed is relatively unimportant. For example, the ON/OFF power switches on most instruments are manually actuated. Likewise manual selector switches are used in instruments for function selection, range changing, introduction of specific circuit elements, and many other control operations.

Manual switches can be classified in a variety of ways. However, it is quite common to classify switches by their actuating mechanism and by their number, type, and arrangement of switch contacts.

Several different types of manual actuating arrangements for switches are illustrated in Fig. 2–74. The choice of which type of actuating mechanism to use for a particular application depends on many factors, including ease of operation, whether repeated actuation is required, whether the switch must interconnect multiple contacts, etc.

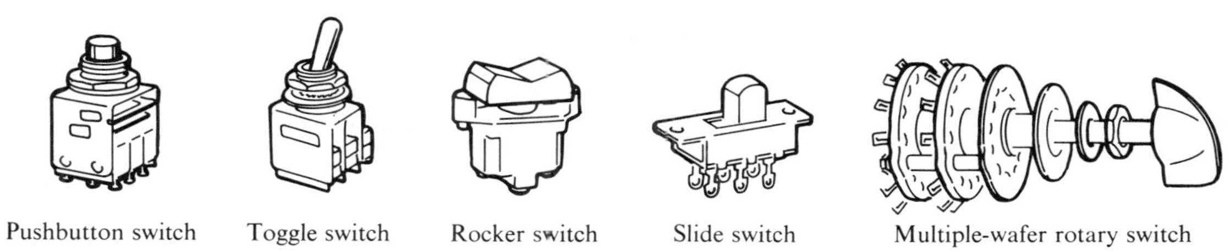

Fig. 2–74 Manually actuated switches showing different actuating arrangements.

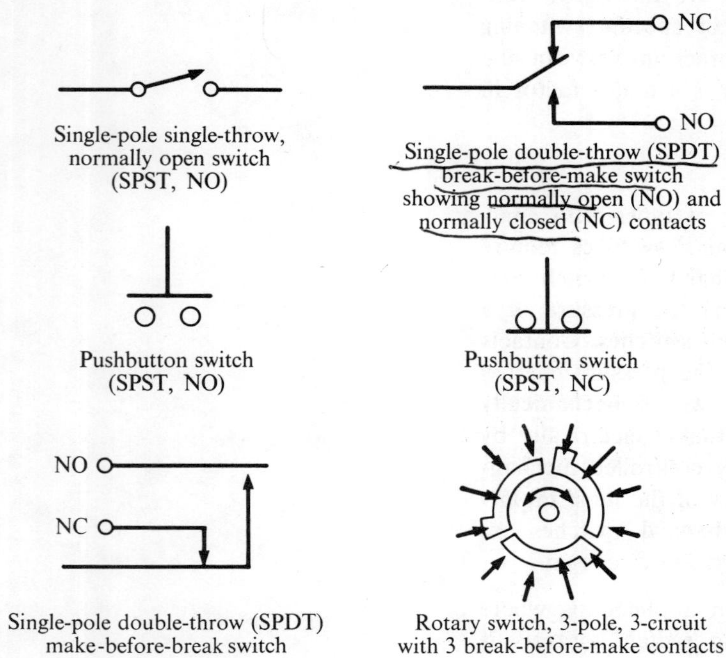

Fig. 2–75 Typical manual switch contact arrangements.

A very important consideration in switches is the switch contacts themselves. Metal-metal contacts are often silver or gold plated for applications where the contact quality is important. Some switches employ metal-liquid contacts, such as mercury-to-dry-metal contacts or mercury-to-wet-metal contacts. Others employ liquid-liquid contacts as in mercury-mercury switches. Mercury switches are actuated by a tilting motion which moves the mercury to make or break the contact.

Some contact arrangements and circuit symbols for typical manual switches are illustrated in Fig. 2–75. Specific circuit applications usually dictate the most appropriate arrangement of contacts. Break-before-make contacts are used when it is permissible to have a momentary open circuit and undesirable to have two circuit components connected together even momentarily. Make-before-break contacts are used when such a momentary open circuit would be undesirable, as in the switch selection of OA feedback resistors for a current follower, where a momentary open circuit would drive the OA to limit.

Manual Switch Characteristics. There are many important characteristics of switches which should be considered for specific applications. Manufacturers of switches usually list **load-life characteristics, contact re-**

sistance, insulation resistance, and **dielectric strength** of the insulation in their ratings of switches as well as contact current and voltage ratings. Load and life ratings are given for stringent test conditions established for the industry. These, of course, apply rigidly only to the test conditions. A switch may have a shorter or longer useful life in different applications, depending on the demands made on the switch. All switches deteriorate with age, and the rate at which deterioration takes place varies greatly with switch design.

The contact resistance of a mechanical switch is the resistance between a pair of closed contacts, or the ON resistance. Typical values are in the milliohm range for new switches. However, the contact resistance increases with age and is greatly influenced by the voltage, current, frequency, and environment of the switch.

The insulation resistance is the resistance between open contacts, or the OFF resistance. Typical values for new switches are in the thousand megohm range. Insulation resistance decreases with age, because of surface contamination.

The dielectric strength is a measure of the insulator's ability to withstand high voltages without breakdown. For new switches a typical value might be 1000 V ac.

Current and voltage contact ratings vary widely for different switch designs. Signal switching applications which involve low power switching can be carried out with miniature or subminiature switches, whose nominal ratings might be 5 A at 125 V ac and 24 V dc. Heavy duty power switches for ac power line switching can have typical ratings as high as 20 A at 125 V ac.

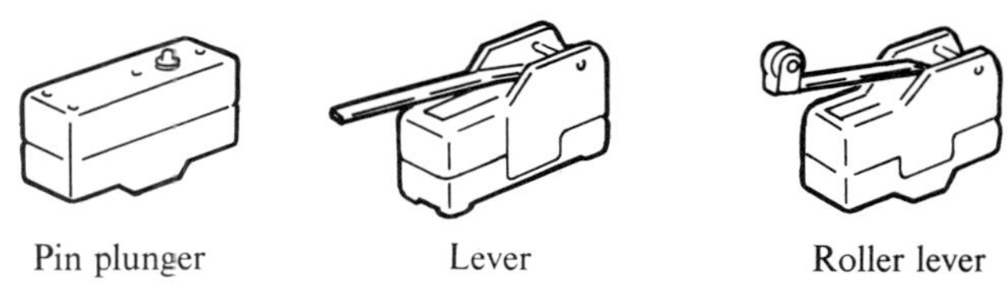

Fig. 2–76 Typical actuating mechanisms for mechanically actuated switches.

Mechanically Actuated Switches. Switches whose contacts are meant to be operated mechanically are also available in a wide variety of forms. Many of the contact considerations and characteristics are similar to those of the manual switches just described. Devices which transfer the actuating pressure to the switching mechanism are called **actuators,** and are available in many forms, as illustrated in Fig. 2–76.

In addition to contact ratings and load-life ratings, there are other important characteristics for mechanically actuated switches. The **operate**

force is the force required (in ounces) to actuate the switch. The **release force** is the force at the actuator at the instant of the return of the contacts to their normal position. **Pretravel** is the distance from the normal position to the operate position. **Overtravel** is defined as the limit of plunger travel beyond the operate position. Finally, **differential travel** is the distance the actuator moves from the operate position to the release position during its return stroke.

Relay Switch Types. Relays are widely used in electronic circuits as remotely controlled mechanical switches to turn a sequence of events ON or OFF. They are not so fast as transistor or optoelectronic switches, but they have higher open circuit resistance, lower contact resistance, and can generally switch higher loads. The electromagnetic relay is a four-terminal device in which the actuating terminals are electrically isolated from the switched signal terminals. Reed relays can be activated by a permanent magnet or a coil.

Electromagnetic relays utilize a current through a coil to provide a magnetic field that moves the switch contacts, as illustrated in Fig. 2–77 for an armature relay. An armature relay operates by energizing an electromagnet (with a suitable current) which attracts a pivoted lever of magnetic material to a fixed pole. The pivoted lever is called the armature. A switch contact moves with the armature to provide a movable contact. If the current in the coil (and, therefore, the magnetic force) is sufficient, the armature moves the movable contact until it touches the stationary contact. A spring holds the movable contact (and armature) in the open position when the coil is not energized.

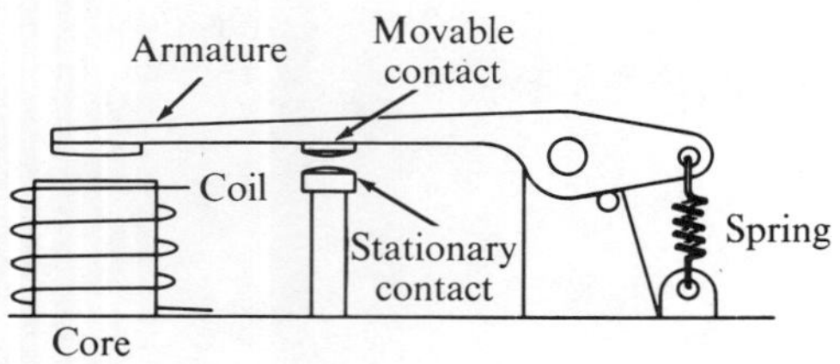

Fig. 2–77 Schematic diagram of single-pole, single-throw, normally open (SPST, NO) relay.

There are many types of electromagnetic relays. With a plunger relay, movable contacts are attached to a plunger that moves within a tubular magnetic coil (solenoid). Rotary relays often utilize the rotation of a motor shaft to move switch contacts.

Relays are constructed according to many designs to perform a wide range of functions. Some of the types of relays are interlock, stepping, sequencing, time delay, latch-in, polarized, differential, and general purpose.

The reed relay is a relatively new type of switch. Two or more metal reeds are enclosed in a hermetically sealed glass capsule. A normally open SPST reed relay is shown in Fig. 2–78. The overlapping reeds can be closed or opened by positioning a permanent magnet near or away from the reed contacts. The reed contacts can also be switched by actuating an electromagnet. Some typical arrangements for the use of the reed relay are illustrated in Fig. 2–79.

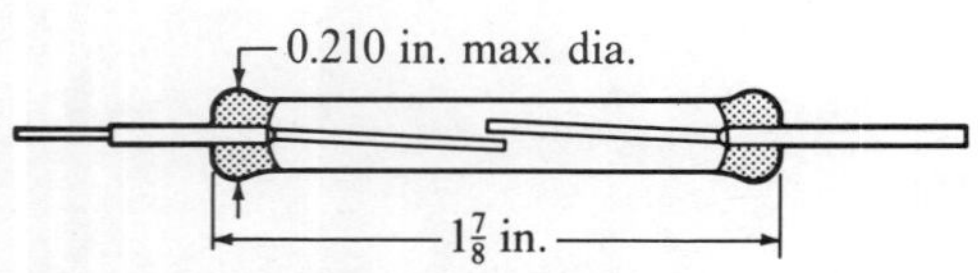

Fig. 2–78 Reed relay, with silver alloy contacts. (Courtesy of Hamlin, Inc.)

Relay Contacts and Characteristics. The number of relay contacts and contact arrangements is determined by the application. The nomenclature

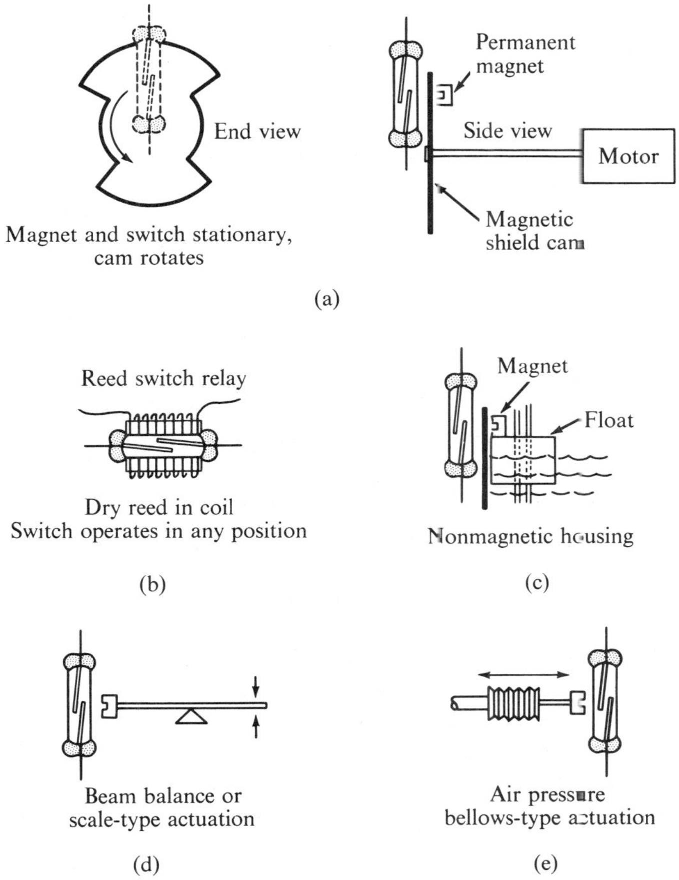

Fig. 2–79 Typical reed switch arrangements. (a) Switch operates whenever cut-out portion of shielding metal is between magnet and switch. (b) Switch located in cylindrical coil. (c) Solution level control. (d) Beam position indicator. (e) Magnet positioned by air-driven bellows. (Courtesy of Hamlin, Inc.)

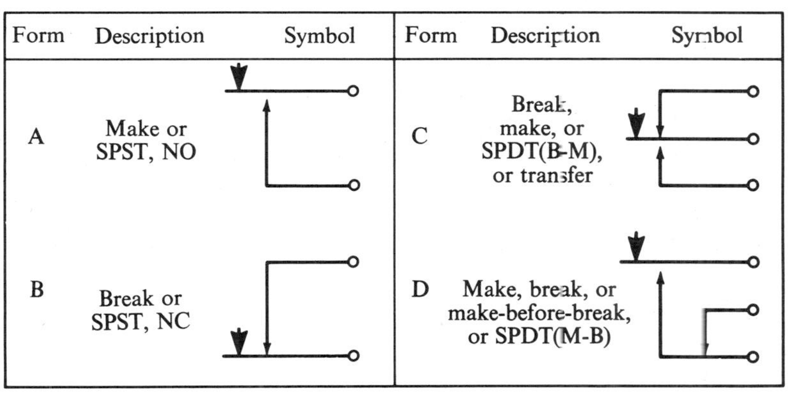

Form	Description	Symbol	Form	Description	Symbol
A	Make or SPST, NO		C	Break, make, or SPDT(B-M), or transfer	
B	Break or SPST, NC		D	Make, break, or make-before-break, or SPDT(M-B)	

Fig. 2–80 Four common forms of relay contacts with designations.

for the basic contact forms used to make up specific contact arrangements is illustrated in Fig. 2–80 for four of the most common forms. Many other contact forms are available.

A combination of two stationary contacts and one movable contact which engages one stationary contact when the coil is energized and the other stationary contact when the coil is not energized is called a single-pole double-throw (SPDT) relay. It is one of the most common contact arrangements and is illustrated as form C in Fig. 2–80.

The current in the relay coil must exceed a certain minimum value for the armature to "pull in" and close the NO contacts. At a somewhat lower current the armature will "drop out" and the NO contacts will open. It is common to design switching circuits to exceed the pull-in current by several times the minimum so as to ensure operation of the relay.

Relays do not operate instantaneously. In fact, most relays require at least a few milliseconds to complete their contact function, although the reed relays have operate times of about 1 msec. The operate time is the time interval from the instant of coil-power application until completion of the last contact function. The release time is the time interval from the instant of coil-power cutoff until the completion of the last contact function. The measurement of the operate and release times gives an appreciation for the limitations of relays as switches. A simple circuit for obtaining this information is shown in Fig. 2–81.

The circuit in Fig. 2–81a provides a voltage V (5 V in the example) across a voltage divider consisting of two 1 kΩ resistors. The voltage across the 1 kΩ resistor R_1 is fed to the input of the oscilloscope. However, at the start the normally closed contacts are connected across the 1 kΩ resistor, thereby shorting the input resistor R_1 and providing zero voltage input to the scope's vertical amplifier. The switch signal S operates the coil of the relay and is also connected to the trigger input of the scope. That is, the +5 V signal triggers the sweep and applies the drive current to the relay coil. There is then a finite time period (while the scope beam is sweeping across the horizontal axis at a known preset velocity) before the movable contact breaks away from the stationary contact. When the contacts do break away, one-half of the voltage V appears across the input to the scope's vertical amplifier, and the trace on the scope is deflected by the input voltage. The magnitude of the deflection depends on the sensitivity setting of the vertical amplifier, as illustrated in Fig. 2–81b.

If the function of the relay in a specific circuit is the breaking of the normally closed contacts, then the time from start of trace to the voltage jump is the operate time.

After the normally closed contacts break, there is a finite transfer time before the movable contact reaches the normally open stationary contact. It is possible to measure this transfer time by connecting the normally

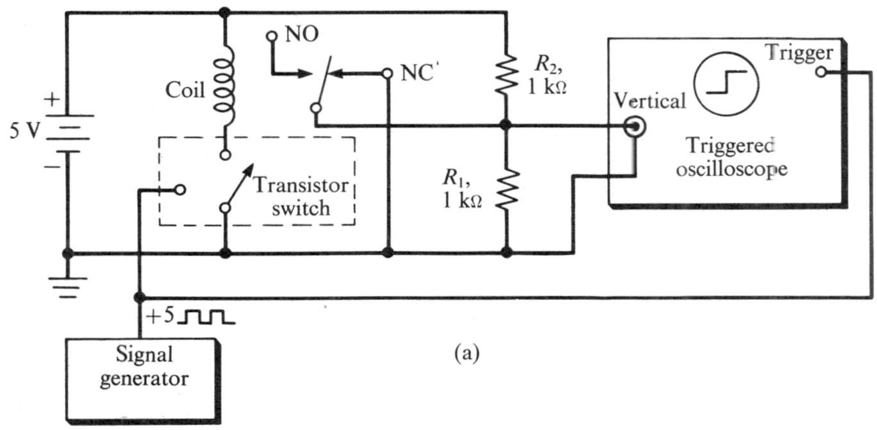

(a)

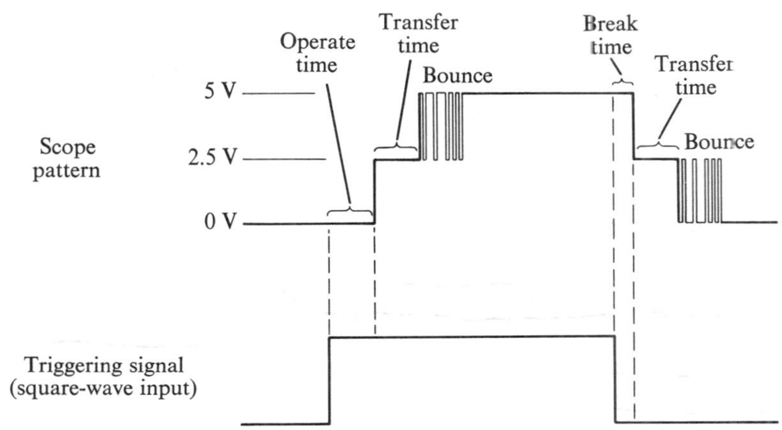

(b)

Fig. 2–81 Circuit for measuring operate, transfer, bounce, and release times.

open contact to the +5 V terminal. Note that the transfer time should not include the bounce time, which is very obvious on the scope trace and of considerable duration.

When the contacts break or make there is always some contact bounce. Although this bounce and its duration can seriously distort the switched signal, it is not included in the operate and release times. The bounce time is also a characteristic of manual switches.

Experiment 2–15 *Relay Characteristics*

The operate, release, and contact bounce times can be critical for suitable operation of some electronic circuits. It is the purpose of this experiment to observe these times and determine their significance in practical circuits. The pull-in and drop-out currents are also determined for a relay.

2–3.3 SOLID STATE SWITCHES

The advent of solid state devices for ON-OFF control of electrical quantities has been an essential part of the modern instrumentation revolution. Because of the various limitations of mechanical switches, which were detailed in the previous section, solid state switches are imperative for the short-time (μsec to nsec) ON-OFF control operations that many modern experiments require. The recent availability of inexpensive diodes, transistors, and thyristors has brought about new short-time measurement and control applications which would have been impossible only a few years ago. In this section the switching characteristics of solid state devices are described, and the graphical analysis of switching circuits is presented in order to provide a basic understanding of the types and properties of modern high speed switches.

The pn Junction Diode

Because of its special current-voltage characteristic the junction diode is often used as a high speed ON-OFF control device. It has the disadvantage of being a two-terminal device, which means that the signal that turns the switch ON and OFF is also the signal that is transmitted. In other words, the actuating signal and the signal to be switched must be the same signal in a junction diode.

Current-Voltage Curves and Load Lines. The circuit of Fig. 2–82 shows the junction diode acting as a switching device in a circuit with a voltage source which alternates between +10 V and −10 V. The switching cir-

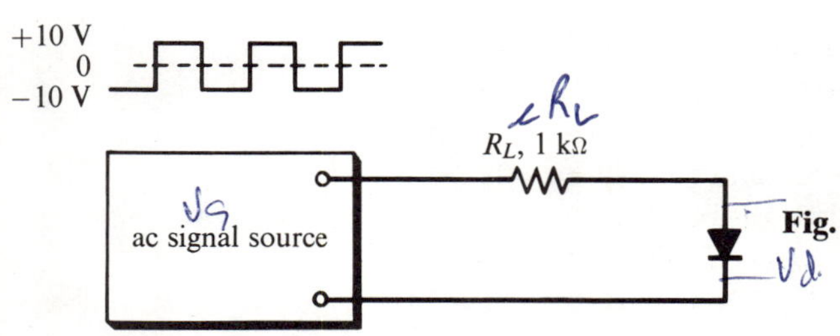

Fig. 2–82 Alternating current switching circuit.

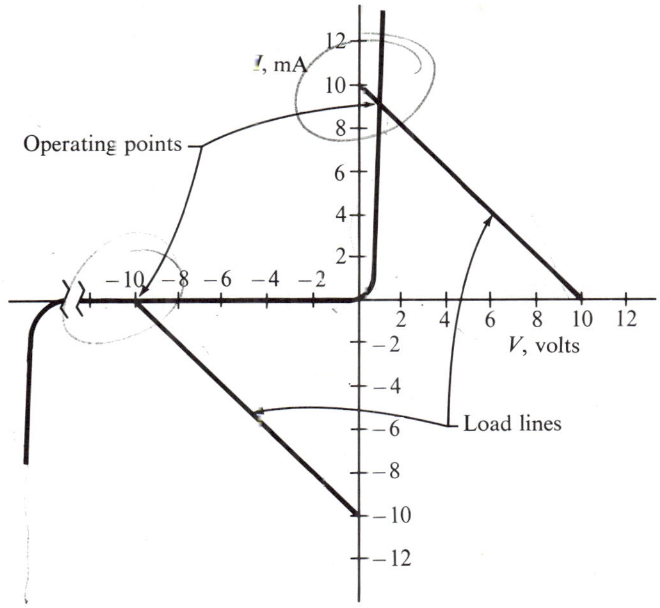

Fig. 2–83 Operating points for circuit of Fig. 2–82.

cuit can be graphically analyzed by the technique introduced in Section 2–3.2. The load line for the circuit is superimposed on the diode current-voltage curve and the operating points are established graphically. The equation for the load line is, as before, $v_d = v_s - iR_L$, where v_d is the voltage across the diode, v_s is the source voltage, and iR_L is the voltage drop across the load resistor. Since v_s has two values in the circuit of Fig. 2–82 (+10 V and −10 V), there are two load lines: one for each value of v_s, as shown in Fig. 2–83. The two operating points are found from the intersections of the load lines and the current-voltage curve. From the two operating points shown in Fig. 2–83, it can be seen that the voltage drop across the load resistor will be $\approx 9.4 \text{ mA} \times 1 \text{ k}\Omega = 9.4$ V when $v_s = +10$ V and ~0 V when $v_s = -10$ V. Hence from the operating points it is clear that a diode can be a switch which is OFF or ON depending on the polarity of the applied signal. The voltages across both the diode and the load for each value of v_s can be obtained by the above technique.

If the signal source is a sinusoidally varying ac voltage as shown in Fig. 2–84a, a graphical analysis can still be made although it is somewhat cumbersome, since load lines must be drawn for each value of v_s. As v_s varies with time, beginning with $t = 0$, the load line shifts outward horizontally during the first quarter-cycle until v_s reaches its peak value, V_p, as illustrated in Fig. 2–84b. During the second quarter-cycle, the load line moves

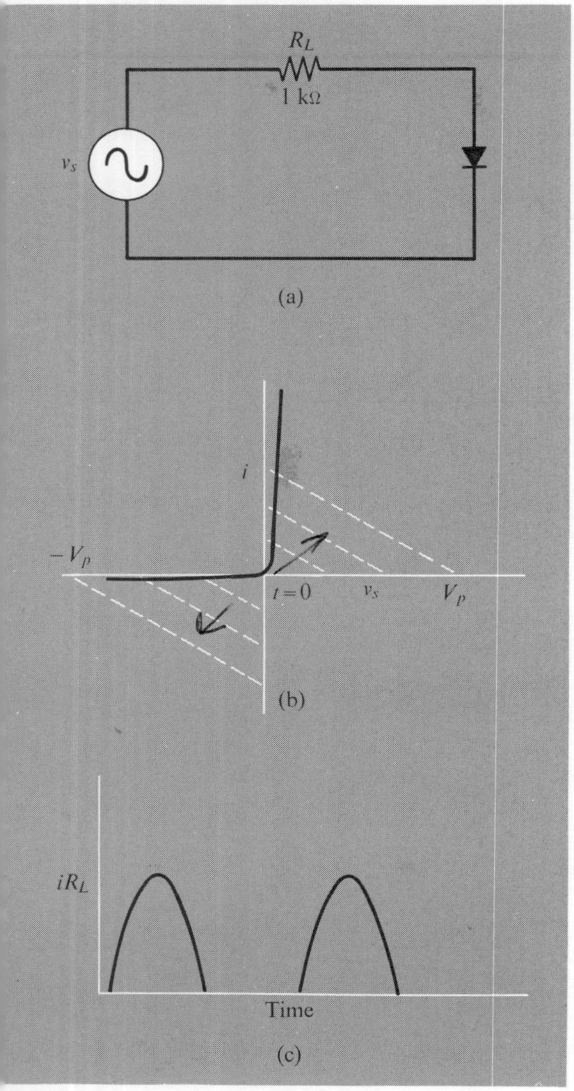

Fig. 2–84 Diode switch with sine wave source: (a) switching circuit; (b) graphical analysis; (c) load voltage vs. time.

inward horizontally as v_s decreases to zero. When v_s goes to negative values, the diode is reversed biased and the load line is drawn in the third quadrant. Again the load line moves outward horizontally until the peak negative value $-V_p$ is reached and moves inward horizontally until the signal source reaches zero, at which point the cycle repeats itself. The operating points for each value of v_s can be determined from the intersection of the *i-v* curve and the varying load line. The value of current at each operating point can be read from the graph, and the voltage drop across the load can be calculated at each instant in time and plotted, as shown in Fig. 2–84c.

Circuit Models. In many situations, particularly those in which the signal source varies continuously with time, the graphical analysis technique described above is rather cumbersome, and it becomes easier to work with a linear circuit model for the diode. For switching applications in which the signal source varies between two fixed voltage values, as in Fig. 2–82, the circuit model used is usually the **static model** of the diode. In the static model, a forward biased diode is modeled as a simple resistor of resistance R_{fs}, the static forward resistance. The value of R_{fs} is found from the ratio of diode voltage to diode current at the ON operating point. Likewise the reversed biased diode is modeled as a simple resistor of resistance R_{rs}, where the reverse static resistance is found from the OFF operating point. Hence the complete linear model for the circuit of Fig. 2–82 consists of two circuits: one for the +10 V condition of the signal source and the other for the −10 V condition, as shown in Fig. 2–85a and b. Since these circuits give the same electrical quantities at the two values of v_s as does the diode, they are often known as **equivalent circuits.**

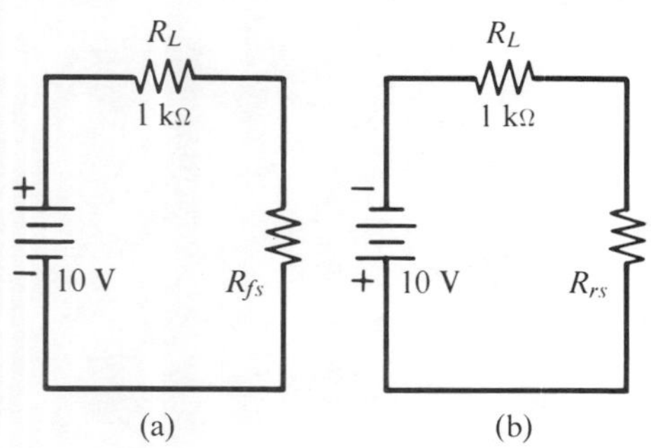

Fig. 2–85 Static circuit models for the diode switch of Fig. 2–84: (a) forward biased equivalent circuit; (b) reversed biased equivalent circuit.

Fig. 2–86 Piecewise-linear model of diode: (a) current-voltage characteristic; (b) reverse bias model; (c) forward bias model; (d) simplified model neglecting reverse current.

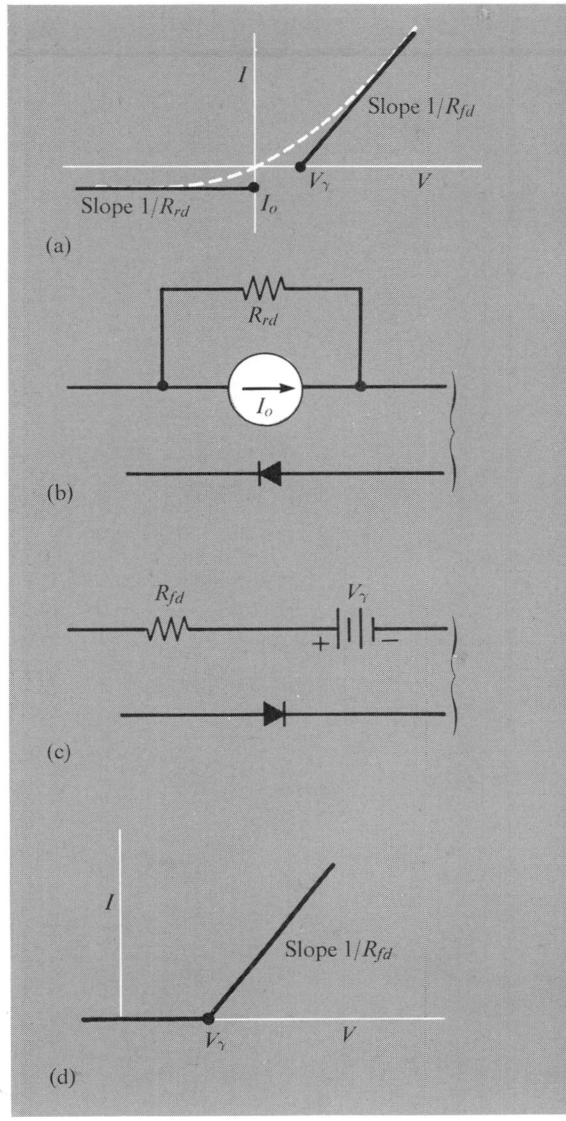

The static model of the diode given above is not particularly useful for continuously varying signals such as sine waves, or for two-state signals when the values of v_s are unknown. Instead a linear model consisting of two straight-line segments is used. Such models are known as **piecewise-linear models.** The most complex piecewise-linear model that is used for the pn junction diode is illustrated in Fig. 2–86. In Fig. 2–86a the diode in reverse bias is approximated by a straight-line current-voltage characteristic whose slope is the reciprocal of the reverse dynamic resistance, R_{rd}. The reverse dynamic resistance is the reciprocal of the slope of the diode I-V curve in the region of the reverse operating point ($R_{rd} = dV/dI$). The resulting circuit model of the diode in reverse bias is shown in Fig. 2–86b and consists of a constant current source and a shunt resistance of value R_{rd}. In forward bias the diode current-voltage characteristic is approximated by a second straight-line segment whose slope is now the reciprocal of the forward dynamic resistance R_{fd}. The forward dynamic resistance is the reciprocal of the slope of the diode I-V curve in the region of the forward operating point ($R_{fd} = dV/dI_m$). The straight-line characteristic intersects the voltage axis at voltage V_γ, which is sometimes called the **threshold voltage or cut-in voltage.** For forward currents of up to 10 mA, V_γ is approximately 0.2 V for germanium diodes and 0.6 V for silicon diodes. For larger currents, up to 50 mA, better values of V_γ are 0.3 V for germanium and 0.65 V for silicon. The circuit model which results in the forward bias region is shown in Fig. 2–86c and consists of replacing the diode with a constant voltage source of voltage V_γ and a series resistor R_{fd}.

Because I_i, the reverse current of the diode, is usually very small (of the order of nanoamperes for a silicon diode), it is often completely neglected, as shown by the simplified piecewise-linear model of Fig. 2–86d. Here the diode is considered an open circuit of infinite back resistance for values of $v_s < V_\gamma$; for $v_s > V_\gamma$ the circuit model of Fig. 2–86c applies. This simplified model is most often used in applications where the reverse current can be neglected.

A useful relationship for estimating the dynamic resistance of a forward biased diode can be obtained from the basic diode current-voltage

(I-V) equation

$$I = I_i(e^{V/\eta V_T} - 1), \qquad (2\text{–}100)$$

where V_T, *the voltage equivalent of temperature,* is equal to $(kT/Q_e) = T/11{,}600$, and η is a characteristic of the diode. For germanium $\eta \simeq 1$ and for silicon, $\eta \simeq 2$.

Forward bias corresponds to positive values of V in Eq. (2–100). For values of V greater than 0.25 V for silicon or 0.12 V for germanium, the exponential term is more than 100, so the quantity (-1) is negligible and Eq. (2–100) is approximately

$$I = I_i e^{V/\eta V_T}, \qquad (2\text{–}101)$$

showing an exponential increase in current as a function of the voltage. Taking the logarithm of both sides of Eq. (2–101) and evaluating at 300°K,

$$\log I - \log I_i = V/(2.3\eta V_T) = V/(0.059\eta).$$

Thus the voltage increases by 59η mV for every tenfold increase in current at room temperature. For real diodes, a plot of log I vs. V often gives a straight line over several orders of magnitude.

If Eq. (2–101) is differentiated with respect to V, Eq. (2–102) results:

$$\frac{dI}{dV} = \frac{1}{\eta V_T} I_i e^{V/\eta V_T} = \frac{1}{\eta V_T} I, \qquad (2\text{–}102)$$

or inverting,

$$\frac{\eta V_T}{I} = \frac{dV}{dI} = R_{fd} = \frac{0.026\eta}{I}. \qquad (2\text{–}103)$$

For example, according to Eq. (2–103), if $I = 0.1$ mA, then the dynamic resistance is 260 Ω for germanium and 520 Ω for silicon, but at 1 mA it is about 26 Ω and 52 Ω, etc. At lower currents, the ohmic resistance of the semiconductor becomes significant.

Classes of Diodes. The two general classes of diodes are signal diodes and rectifier diodes. The signal diodes are used for many functions in digital instrumentation and detector circuits where the forward current and reverse voltage requirements are not high (e.g., 50 mA and 50 V) but where fast response times, low capacitance, and low leakage currents are desirable characteristics.

The rectifier diodes are capable of high current, starting at about 100 mA and ranging to several hundred amperes. The reverse voltage ratings typically vary from about 50 to 600 V, but some newer designs exceed 1000 V. All pn junctions can be damaged by excessive reverse voltages unless the current is limited to a safe value, and consequently they have peak inverse voltage (PIV) ratings.

Diode Switching Times. Most electronic switching devices operate inherently many orders of magnitude more rapidly than mechanical switches. The development of signal switching techniques operable on the microsecond or lower time scale has been a major breakthrough for scientific instrumentation. As we continue in our efforts to measure time more precisely and to resolve events separated by shorter and shorter times, the problem of switching time continues to limit the accuracy and speed of our measurements.

A reverse biased junction diode has very few charge carriers in the junction region. If the diode is suddenly forward biased, the resistance will remain high until majority carriers, accelerated by the new potential gradient, cross the junction boundary. This requires only a few nanoseconds. Having crossed the junction, they are now minority carriers and are neutralized eventually by majority carriers. This reduces the majority carrier concentration near the junction, causing a net diffusion of majority carriers toward the junction from each side. Once established, this diffusion maintains the supply of charge carriers at the junction necessary for the flow of charge across the junction. But until the concentration gradient is established, an electric potential gradient is required to maintain the necessary rate of transport of majority carriers to the junction. Thus, even though forward current is flowing, the forward potential drop across the diode will be higher than its steady-state value. The attainment of steady state is usually accomplished within a fraction of a microsecond. In most applications the turn-on time would be just the time required for the forward current to reach a particular fraction of the steady-state level after the forward bias was applied.

The forward biased junction diode has relatively large concentrations of minority carriers near the junction. These carriers have just crossed the junction but have not yet been neutralized by the majority carriers. When a reverse bias is suddenly applied to a forward biased diode, the minority carriers in the junction region will be attracted back across the junction by the new gradient. That is, electrons near the junction in the p region will be accelerated back toward the nonpositive n region, and similarly for the holes near the junction in the n region. In other words, the junction is still conducting. This gives rise to a substantial reverse current until the minority carriers in the junction region are depleted to their usual low value for the reverse biased diode. This is called the **stored-charge effect** because of the excess of minority charge carriers which are stored in the junction region during conduction and which need to be swept out of the junction region before the diode is OFF. This effect will be discussed again in more detail when the junction transistor is described.

The diode current behavior following the sudden application of a reverse bias is shown in Fig. 2–87. The storage time, t_S, is the time required

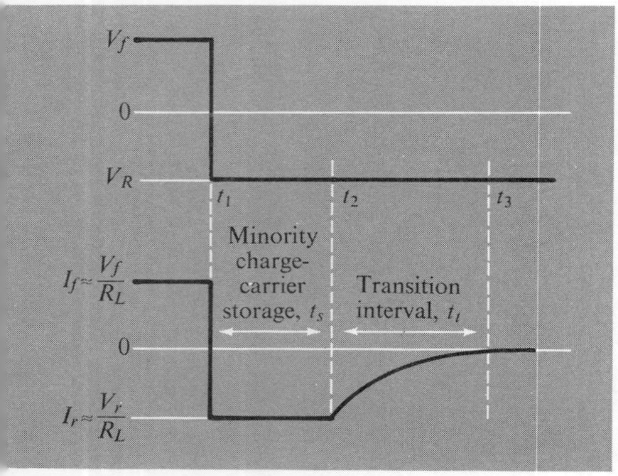

Fig. 2–87 Diode current behavior after application of step reverse bias.

to sweep the minority carriers from the junction. During the transition interval, the diode resistance is increasing. This occurs gradually because excess minority carriers farther from the junction are still crossing the junction and because the capacitance of the reverse biased pn junction must charge through R_L to the reverse bias potential $-V_r$.

Schottky Barrier Diodes. Special high speed switching diodes, known as Schottky diodes, are made which greatly reduce the amount of the stored charge and the width of the junction region. Such devices in appropriate circuits have achieved nanosecond switching times. The Schottky diode is a metal-semiconductor junction rather than a normal pn junction. Since there are no mobile positive charge carriers in the metal, the forward current is essentially all carried by electrons flowing from the semiconductor into the metal. These electrons rapidly reach equilibrium with other electrons in the metal, which eliminates the stored charge effect. Schottky diodes have a lower forward voltage drop across them for a given current than a pn junction diode.

Experiment 2–16 *Characteristic Curves and Resistance of a Diode*

The characteristic curve for a silicon diode is obtained with an oscilloscopic curve tracer, and the significant diode parameters are measured from the curve. The values obtained from the forward biased diode characteristic curve are used to plot log I vs. V so as to determine the logarithmic relationship over a few decades. The dynamic resistance of the diode is also determined and compared with the static resistance at a current of 1 mA. The problems of using an ohmmeter to measure diode resistance are determined.

Reverse Breakdown and Zener Diodes

In Fig. 2–83 it was noted that when the reverse voltage exceeds a certain value the reverse current suddenly increases, and this is called the breakdown voltage. The large reverse bias produces an electric field (greater than 10^6 V/cm) that is sufficient to break the covalent bonds and produce free mobile electron-hole pairs, by the process of tunneling. It occurs in abrupt junctions between highly doped regions. Another type of breakdown can occur called avalanche breakdown, which is caused by carriers that gain sufficient kinetic energy to ionize other atoms with which they collide, thus producing additional mobile electron-hole pairs. The mechanism of breakdown depends on the voltage region of breakdown, and both mechanisms can be present in the same diode. Silicon diodes that break down above 8 V probably depend on the avalanche mechanism, and those in the region below 4 V on the tunneling mechanism. For those diodes that break down between 4 and 8 V both mechanisms are involved, and the dominant mechanism depends on the impurity distribution at the junction. Even though the junction breakdown mechanism originally proposed by C. Zener was the tunneling mechanism, all diodes designed specifically for operation in the reverse breakdown region are called **Zener diodes.** Two common symbols for Zener diodes are shown in Fig. 2–88. The symbol is meant to suggest the shape of the *I-V* curve.

Fig. 2–88 Zener diode symbols.

In a similar manner, the reverse bias region before breakdown is sometimes used as a source of very small constant currents. A very high resistance is a desirable characteristic in constant current sources. This application of the diode is not very common, however, because of the large effect of temperature on the reverse current I_i and the limited range of possible currents.

The use of a Zener diode for regulating the output voltage of a power supply is described in Section 2–4.

Experiment 2–17 *Reverse Breakdown and Zener Diodes*

In this experiment the reverse characteristic curve for a Zener diode is observed and the significant parameters of the Zener diode are measured from the curve.

Microcircuit Applications of pn Junctions

The development of microcircuits consisting of hundreds of components (diodes, transistors, capacitors, and resistors) fabricated in a silicon chip has been at the heart of the instrument/automation revolution. Many components can be interconnected right on the chip to form complex circuits called **integrated circuits.** The reverse biased pn junction finds many applications in such microcircuits.

Figure 2–89 illustrates a pn junction and the depletion layer associated with it. As was pointed out in Section 2–3.1, the depletion layer contains very few mobile charge carriers. When the junction is reverse biased, a negligibly small current flows across the junction. Hence a reverse biased pn junction is an excellent insulator and is extensively applied to provide isolation between the p- and n-type semiconductor materials across the junction. In most present integrated circuits the various active components are isolated from each other inside the tiny silicon chip by means of such reverse biased pn junctions. Components are then connected by means of a metal, such as aluminum, which is deposited on the chip surface to form

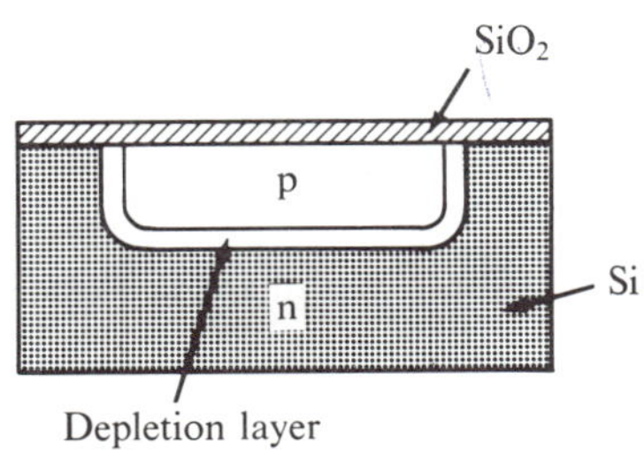

Fig. 2–89 A pn junction used for isolation.

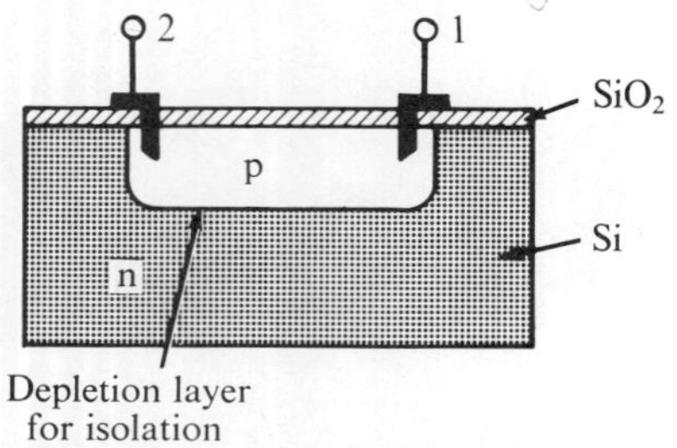

Fig. 2–90 A diffused microresistor with pn junction isolation.

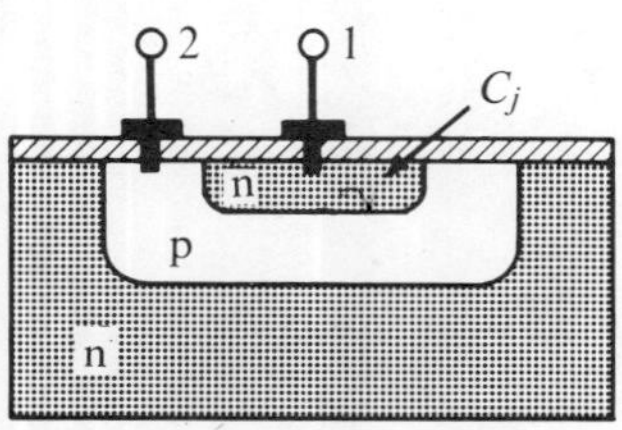

Fig. 2–91 A microcapacitor from a pn junction.

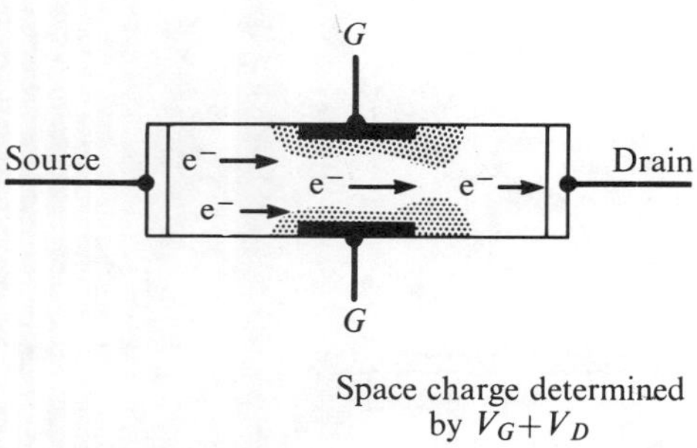

Fig. 2–92 Illustration of channel space charge.

the interconnection pattern. Because there is capacitance associated with the pn junction, the degree of isolation of components by reverse biased junctions falls off at high frequency.

Another common use of the reverse biased pn junction for isolation in microcircuits is illustrated in Fig. 2–90, where a diffused microresistor is shown. The p-type semiconductor is used as the resistor and the reverse biased junction isolates the resistor from any other components on the chip. Terminals 1 and 2 provide the electrical contacts for the resistor. Resistance values ranging from 100 Ω to about 30 kΩ can be obtained in this manner.

Reverse biased pn junctions are also occasionally used as microcapacitors for integrated circuits as illustrated in Fig. 2–91. Here two junctions are shown. One of the junctions provides the desired capacitance C_j, while the other isolates the capacitor from any other components on the chip. Electrical connections to the capacitor are provided by the terminals 1 and 2. Unfortunately the small size of integrated circuits prevents the achievement of large values of capacitance in this manner, and typical maximum capacitances are in the range of tens of picofarads.

Field-Effect Transistor Switches

The field-effect transistor (FET) was briefly introduced in Module 1 with respect to its application as the input stage of a voltmeter. In this application the high input resistance of the FET prevents loading of voltage sources that are measured with the FET voltmeter. Another very important application of the FET is as a high speed switch with very high OFF/ON resistance ratio.

Junction Field-Effect Transistors. Recall that the operation of the unipolar FET as a variable resistance depends on the formation of a "channel" of variable thickness through which the electrons must pass in going from the source to the drain. This is illustrated in Fig. 2–92 for an n-type semiconductor. The channel is formed by introducing p-type impurities into opposite sides of the semiconductor bar, and this type of FET is called a junction FET, abbreviated J-FET. A space charge is developed at the p-type "gate" regions which provides a certain effective channel thickness through which electrons flow. If a reverse bias is applied between the source and gate, the space charge is increased and the effective channel thickness for electron flow is decreased.

When the voltage V_D is applied as shown in Fig. 2–93a, the drain is made positive with respect to the source, and electrons flow from the source through the channel to the drain. The gate voltage V_G controls the current by controlling the effective channel thickness. As would be ex-

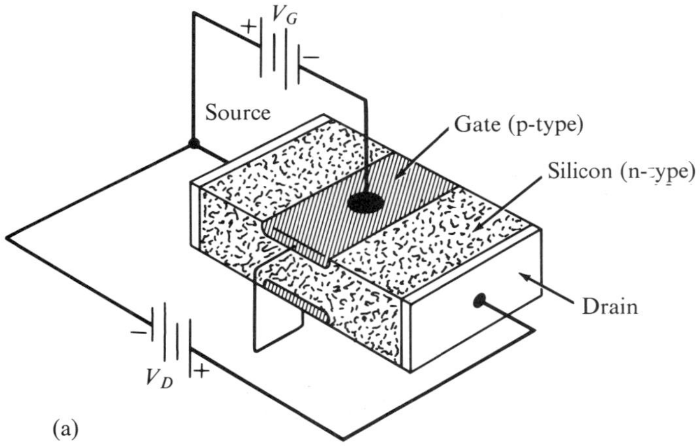

(a)

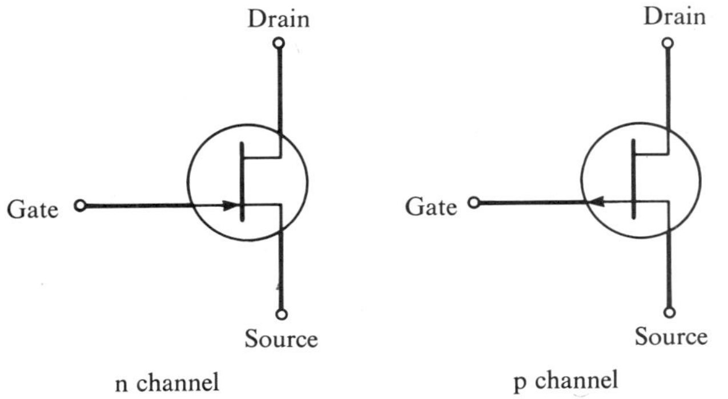

(b)

Fig. 2–93 The junction FET; (a) pictorial; (b) symbols.

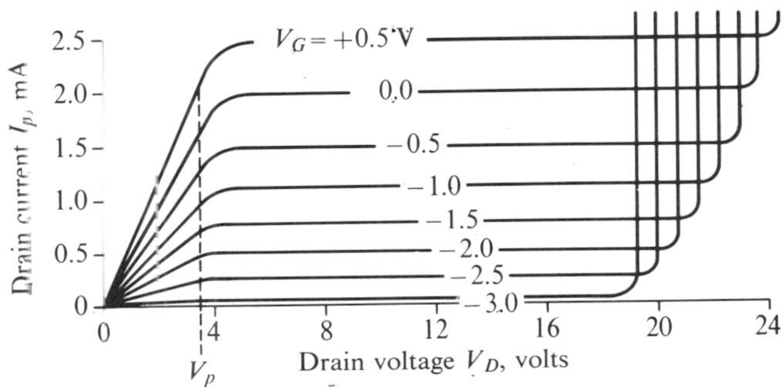

Fig. 2–94 Characteristic curves for junction FET.

pected, there is a critical drain voltage above which there is no increase in drain current. This is called the "pinch-off voltage" V_p and is dependent on the mobility of the electrons in the silicon bar. As shown in Fig. 2–94 a plot of drain current vs. drain voltage provides a family of characteristic curves for different values of gate voltage V_G. The circuit symbols for n- and p-channel FET's are shown in Fig. 2–93b.

When the FET is used as a switch, the gate bias V_G is made sufficiently large to turn the carrier flow OFF. The OFF resistance is very high, typically greater than $10^9\ \Omega$. The ON resistance for zero gate bias can be estimated at worst to be $R_{ON} = V_p/I_{DSS}$, where I_{DSS} is the current determined from the point where the pinch-off voltage V_p crosses the characteristic curve for $V_G = 0$ in Fig. 2–94.

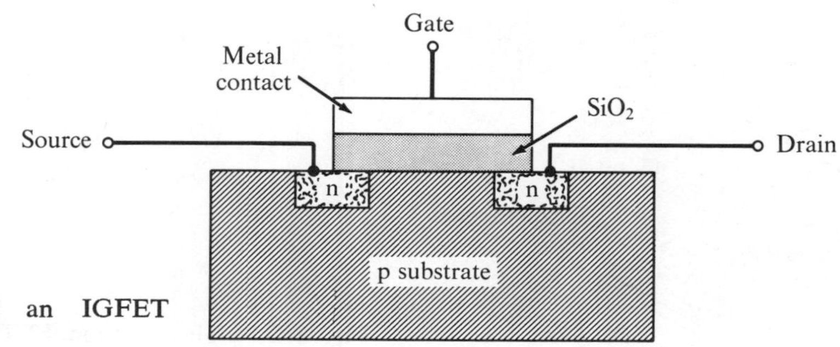

Fig. 2–95 Representation of an IGFET (MOSFET).

Insulated Gate Field-Effect Transistor (IGFET or MOSFET). It is possible to construct the FET in several ways. The junction FET of Fig. 2–93 is frequently used, but it does not have as high an **OFF** resistance as the insulated gate FET (IGFET).

The IGFET can be constructed so that essentially no drain current flows under zero bias conditions, as illustrated in Fig. 2–95. The pn junctions in the substrate very effectively isolate the source from the drain under zero bias on the gate. The drain-source voltage reverse biases the drain junction so that drain current is essentially zero. Typical resistances are $10^{14}\ \Omega$ or more.

A thin layer of metal oxide (e.g., SiO_2) is deposited on the surface over the channel, and this is covered with a thin layer of deposited metal, forming a capacitor between the gate and substrate. The application of a positive voltage between gate and source will induce a negative charge in the channel so that it becomes in effect an n-type material and current flows between source and drain. The gate acts to enhance the number of current carriers in the channel. It is from the use of the semiconductor with metal and oxide structure that the name MOSFET, frequently used to identify the device, originates. The abbreviation MOST is also used to identify this type of transistor. In those cases where the **insulated gate** is formed by a chemical compound other than the oxide, the appropriate name is IGFET.

The MOSFET can also be fabricated so that with $V_G = 0$ there is a conductive path between source and drain. A negative voltage applied to the gate will then induce a positive charge in the channel which causes a depletion of carriers and can turn the MOSFET **OFF**.

The symbols for the MOSFET are shown in Fig. 2–96.

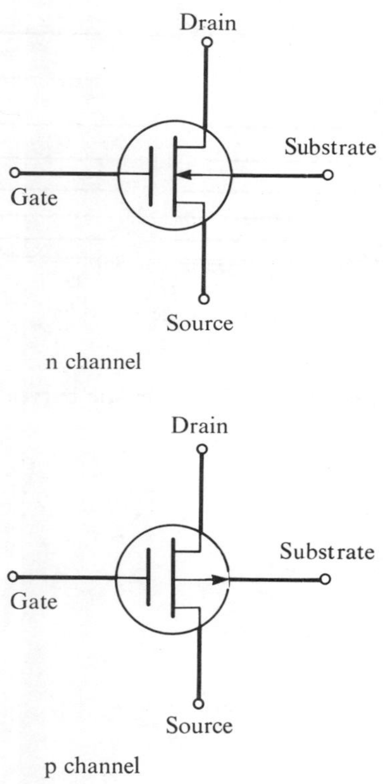

Fig. 2–96 MOSFET symbols.

Experiment 2–18 *Field-Effect Transistor (FET) Characteristics*

The characteristic curves for a FET are obtained from a curve tracer and the load line and ON-OFF characteristics are measured.

Junction Transistor Characteristics

Seldom has a component in any apparatus received so much general publicity as has the transistor, a word known to every schoolchild. This remarkable device has brought about major advances in electronics and has pointed the way to the development of other devices and the new integrated circuits.

One of the important applications of the transistor is as a switch to control electrical current and voltage. The transistor has the advantage that it is a three-terminal device and hence offers some isolation between the information signal to be switched and the electrical actuation signal.

The operation of the junction transistor is based on many of the principles described in Section 2–3.1 for the pn junction. In fact, the transistor is essentially made up of two pn junction diodes coupled by a very thin common base, either of p-type or n-type semiconductor material. Therefore, there are two types of junction transistors depending on the material of the base, the npn and the pnp transistors, as shown by the pictorial representation in Fig. 2–97. Because of the two pn junctions, this type of transistor is often called a **bipolar junction transistor** and referred to as a BJT.

In Fig. 2–97 a lead is shown attached to the base. Leads are also shown connected to the two regions of semiconductor material that are adjacent to and of opposite type to the base material. These regions are called the **emitter** and the **collector.** Although it would appear from the figure that the emitter and collector are identical, the fabrication techniques provide each with certain practical characteristics described later in this section. However, in principle, the emitter-base and collector-base junctions are of similar type, and the operation of the transistor depends on the applied voltages between the three terminals.

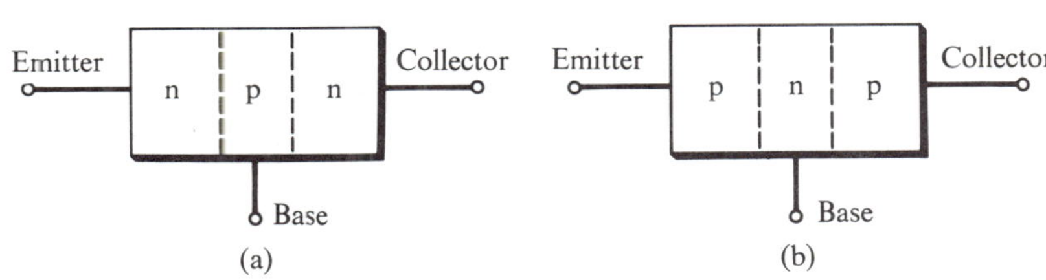

Fig. 2–97 Basic transistor types; (a) npn; (b) pnp.

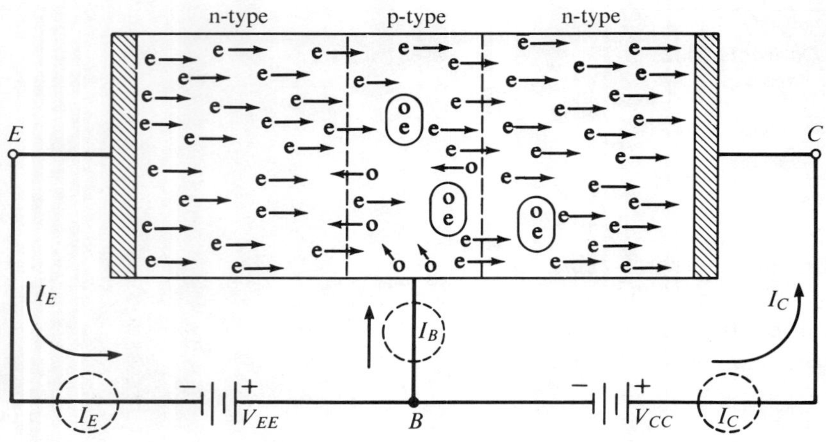

Fig. 2–98 Representation of npn transistor in operation with forward biased emitter-base junction and reverse biased collector-base junction (e = electrons, o = holes, and oe = recombination of holes and electrons).

Principles of Transistor Operation. The internal operation of the npn transistor is represented schematically in Fig. 2–98. It should be noted that the base is very thin, only about 10^{-3} to 10^{-4} cm for a typical transistor. If the emitter-base junction is forward biased the energy barrier is reduced, so that the majority carriers in the emitter (electrons for the npn transistor) cross the barrier and are emitted or ejected into the base, where they become minority carriers. The emitted electrons that cross the junction rapidly diffuse through the very thin base region. When they reach the region of the base-collector junction they are rapidly accelerated toward the collector. Note that because the collector-base junction is reverse biased the n-type collector is positive with respect to the p-type base. The electrons that reach the collector junction are, therefore, attracted to the collector and into the external circuit. In effect, then, the movement of minority carriers through the base (i.e., electrons in the p-type base) primarily determines the collector current I_C.

Some of the electrons combine with holes as they diffuse through the base, and thus a small base current is required to maintain the hole concentration. Also contributing to the base current is the need to compensate for the difference between hole currents across the two junctions. The holes in the base cross the forward biased emitter-base junction, but the number of these holes is much smaller than the number of electrons crossing from emitter to base, because the emitter has a much higher impurity content (charge carrier density) than the base. The number of holes that cross the collector-base junction is usually very small because the collector is heavily doped and the junction is reverse biased. Therefore, the base current I_B is primarily a result of the emission of holes from base to emitter and the electron-hole recombination that occurs in the base.

Because $I_E = I_B + I_C$ from applying Kirchhoff's Law to junction B of Fig. 2–98, and because I_B is usually small compared to I_E, the collector current I_C will be just a little less than I_E. Typical relative values might be $I_C = 99$, $I_E = 100$, and $I_B = 1$. For the configuration shown in Fig. 2–98 it is useful to define a forward current gain α_N as

$$\alpha_N = I_C/I_E, \tag{2–104}$$

the fraction of the emitter current that appears in the collector circuit.

The collector current I_C is assumed here to be a fraction θ of the injected electron current I_n,

$$I_C = \theta I_n, \tag{2–105}$$

and because the emitter current I_E is the sum of the electron current I_n and the hole current I_p,

$$I_E = I_n + I_p, \tag{2–106}$$

it follows from Eqs. (2–104), (2–105), and (2–106) that

$$\alpha_N = \frac{I_C}{I_E} = \frac{\theta}{1 + I_p/I_n}. \tag{2–107}$$

Therefore, even if there is no recombination in the base, so that $\theta = 1$, it can be seen from Eq. (2–107) that α_N would be less than unity. As noted previously, the fabrication procedures for the emitter and base regions of the npn transistor result in very few holes crossing the emitter-base junction as compared to the number of electrons that cross, so that the ratio I_p/I_n is small, and from Eq. (2–107) it follows that α_N is approximately equal to θ.

In the discussion above it was assumed that the collector current I_C was a fraction θ of the emitter electron current I_n. However, it is possible that the kinetic energy of the electrons as they are accelerated through the base-collector junction will be sufficient to cause additional electrons to be released from the atoms in the collector. The collector current I_C in the external circuit will thus exceed the rate of electron flow into the collector through the base-collector junction. Therefore, θ can be expressed as the product of two terms: there is a term M to account for any collector-current multiplication, and a term β_t (called the "base transport factor") that is the ratio of carriers which flow from base into collector to the number of carriers entering the base from the emitter, so that $\theta = \beta_t M$. The current gain relationship of Eq. (2–107) can thus be written

$$\alpha_N = \frac{1}{1 + I_p/I_n} \beta_t M = \epsilon \beta_t M. \tag{2–108}$$

In practice, transistors are generally designed to keep α_N close to unity.

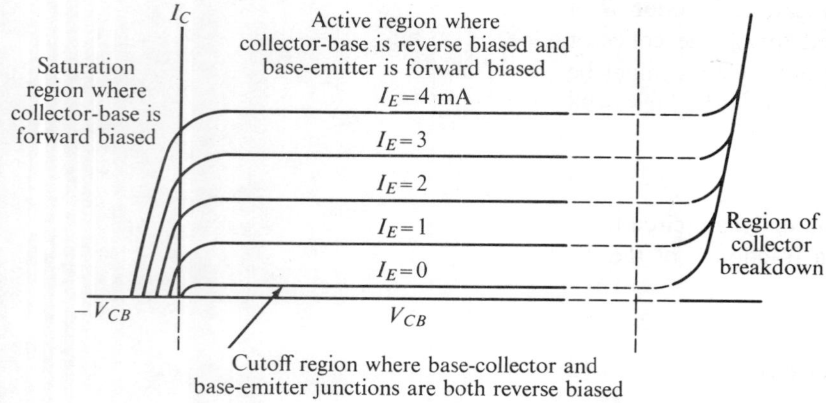

Fig. 2–99 Collector characteristics for common-base transistor configuration.

Common-Base Collector Characteristics. In Fig. 2–98 the connection of one lead of V_{EE}, one lead of V_{CC}, and the base lead to a common point provides the so-called **"common-base"** circuit configuration. If the collector current I_C is plotted against the collector-base voltage V_{CB} for fixed values of emitter current I_E, the characteristic curves of Fig. 2–99 are obtained for the npn transistor described above.

It can be seen from Fig. 2–99 that when $I_E = 0$ there is a collector current, which is a result of collector-base reverse leakage current I_{CO}. The collector current can be written

$$I_C = \alpha_N I_E + I_{CO}. \tag{2–109}$$

If the base-emitter and collector-base junctions are both reverse biased, the collector current can be decreased below I_{CO}. In this case the sign of I_E in Eq. (2–109) is negative. If the collector-base and emitter-base junctions are both forward biased, then the operating characteristics are in the region left of the vertical axis, and this is the saturation region for the common-base connection.

In Fig. 2–99, in the region that is bounded by the two vertical lines and the line for $I_E = 0$, the base-emitter junction is forward biased and the collector-base junction reverse biased. In this region the collector-base voltage V_{CB} has very little effect on the collector current I_C. This is often called the "active" region, which is utilized for small-signal linear circuits.

As V_{CB} is increased, the depletion layer of the collector-base junction increases so that the effective width of the base decreases. This decrease of effective base width means that for the npn transistor the electrons from the emitter need to move an even shorter distance before they enter the collector. Therefore, there is less chance for recombination and less loss of electrons in the base. When the value of V_{CB} gets too high, an effect called

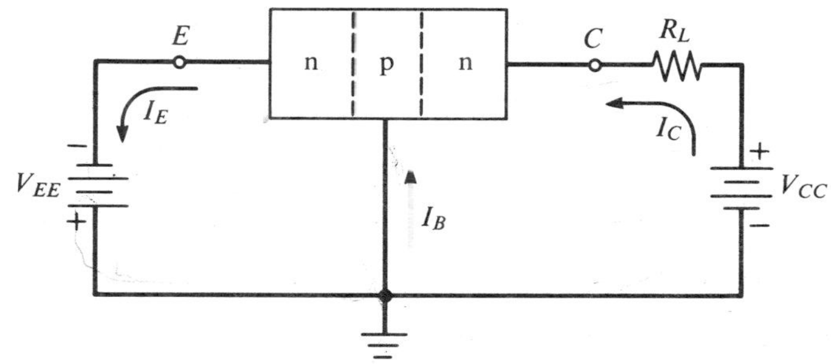

Fig. 2–100 Effect of voltage drop across load resistor on the transistor operation.

punchthrough occurs. The large collector-base voltage causes the effective base width to approach zero so that the emitter electrons (for the npn transistor) move directly to the collector and breakdown occurs. In this case the value of M in Eq. (2–108) increases rapidly and/or Zener breakdown occurs. The breakdown region is shown in Fig. 2–99 to be to the right of the dotted vertical line.

Saturation. The saturation effect is very important for switching circuits. If a load resistor R_L is connected in the collector circuit as shown in Fig. 2–100, then the collector-base voltage $V_{CB} = V_{CC} - I_C R_L$. If the voltage drop across the load, $I_C R_L$, is less than V_{CC}, then the collector-base junction remains reverse-biased, and the emitter electrons which cross the collector junction flow into the external circuit. However, when the emitter current I_E is increased sufficiently, the collector current I_C increases proportionately, and the voltage drop $I_C R_L$ across the load can exceed V_{CC}. In this case the collector-base junction becomes forward biased. Now the collector can inject electrons into the base, although at the same time it is receiving emitter-injected electrons to provide the collector current I_C. The net result is that there is a buildup of electron concentration at the collector-base junction, and the transistor is said to be in **saturation.**

If the emitter current is suddenly decreased, the collector current continues to flow for a short time because of the electrons stored at the collector junction. The collector current remains relatively constant for a brief period after cutoff of I_E until the electron concentration at the collector junction returns to zero. This time delay between emitter and collector current changes can be very important in switching circuits, and is often called the **storage-time delay.** It is discussed later in this section. The storage time can be reduced if the collector can be made to act as a poor emitter when forward biased. If the collector is fabricated from a high resistivity material, its electron injection efficiency is low, and this decreases the storage-time delay.

Common-Emitter Collector Characteristics. When the transistor emitter lead is in common with the base and collector voltage sources, a so-called **"common-emitter"** circuit exists as illustrated in Fig. 2–101. The base-emitter junction is shown forward biased so that electrons that diffuse to the collector junction are accelerated into the collector and move through the external circuit. Again, from Kirchhoff's Law and the direction of the electrode currents shown in Fig. 2–101,

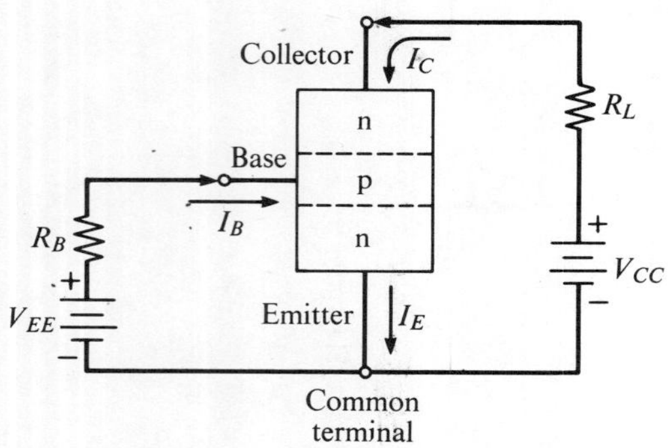

Fig. 2–101 Common-emitter connections for npn transistor.

$$I_E = I_C + I_B. \tag{2–110}$$

Typical collector characteristic curves are shown in Fig. 2–102 for the common-emitter connection of the transistor. These curves illustrate the relationship of I_C, V_{CE}, and I_B, and are useful in evaluating several dc parameters and for circuit analysis.

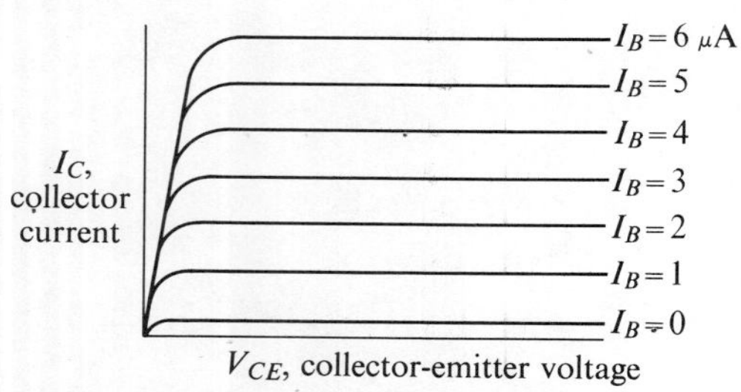

Fig. 2–102 Collector characteristic curves for the common-emitter connection of npn transistor.

The voltage from collector to emitter, V_{CE}, is equal to the sum of V_{CB} and V_{BE}, as illustrated in Fig. 2–103. If the transistor goes into saturation, the polarity of V_{CB} reverses but V_{CB} always remains smaller than V_{BE}. Since the voltage across a forward biased silicon junction is about 0.6 V, it would be expected that a saturated silicon transistor would have a collector-emitter voltage $V_{CE(\text{sat})}$ less than 0.6 V; it is typically 0.1–0.2 V. The exact value of $V_{CE(\text{sat})}$ depends on both the base current and collector current and can vary from a few millivolts to over one volt.

It should be noted that the relatively high ohmic resistance R_C in the collector material can cause a voltage drop, $I_C R_C$, so that a more accurate expression for the collector-emitter voltage in saturation is

$$V_{CE(\text{sat})} = V_{CB} + V_{BE} + I_C R_C. \tag{2–111}$$

Methods for fabricating transistors are discussed in Appendix B.

Transistor Circuit Symbols and Types. The junction transistor symbols used in schematic diagrams are shown in Fig. 2–104. The arrowhead indicates the direction of conventional collector current flow through the transistor and is always located on the emitter. The symbols may or may not be enclosed in circles, but the circle has no electronic significance.

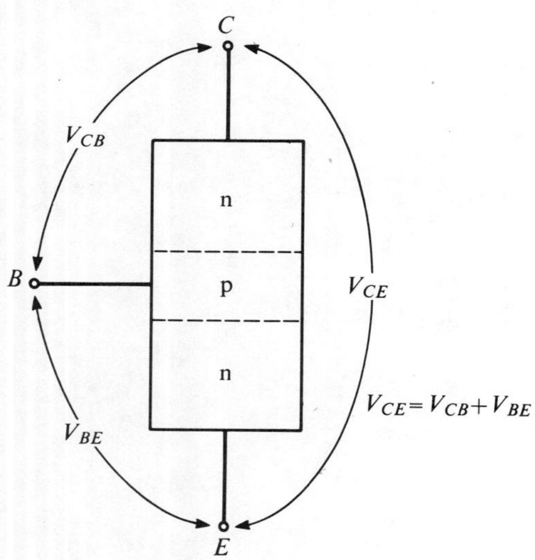

Fig. 2–103 Voltage relationship for common-emitter connection of npn transistor.

The common-base and common-emitter connections for the transistor were illustrated in Figs. 2–98 and 2–101 respectively. Another method of connecting the transistor is to make the collector lead common to the input and output. The common-collector configuration is usually called an **emitter-follower** circuit. The three basic configurations are illustrated in Fig. 2–105.

If the common connection in each of the circuits is grounded (as indicated by the dotted ground symbol), the circuits in Fig. 2–105 are referred to as grounded-base, grounded-emitter, and grounded-collector circuits respectively.

Switching Circuit Analysis. In the first part of this section, it was shown that a diode could be operated as a switch that is either ON or OFF, depending on the polarity of the applied signal. A junction transistor can act as a switch that is ON or OFF depending on the input current. Since there are three terminals, only one terminal needs to be common to the actuating and switched circuits, as illustrated in Fig. 2–106.

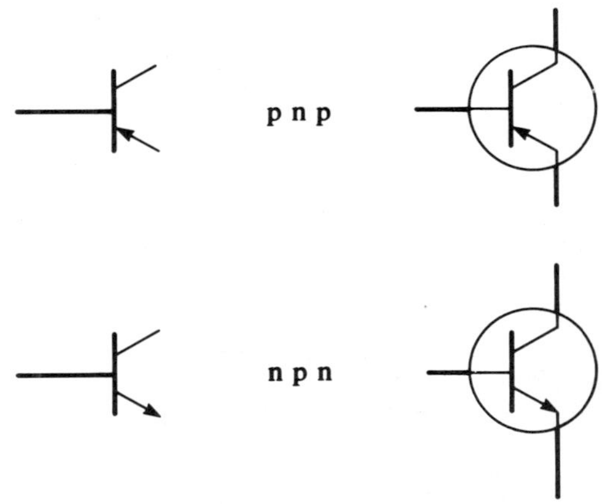

Fig. 2–104 Symbols for pnp and npn transistors.

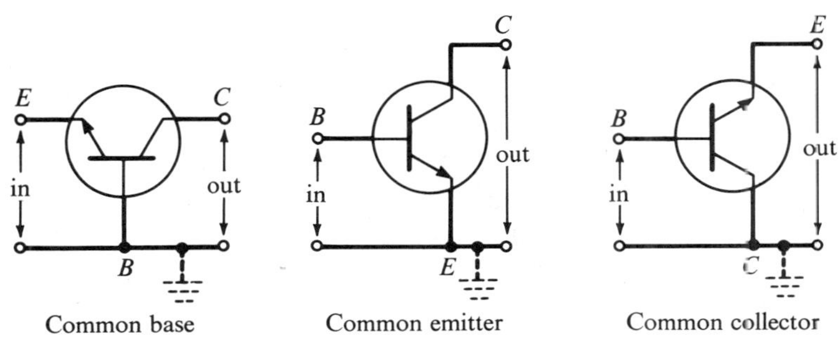

Fig. 2–105 Types of circuit connections for the transistors.

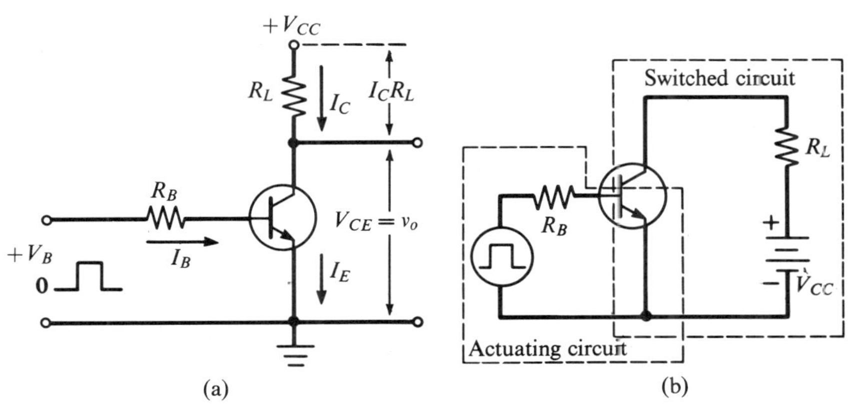

Fig. 2–106 A common-emitter npn transistor switch circuit: (a) typical schematic; (b) outline of actuating and switched circuits.

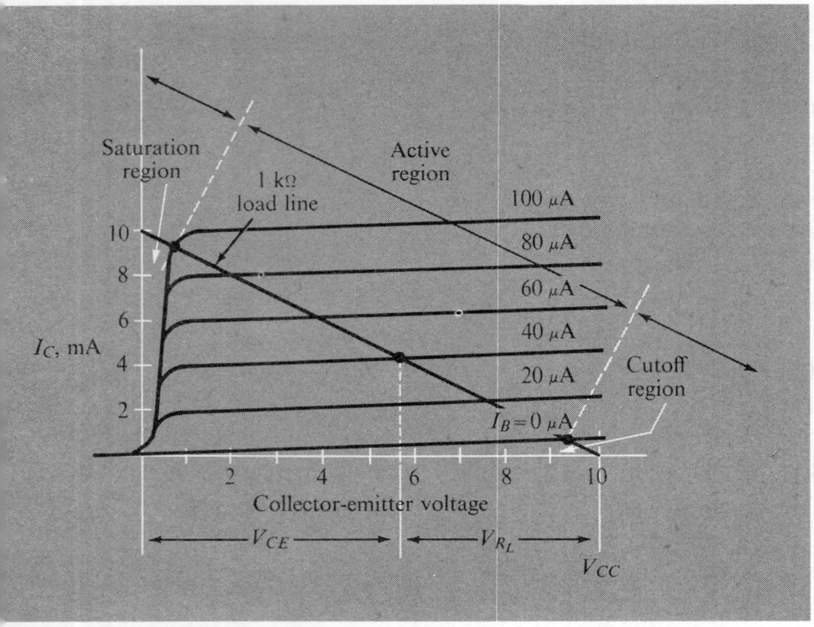

Fig. 2–107 Transistor characteristic curves.

The dc analysis of the common-emitter switch circuit of Fig. 2–106 is done as outlined in Section 2–3.2 by superimposing the load line and *I-V* curves for the collector-emitter circuit (I_C vs. V_{CE}). This is shown in Fig. 2–107 for a typical transistor with a 1 kΩ collector load resistor R_L and $V_{CC} = 10$ V. In the transistor case, there is a separate *I-V* curve for each value of the base current I_B. It is customary to show a few representative curves and to interpolate if any intermediate curves are required. By varying I_B, the operating point can be placed almost anywhere along the load line. In the "active" region, a small change in I_B results in a much larger change in I_C. This is the mode of operation for amplifier circuits. The ratio I_C/I_B is the dc current gain, sometimes called the "dc forward current transfer ratio" or the "dc beta" for short. The symbol for this characteristic is h_{FE} or β_N, where

$$\beta_N = h_{FE} = I_C/I_B. \qquad (2\text{–}112)$$

Because $\alpha_N = I_C/I_E$ and $I_B = I_E - I_C$, it follows that, by substitution into Eq. (2–112), $\beta = I_C/I_B = \alpha_N/(1 - \alpha_N)$, which can be a useful relationship if only β_N or α_N is given. For example, if $\alpha_N = 0.99$, then $\beta_N = 0.99/0.01 = 99$.

In switching circuits one is more interested in the regions of maximum and minimum conductance than in the active region. If the ON transistor were a perfect conductor, that is, if $V_{CE} = 0$, the collector current I_C would have the maximum possible value,

$$I_{C(\max)} = V_{CC}/R_L. \tag{2–113}$$

If the OFF transistor were a perfect insulator the collector current $I_C = 0$, and

$$V_{CE(\max)} = V_{CC}. \tag{2–114}$$

Two points ($V_{CE} = 0$, $I_C = V_{CC}/R_L$) and ($I_C = 0$, $V_{CE} = V_{CC}$) are used to draw the load line. Since $I_C R_L + V_{CE} = V_{CC}$, the collector current can be expressed as

$$I_C = -\frac{1}{R_L} V_{CE} + \frac{V_{CC}}{R_L}. \tag{2–115}$$

That is, the collector current I_C can be determined from the slope $(-1/R_L)$ and intercept (V_{CC}/R_L) of the load line. As shown in Figs. 2–106 and 2–107, the sum of V_{CE} and V_{R_L} equals V_{CC} at any point along the load line.

If a voltage of V_B is applied to the input of the circuit in Fig. 2–106a, the base current I_B will flow, and the switch is turned ON if I_B is sufficiently large. Whether the switch can be considered ON can be determined by the point where a specific value of I_B crosses the load line in Fig. 2–107. Usually it is desired that the switch go into saturation, and thus V_{CE} is very small (about 0.1 to 0.2 V).

To ensure that the transistor goes into saturation and the switch is truly ON, the base network should provide a turn-on base current I_{B1} such that

$$h_{FE} I_{B1} > I_{C(\max)}, \tag{2–116}$$

where

$$I_{C(\max)} = \frac{V_{CC} - V_{CE(\text{sat})}}{R_L} \tag{2–117}$$

and

$$I_{B1} = \frac{V_B - V_{BE(\text{ON})}}{R_B}. \tag{2–118}$$

The voltage $V_{BE(\text{ON})}$ is the voltage drop across the forward biased base-emitter junction.

To ensure that the transistor is cut off ($I_E = 0$) and the switch is truly OFF, the base network must prevent the base-emitter junction from being forward biased. Referring to Fig. 2–106, in a cut-off transistor, the base current I_B is equal to the reverse bias collector-base leakage current I_{CO}.

This leakage current is in the opposite direction to the I_B arrow shown in the figure. The actual base-emitter voltage will be equal to the applied actuating voltage (OFF value) plus $I_{CO}(R_B + R_s)$, where R_s is the voltage source resistance. If the OFF value of the actuating voltage is 0 volts as shown and $(R_B + R_s)$ is 10 kΩ, and $I_{CO} = 10$ μA, the base-emitter voltage will be +0.1 V. In some applications, this amount of forward bias may produce an unacceptably large emitter current for the OFF state. To avoid this problem, it is necessary to apply a negative value of actuating voltage such that the resulting base voltage is actually slightly negative. Obviously high values of R_B and R_s are to be avoided. If the base circuit is left open ($R_s = \infty$), the charge leaking into the base from the collector raises the base voltage until the base-emitter junction is forward biased enough to conduct current from the base at the rate I_{CO}. This is an effective base-emitter current source of I_{CO} which results in a collector current of $\beta_N i_B = \beta_N I_{CO}$ according to Eq. (2–112).

In Fig. 2–106 it can be seen that $v_o = V_{CC} - I_C R_L$. When the transistor is cut off the output voltage v_o is at its maximum of nearly $+V_{CC}$ volts. When the base input signal goes to its maximum of $+V_B$ volts so that the transistor saturates, then v_o drops to a minimum value of $V_{CE(\mathrm{sat})}$, about 0.1 to 0.2 V. In other words, the switch of Fig. 2–106 inverts the input signal. It is, therefore, often called an inverter. The grounded-emitter switch always provides an output voltage change of opposite direction with respect to the input voltage change.

Experiment 2–19 *Transistor Characteristics*

The collector characteristic curves for an npn transistor that is connected in a common-emitter circuit are obtained with an oscilloscopic curve tracer. If a scope camera is available, the curves should be photographed. The recorded curves are to be used to evaluate several dc parameters. The values of $V_{CE(\mathrm{sat})}$ and $V_{CE(\mathrm{OFF})}$ are measured and compared with the expected values. The dc beta β_N, or h_{FE}, is determined from the characteristic curves and compared with typical values for the specific type of transistor.

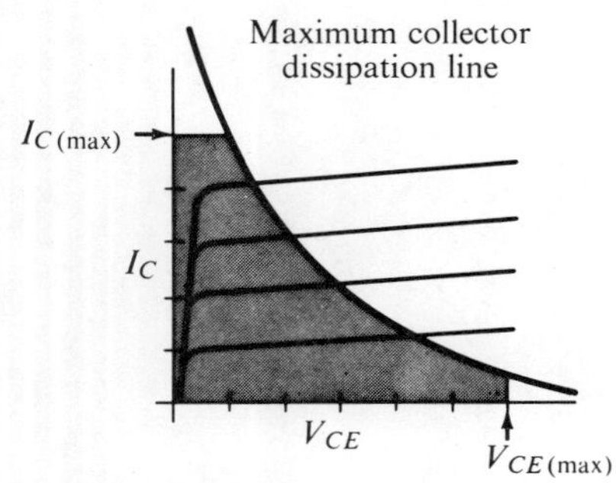

Fig. 2–108 Limiting regions for a transistor.

Voltage and Current Limitations in Transistors. Transistor switching circuits are routinely used in switched circuits where the supply voltage is hundreds of volts or where the current switched can be several hundreds of amperes. Despite this great versatility, there are a number of ways of destroying a transistor through careless circuit design. Three limitations encountered under normal bias conditions are shown in Fig. 2–108. The maximum collector dissipation in watts is $I_C V_{CE}$. The safe (with adequate cooling) region is the shaded area. The load line should normally stay within the safe boundaries. It is possible, in switching applications, for the load line to go through the area outside the maximum dissipation line if both operating points are well within the maximum limitations and the transition time is very short.

In addition to the above limitations, there are two breakdown voltages that must not be exceeded unless there is some protection against the damaging currents which could result from the breakdown. The breakdown voltages are BV_{CBO}, the maximum inverse collector-base voltage, and BV_{CEO}, the maximum collector-emitter voltage.

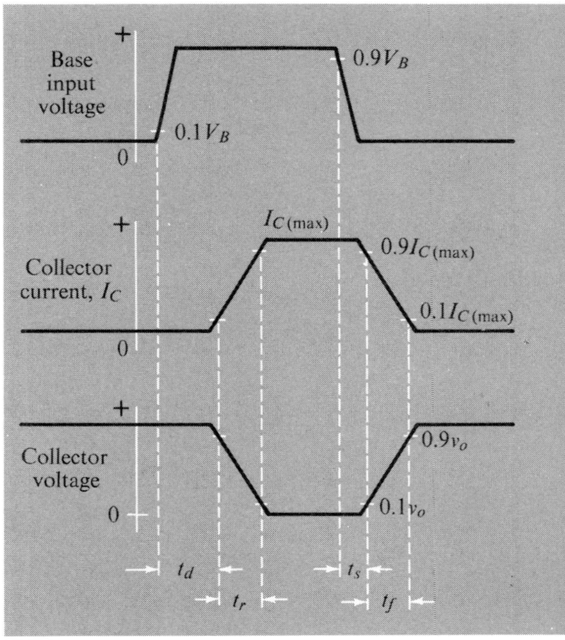

Fig. 2–109 Transistor switching time.

Transistor Switching Time. The transistor switch cannot respond instantaneously to a turn-on or a turn-off actuating signal. In many applications it is important to be aware of the magnitude of possible errors caused by a time delay in the switching circuit and how to minimize the time lags. In Fig. 2–109 the response of a transistor switching circuit is shown when a turn-on and turn-off actuating signal V_B is applied to the input circuit of Fig. 2–106.

Each segment of the turn-on and turn-off times illustrated in Fig. 2–109 will be considered briefly. First, there is the delay time, t_d. The base-emitter input capacitance must charge through R_B before the base-emitter junction is actually forward biased; there is a finite transit time for the first minority carriers to cross the base region; and some time is required for the collector current to reach 10% of its final value.

Next, there is the rise time, t_r. This is the time needed to establish the concentration of minority carriers in the base region (i.e., the electron density in the p-type base of the npn transistor) which is required to carry 90% of the final value of I_C. The combination of t_d and t_r is the turn-on time, t_{ON}. The rise time can be substantially reduced if the base current is larger than the minimum needed for saturation. The collector current rises toward $h_{FE}I_B$ with a characteristic time constant, but cannot, of course, exceed $I_{C(\max)}$. However, increasing I_B greatly beyond the current required to saturate the transistor substantially increases the steady-state excess minority carrier concentration in the base region during saturation.

Assuming now that the actuating signal was of sufficient duration and magnitude to saturate the transistor, it is important to consider the turn-off behavior. First, there is the storage time, t_s. This is the time necessary to clear the collector-base junction of excess minority carriers, as discussed under saturation earlier in this section, and to decrease I_C to 90% of its maximum value. Since the saturation of the transistor occurs as soon as the collector-base junction becomes forward biased, any further increase of base current causes excess charge to accumulate in the base region. Then, when the forward bias on the base-emitter junction is suddenly removed, the excess base charge flows into the collector. At the instant the potential gradient at the collector-base junction reaches zero, the collector no longer emits carriers into the base region, and the transistor comes out of saturation.

The fall time, t_f, is the time required for the output current to fall from 90% to 10% of its maximum value, and depends on the time required to discharge the collector-emitter capacitance and the time required for the minority carriers in the base to be collected so that there is zero density gradient at the emitter junction as well as at the collector junction. This is illustrated in Fig. 2–110, where the charge density gradients in the base between the emitter and collector junctions are drawn.

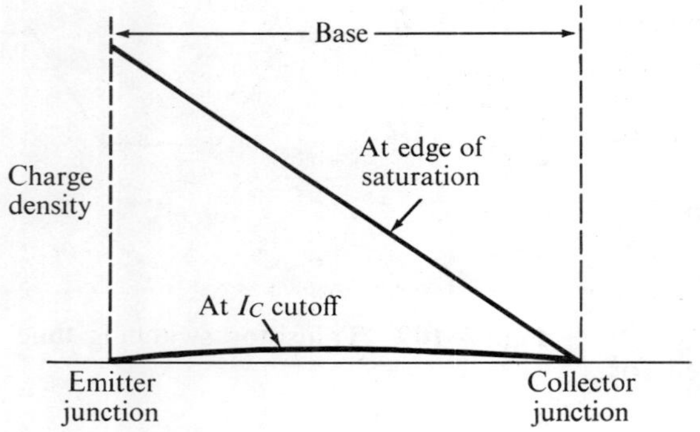

Fig. 2–110 Charge density gradients in the base.

To summarize then, the important switching times are as follows:

t_d: Delay time or "turn-on delay," the finite time that elapses between application of the base input voltage and the start of collector current flow in the transistor. The time t_d is measured at values of input voltage and output current that are 10% of the maximum values.

t_r: Rise time, the time required for the collector current to increase from 10% to 90% of its total change.

t_s: Storage time, the time required for the collector current to decrease to 90% of its maximum value after the input has decreased to 90% of its maximum value.

t_f: Fall time, the time required for the output waveform to fall from 90% to 10% of its maximum value.

t_{ON}: Turn-on time, the sum $t_d + t_r$.

t_{OFF}: Turn-off time, the sum $t_s + t_f$.

It should be noted that the various transistor switching times are usually measured by using the V_{CE} output waveform because it is easy to display on the oscilloscope. This is valid only for cases where the load is resistive.

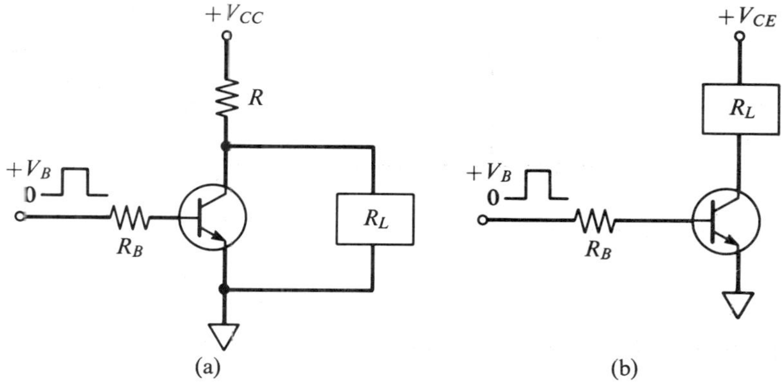

Fig. 2–111 Connection of a load to a BJT switch.

Connection of Load. A basic application of the transistor switch is to control current in an external load. There are two ways to connect the load, as illustrated in Fig. 2–111. When the load is placed in *shunt* with the transistor, a resistance R is connected in series between the parallel network and the supply voltage. When the switch is OFF, current flows through the load and the series resistance R. When the switch is ON, the load is effectively short-circuited and the current is steered from the load into the transistor. When the external load is in series with the transistor, the current through the load is the same as the collector current I_C.

If the external load requires a larger current than can be supplied by one transistor that is controlled by a current-limited base control voltage source, it is possible to cascade transistors as shown in Fig. 2–112.

The base-emitter junction of Q_2 is the shunt load of Q_1. The base current of Q_2 is dependent on the supply voltage V_{CC}, the Q_1 load resistor R_{L_1}, and $V_{BE(ON)}$. The current $I_{B(Q_2)} = (V_{CC} - V_{BE(ON)})/R_{L_1}$.

Whenever Q_1 is ON the base current for Q_2 is cut OFF, and when Q_1 is OFF there is a base current $I_{B(Q_2)}$ for Q_2. The current through the load is equal to the collector current for Q_2 because it is in series with Q_2. If resistor R_{L_1} is chosen such that the base current of Q_2 causes saturation (see Eqs. 2–116 through 2–118), the current in the load $I_{C(Q_2)} = (V_{CC} - V_{CE(sat)})/R_{load}$.

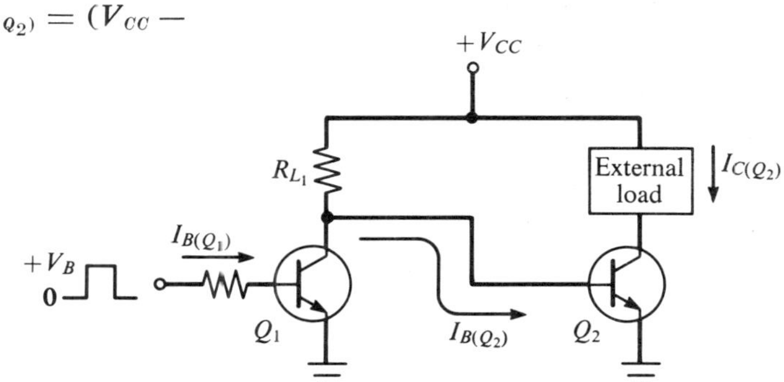

Fig. 2–112 Cascaded transistors for large loads.

Note that with the circuit of Fig. 2–112 the current in the load is OFF when the base voltage of transistor Q_1 is positive, and vice versa. If it is desirable to have the current in the load go ON when the input goes positive, it is necessary to add a transistor which acts as an inverter between the two transistors in Fig. 2–112 or connect the emitter of Q_1 to the base of Q_2. In the latter case, Q_1 acts to switch ON the base current of Q_2.

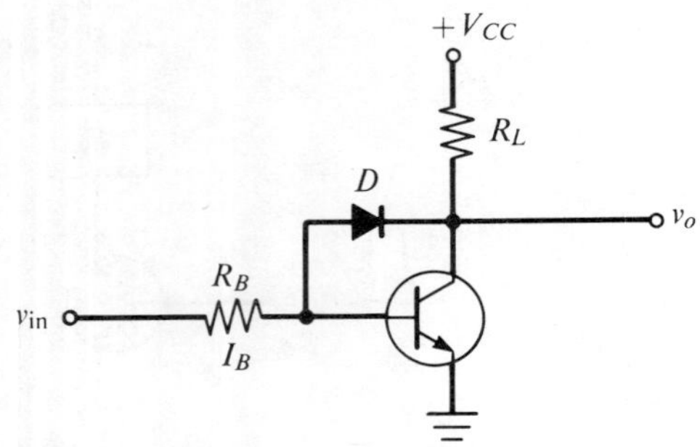

Fig. 2–113 Nonsaturating transistor switch with diode clamp.

Nonsaturating Switching Circuits. The storage time caused by allowing a transistor to saturate significantly limits the switching speed. The problem of storage time does not arise if the transistor is not allowed to saturate, and thus an improvement in switching speed is possible.

One method of providing a nonsaturating transistor switch is to clamp the collector in such a way that the collector-base junction cannot become forward biased. One method of clamping is shown in Fig. 2–113. The diode[5] D becomes forward biased as the transistor begins to leave the active region, at the edge of saturation. Therefore, the collector never becomes sufficiently negative with respect to the base for the collector-base junction to become forward biased, and the transistor cannot saturate. The excess base current is conducted away through the diode. The Schottky diode is especially effective in this application. Its forward voltage drop is less than that of the collector-base junction, thus keeping the transistor farther from saturation. In addition, the Schottky diode has essentially no stored charge itself. For these reasons, the storage time is almost completely eliminated in a Schottky clamped transistor.

Note 5. Clamping Diode.

A germanium or a Schottky diode is best to use because it conducts sooner and ensures that the transistor does not get close to saturation. A silicon diode could be used because the clamping diode carries only a low current compared with the transistor base-emitter junction, and consequently the forward voltage drop across the clamping diode is less than V_{BE}.

Another method of providing nonsaturated switches is to control the collector current. A circuit utilizing transistors operated in the common-base configuration is shown in Fig. 2–114. When the input signal is negative, transistor Q_1 is cut off, but the base-emitter junction of Q_2 is forward biased, and thus Q_2 is ON and the emitter current $I_E \approx V_{EE}/R_E$. Now when the input signal goes positive the transistor Q_1 goes ON and the voltage drop across R_E causes the base-emitter junction of Q_2 to become reverse biased and Q_2 goes OFF. The current is in effect "steered" to either Q_1 or Q_2 by the input signal. The steering depends on both the polarity

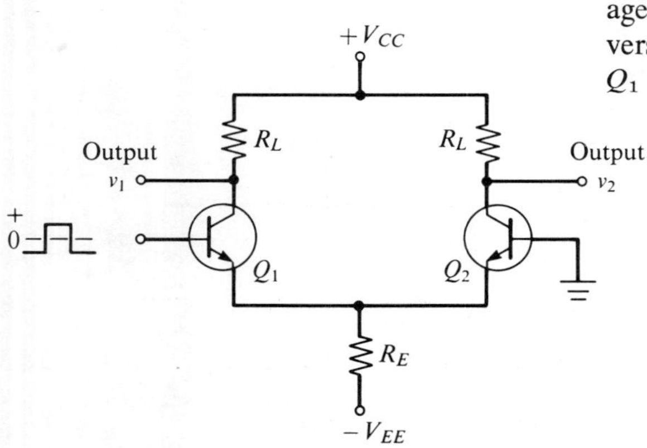

Fig. 2–114 Nonsaturating transistor switches.

and the amplitude of the input signal. The collector-base junction will always be at some value of reverse or zero bias if $I_C R_L \leqslant V_{CC}$, and the transistor will thus stay out of saturation.

The outputs v_1 and v_2 are complementary, with v_2 in phase with the input signal. A change of input signal of only a few tenths of a volt is required to switch states.

Experiment 2–20 *The Basic Bipolar Junction Transistor (BJT) Switch*

The basic BJT switch circuit is connected and tested so that its important characteristics can be observed. The inverting action of the transistor is first observed. If a high speed oscilloscope is available, the switching times are measured. Then the transistor switch is connected to a load so as to control the current in an external load. The nonsaturating transistor switch circuit is also connected and tested and its characteristics are compared with those of the saturated switch.

The Unijunction Transistor

The unijunction transistor (UJT) has some unique features as compared to the two-junction npn and pnp transistors. It is a three-terminal device with only one pn junction. The inherent stability of the characteristic parameters provides advantages for certain circuit applications.

Characteristics and Biasing. The UJT is conveniently considered as a bar or strip of silicon to which a pn junction is made, as illustrated in Fig. 2–115a, although present structures of the UJT deviate from the bar structure for simplicity and economy of production. The leads from the two ends are called base 1 and base 2. The rectifying contact is called the emitter, and it is represented as being between base 1 and base 2 on the bar of silicon. The schematic circuit symbol is shown in Fig. 2–115b. The

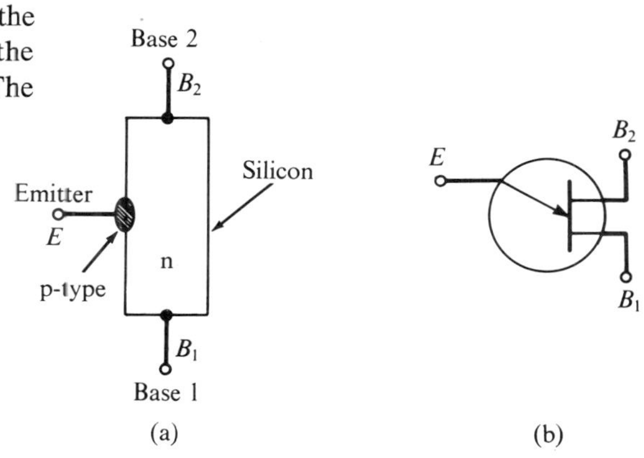

Fig. 2–115 Unijunction transistor: (a) structure representation; (b) circuit symbol.

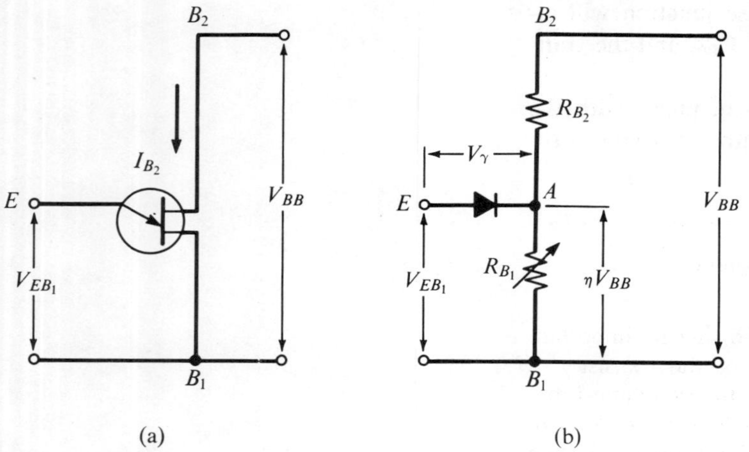

Fig. 2–116 Unijunction transistor: (a) nomenclature; (b) simplified equivalent circuit.

operation of the UJT is readily illustrated by consideration of the equivalent circuit in Fig. 2–116.

If a voltage V_{BB} is applied so that base 2 is positive, a current will flow between base 2 and base 1 as determined by the interbase resistance R_{BB} (typically 7 kΩ) of the silicon bar, so that

$$I_{B2} = V_{BB}/R_{BB}. \qquad (2\text{–}119)$$

The resistance R_{BB} can be considered as consisting of the resistances R_{B1} and R_{B2} in series, acting as a voltage divider. A fraction η (typically 0.6) of the applied voltage (that is, ηV_{BB}) is divided between the emitter contact and base 1. Therefore, the pn junction at the emitter is reverse biased and only a small leakage current normally flows in the emitter lead.

When the voltage V_E that is applied between the emitter and base 1 is increased, a voltage is reached where V_{EB_1} is equal to the sum of the voltage across R_{B1} and the forward voltage drop of the pn junction V_γ. This voltage is known as the peak-point voltage V_p,

$$V_p = V_\gamma + \eta V_{BB}. \qquad (2\text{–}120)$$

For emitter-base voltages greater than V_p the pn junction is forward biased, so that holes are injected from the emitter into the silicon bar. Because base 1 is negative with respect to the emitter, the electric field is such that most holes move toward the base 1 terminal. An equal number of electrons are injected from base 1 to maintain electrical neutrality in the bar. The increase in current carried in the silicon bar decreases the value of R_{B1}. This causes the fraction of voltage between point A and base 1 to decrease, and this in turn causes a further increase of emitter current I_E and a lower resistance for R_{B1}.

A plot of V_E vs. I_E is shown in Fig. 2–117, which shows the region of negative resistance where the voltage across R_{B1} decreases as I_E increases. The region of negative resistance occurs between the peak-point voltage V_p and the so-called valley-point voltage V_v. At the valley-point voltage and higher voltages the density of charge carriers is so high that the lifetime of the carriers is decreased. This decrease counteracts the effect of new carriers being generated, and thus the emitter voltage V_E increases gradually at currents above I_v. The region to the right of the valley point is known as the saturation region, where the dynamic resistance is positive.

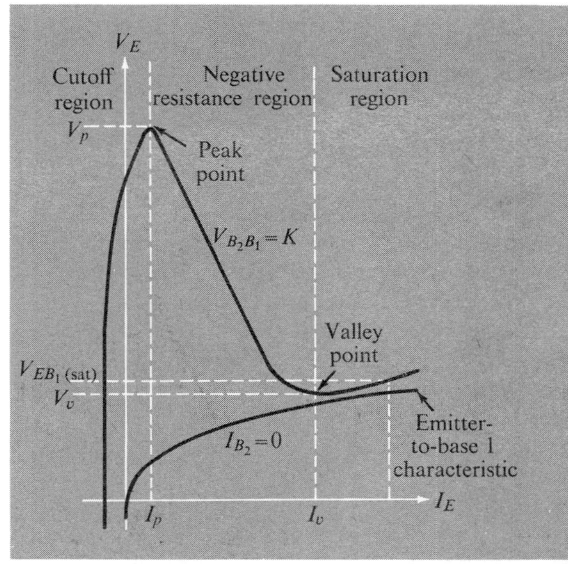

Fig. 2–117 Static emitter characteristic curves for unijunction transistor. (Courtesy of General Electric Co.)

The emitter characteristic curve given in Fig. 2–117 is not drawn to scale, in order better to show the various operating regions. A typical emitter characteristic curve drawn to scale is shown in Fig. 2–118, where only a small portion of the cutoff region is shown. The peak-point voltage is about 16 V and is reached at a forward emitter current of about 10 pA. The voltage then remains constant until about 0.1 μA, where it starts to decrease. The valley-point voltage of 1.6 V is reached at a valley current of about 8 mA. The saturation resistance can be determined by the slope of the *I-V* curve in the saturation region, and is seen to be about 5 Ω.

Peak-Point Voltage. The peak-point voltage V_p is the most important characteristic of the unijunction transistor because it determines the trigger or switch point in its circuit applications. The value of V_p is expressed in Eq. (2–120) as a function of the intrinsic standoff ratio η.

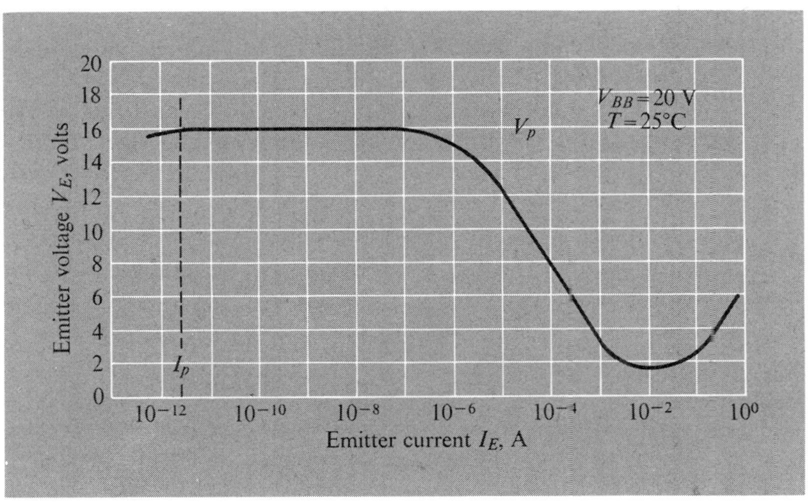

Fig. 2–118 Static emitter characteristics for unijunction transistor. (Courtesy of Motorola Inc.).

Intrinsic Standoff Ratio. A rearrangement of Eq. (2–120) provides the definition of the intrinsic standoff ratio,

$$\eta = \frac{V_p - V_\gamma}{V_{BB}} = \frac{R_{B1}}{R_{BB}}. \tag{2–121}$$

This ratio is not completely independent of temperature but the temperature coefficient is quite small, about 0.06%/°C. The intrinsic standoff ratio is also slightly dependent on the supply voltage V_{BB}, but the variation is generally negligible for practical purposes.

UJT Trigger Circuit. One of the most important uses of the UJT is to provide a precise time-delayed trigger pulse for activating other devices. The basic UJT trigger circuit is shown in Fig. 2–119a. The emitter–base 1 voltage is provided by the voltage across the timing capacitor C. At the start, capacitor C is shorted by switch S_1. When S_1 is opened, the capacitor charges through resistor R from the power supply. The emitter voltage v_E thus increases exponentially toward the supply voltage $+V$ with a time constant RC. When the emitter voltage reaches the peak-point voltage V_p, the UJT turns ON and the capacitor discharges through R_1. When v_E reaches about 2 V, the UJT turns OFF. Since R_1 is usually a small resistance value ($<100\ \Omega$), the discharge of timing capacitor C is usually quite rapid, and a sharp trigger pulse is obtained across R_1 as shown in Fig. 2–119b. If only a single timing pulse is desired, capacitor C must be shorted by closing switch S_1 when the UJT turns OFF. The time delay is approximately given by the RC time constant and can be varied within limits by varying R and/or C. The upper limit on the time delay depends on the desired timing precision, the V_p of the UJT, the leakage current of capacitor C, and the temperature. The upper limit for R is determined by the requirement that the current to the UJT emitter be large enough for it to fire (larger than I_p).

If switch S_1 is omitted or left open, capacitor C will begin to recharge after the emitter voltage falls to about 2 V and the UJT turns OFF. Again C will charge until v_E reaches V_p, at which time discharge through R_1 occurs. This process keeps repeating itself as shown by the waveforms in Fig. 2–119c. Because of the repetitive nature of the waveforms at the emitter and across R_1, this circuit is known as a relaxation oscillator, and has many important applications.

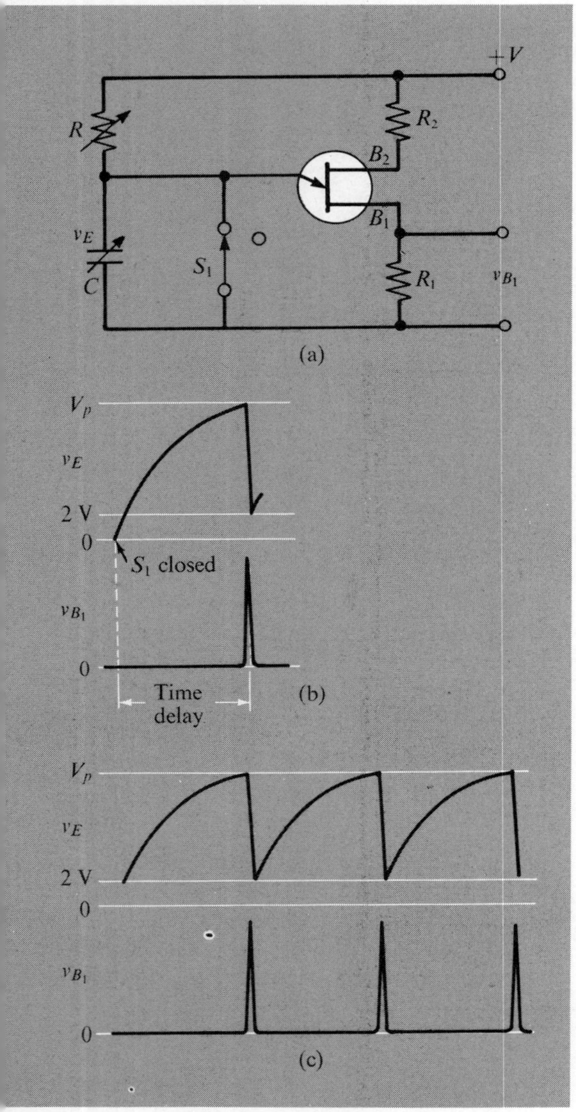

Fig. 2–119 Basic UJT trigger: (a) circuit; (b) waveforms for single pulse; (c) waveforms for repetitive operation (relaxation oscillator).

Experiment 2–21 *Unijunction Transistor (UJT) Trigger Circuit*

A unijunction transistor is combined with an *RC* network to provide a simple time delay and trigger circuit. The range of useful time delay is investigated and the circuit signals are observed.

The SCR, Triac, and Other Thyristors

There are a variety of important semiconductor switches which are pnpn four-layer devices, called **thyristors.** These devices find numerous applications as switches for controlling ac and dc power, as will be discussed in Section 2–3.7. The term thyristor applies to any pnpn semiconductor switch. The most common of these devices is the silicon controlled rectifier (SCR), which is a unidirectional three-terminal switch (current can flow in one direction only). Such unidirectional three-terminal switches are often called **reverse blocking triode thyristors.** Other important reverse blocking triode thyristors are the silicon unilateral switch (SUS), the gate turn-off switch (GTO), the programmable unijunction transistor (PUT), and several light-actuated switches which will be described in Section 2–3.5. There are other unidirectional or reverse blocking thyristors such as the Shockley or four-layer diode, which is classified as a reverse blocking diode thyristor. The silicon controlled switch, which has two control inputs, is a reverse blocking tetrode thyristor. Other thyristors can conduct current in both directions and are known as **bidirectional thyristors.** The most important of these are bidirectional triode thyristors, such as the triode ac switch (triac) and the silicon bilateral switch (SBS). The characteristics and general mechanisms for switching thyristors are first considered. Then specific pnpn devices are briefly described.

Two-Transistor Representation of pnpn Devices. In Fig. 2–120a the pnpn device is shown as a four-layered pellet of alternating p- and n-type semiconductor material (usually silicon). The four-layer diode has leads

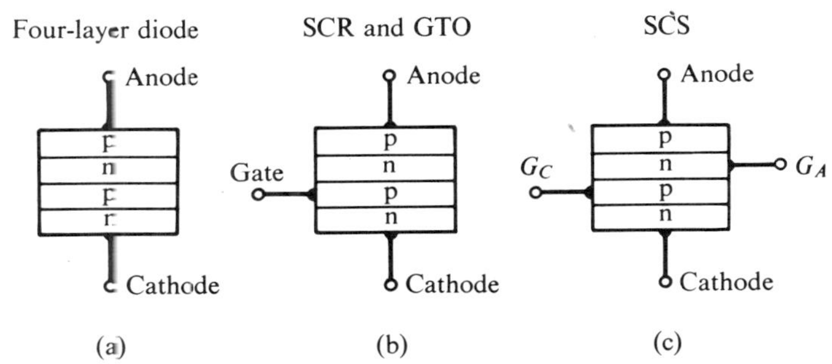

Fig. 2–120 Illustration of external lead connection for various pnpn devices.

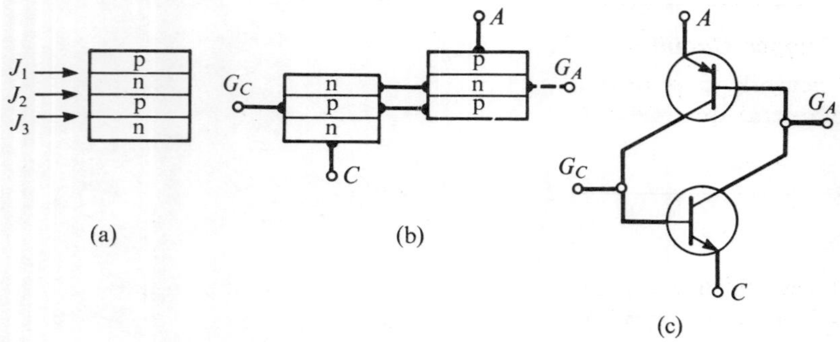

Fig. 2–121 Two-transistor representation of pnpn device. (a) Junctions indicated. (b) Hypothetical separation into two transistors. (c) Schematic of two-transistor equivalent for pnpn device.

connected only to the end layers of semiconductor. The SCR has three layers accessible by external leads, and the SCS has all four layers accessible, as shown in Fig. 2–120b and c respectively. Each layer interacts with its adjacent layers as in the transistor. It can be seen in Fig. 2–120 that, starting from one end of the stack, three adjacent layers form a pnp transistor, and from the other end an npn transistor. This two-transistor concept is illustrated in Fig. 2–121, where it is apparent that the pnpn structure can be considered as a complementary (npn-pnp) transistor feedback pair. The collector of each transistor is direct-coupled to the base of the other, providing a positive feedback loop. That is, a small change in base current of the npn transistor causes a larger change in its collector current, which is the base current of the pnp transistor. The collector current of the pnp transistor also contributes to the base current of the npn transistor. Therefore, what begins as a small change in the base current of the npn transistor becomes a large change because of positive feedback.

If the cathode of the pnpn device is connected to the negative terminal and the anode to the positive terminal of the power supply, the center pn junction (J_2 of Fig. 2–121a) is reverse biased. Therefore, the device does not conduct unless the current gain around the feedback loop is approximately unity. This is seen more clearly by considering the equations expressing the total anode-to-cathode current I_A.

If the current gains of the npn and pnp transistors are h_{FE1} and h_{FE2} respectively, then the current gain $\mathcal{G}_l$ of the internal feedback loop is

$$\mathcal{G}_l = h_{FE1} h_{FE2}. \tag{2–122}$$

With the base leakage current of the npn designated as I_{CO1} and that of the pnp as I_{CO2}, it follows that

$$I_{C1} \text{ (for the pnp)} = h_{FE1}(I_{C2} + I_{CO1}) + I_{CO1} \tag{2–123}$$

and

$$I_{C2} \text{ (for the npn)} = h_{FE2}(I_{C1} + I_{CO2}) + I_{CO2}. \tag{2–124}$$

The total anode-to-cathode current I_A is

$$I_A = I_{C1} + I_{C2}. \tag{2–125}$$

Combining Eqs. (2–123), (2–124), and (2–125), we obtain

$$I_A = \frac{(1 + h_{FE1})(1 + h_{FE2})(I_{C01} + I_{C02})}{1 - h_{FE1}h_{FE2}}. \tag{2–126}$$

When the anode is positive with respect to the cathode, the center section is reverse biased and h_{FE1} and h_{FE2} are very small compared to 1, and thus $I_A \approx I_{C01} + I_{C02}$, that is, the sum of only the junction leakage currents. Therefore, the device is in its high resistance OFF state. However, it can be seen from Eq. (2–126) that as $G_l \to 1$, $I_A \to \infty$. In other words, as the loop gain approaches unity the circuit feedback becomes sufficient for each transistor to drive the other into saturation, and the device goes to its low resistance or ON state. When the device is ON, the anode current is limited only by the external circuit.

Mechanisms for Switching pnpn Devices. Any mechanism that increases h_{FE} so that $G_l \to 1$ will turn the device ON. These mechanisms are all based on the emitter-current dependence of h_{FE} as illustrated in Fig. 2–122. When the emitter current is very low the gain h_{FE} is low, but if anything temporarily increases I_E sufficiently, then the pnpn device is rapidly switched ON. Some important mechanisms for producing the necessary emitter current to switch the device ON include anode-to-cathode voltage, rate of voltage change, temperature, transistor action, and radiant energy.

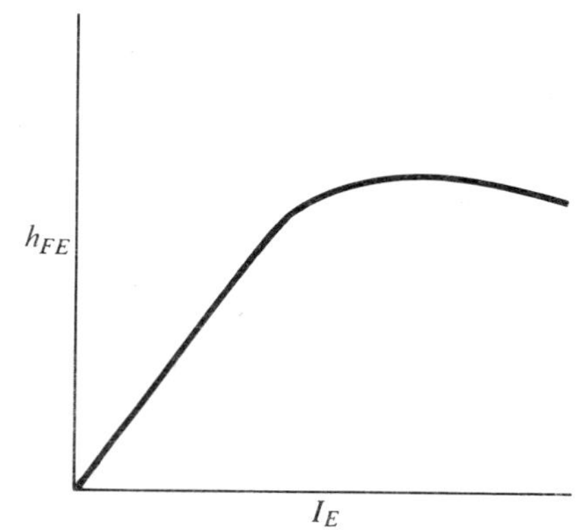

Fig. 2–122 Dependence of h_{FE} on emitter current in a silicon transistor.

When the applied voltage across the device gets high enough there is avalanche breakdown, and when the avalanche current causes $G_l \to 1$ the device is switched ON. This is the normal method used to switch the pnpn diodes into conduction. These diodes can be constructed to switch ON at certain selected anode-to-cathode voltages.

As previously discussed, any pn junction has capacitance. Therefore, when a step function is impressed across the pnpn device a charging current $i = C(dV/dt)$ will flow from anode to cathode to charge the device capacitance. If the current i is sufficient for $G_l \to 1$ the device switches ON; this mechanism is known as "the dV/dt effect."

Because the leakage current in a silicon pn junction doubles for every 8°C increase in temperature, it is possible for $G_l \to 1$ and the device to switch ON from high junction temperature.

A relatively small base current can turn a transistor ON. Likewise, for the pnpn device the injection of current carriers into either "gate" (base region) can switch the device ON. This is the normal method of switching the SCR and the SCS.

The pnpn devices can be turned OFF by the removal of the anode-to-cathode dc supply voltage, or if the supply voltage is ac the device will turn OFF when the voltage is decreased sufficiently. To maintain the device in conduction after a gate signal is removed, it is necessary to exceed a certain load current called the **holding current.**

The junctions in the SCR are all forward biased when the device is conducting, and the base regions are heavily saturated with current carriers. Therefore, to turn the SCR OFF in a minimum time, it is necessary to apply a reverse voltage. This causes the electrons and holes in the region of the two end junctions, J_1 and J_3 in Fig. 2–121, to diffuse to these junctions so as to cause a reverse current in the external circuit. Therefore, the voltage across the SCR will remain at about 0.7 V. When the holes and electrons are removed from the region of the junctions the reverse current ceases and the junctions are again in the blocking state. The reverse voltage across the SCR is now determined by the applied supply voltage. However, the recovery of the SCR is not complete until the center junction is cleared of excess charge carriers, and this depends primarily on the recombination process. This process is quite independent of external applied voltage. When the excess charge carriers at the center junction have been reduced to a low value, it is again safe to apply a forward voltage to the SCR without danger of turn-on. It usually requires about 10 μsec after cessation of the forward current flow before the forward voltage may be safely reapplied. This delay time is known as the **turn-off time.**

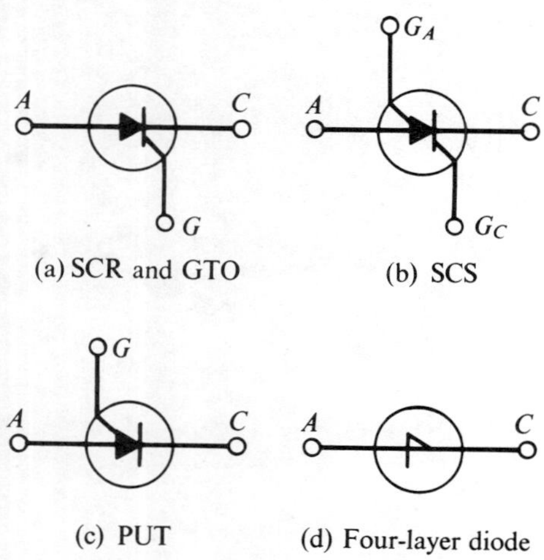

Fig. 2–123 Circuit symbols for common reverse blocking thyristors: A = anode, C = cathode, G = gate.

Reverse Blocking Thyristors. In Fig. 2–123 the circuit symbols are shown for the common types of unidirectional thyristor switches. All of these devices have current-voltage characteristics which are quite similar to that shown in Fig. 2–124 for the SCR. As can be seen from the characteristic curve, the reverse blocking type thyristors conduct in only one of the four quadrants unless the reverse voltage exceeds the reverse breakdown voltage, which is highly undesirable. The major differences between the devices are in the number of control gates (number of terminals) and in the gate construction, which can govern the turn-off characteristics.

As can be seen from Fig. 2–123d the four-layer diode (Shockley diode) has only two terminals and must be actuated by a signal applied at the anode. As such it is not nearly as useful as other pnpn devices, which can be actuated by signals applied to control gates.

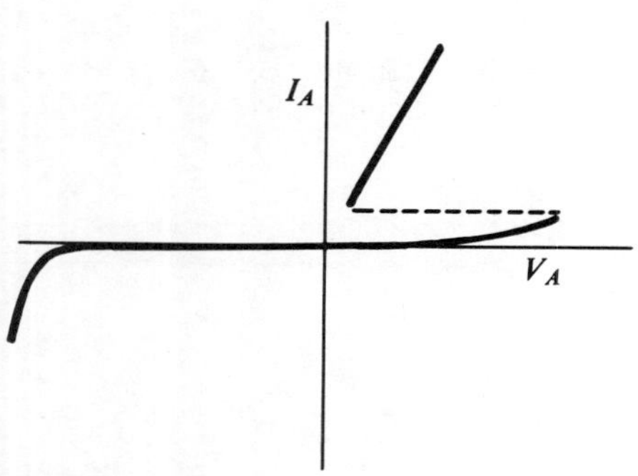

Fig. 2–124 Anode current-voltage characteristic for reverse blocking thyristors.

The SCR has one cathode control gate, as does the gate turn-off switch (GTO). The GTO has a special gate construction that allows it to be turned OFF by a small reverse gate current. The GTO can also be turned OFF in the same way as a conventional SCR. For dc application the GTO enables the turn-off circuitry to be simpler than that for the SCR. However, when GTO's are conducting high currents, there is difficulty in maintaining turn-off control by the gate. Consequently GTO devices are operated at much lower currents than SCR's. In recent years GTO devices have been replaced in many applications by high voltage–high gain silicon power transistors.

The silicon controlled switch (SCS) is a four-terminal device as shown in Fig. 2–120c and Fig. 2–123b. The control gate connections are made to both the lower p-base (G_C) and the upper n-base (G_A). Hence the SCS can be turned ON by either injecting current at G_C or withdrawing current at G_A. Triggering a thyristor by withdrawal of current at the anode gate requires more current than triggering a conventional SCR because the upper n-base area is constructed of high resistivity silicon. Also n-base triggering necessarily removes current from the loop and reduces the loop gain $\mathcal{G}_L$.

The programmable unijunction transistor (PUT) is a thyristor with an anode gate rather than the usual cathode gate of the SCR. The PUT conducts when the anode voltage exceeds the anode gate voltage by the forward voltage drop across one pn junction. At this voltage the peak-point voltage of the PUT is reached and conduction occurs. The major difference between the PUT and the normal UJT is that the peak-point voltage can be controlled in the PUT by controlling the anode gate voltage, whereas the UJT turns ON at a constant fraction η of the supply voltage. PUT's have important applications in time-delay circuits, trigger circuits for high current SCR's, and relaxation oscillators. One important characteristic that makes the PUT more suitable for triggering SCR's than normal UJT's is the higher peak pulse current output of the PUT. Switching times are normally in the 10 nsec region, which makes the PUT quite valuable.

The final reverse blocking thyristor to be discussed in this section is the silicon unilateral switch (SUS). The circuit symbol for the SUS is shown in Fig. 2–125, along with an equivalent circuit and the current-voltage characteristic. The SUS is a miniature SCR with an internal low voltage avalanche diode between the gate and cathode and an anode gate.

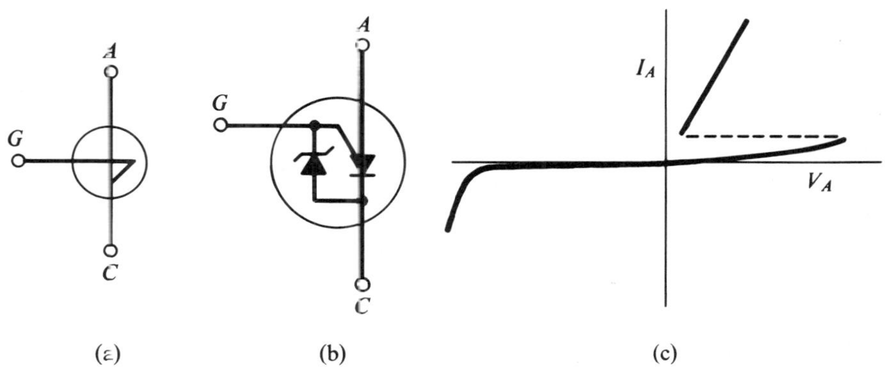

Fig. 2–125 Silicon unilateral switch (SUS): (a) circuit symbol; (b) equivalent circuit; (c) current voltage characteristic.

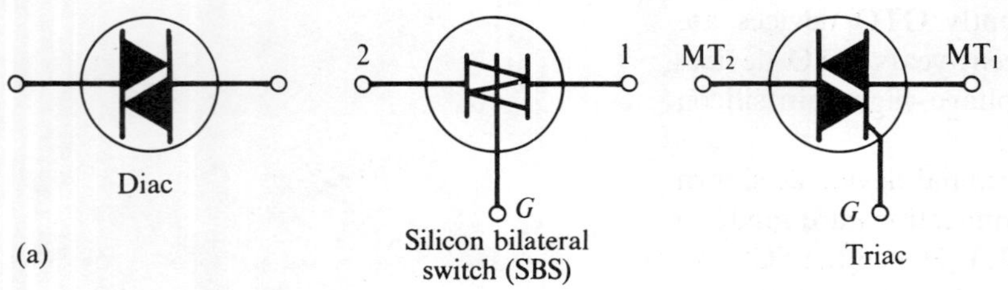

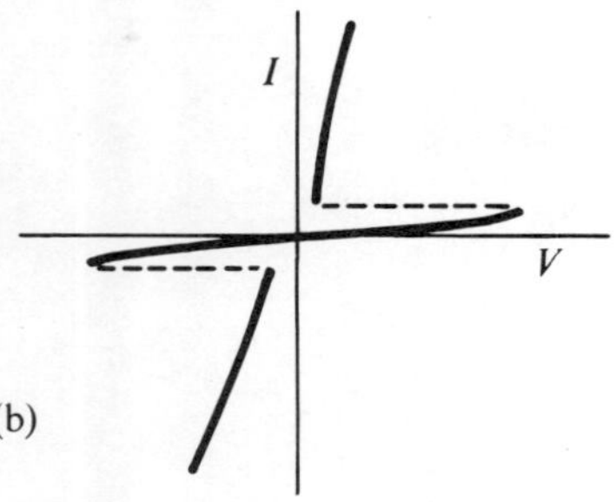

Fig. 2–126 Common bidirectional thyristors: (a) circuit symbols; (b) current-voltage characteristic.

The SUS is usually used in relaxation oscillator circuits. The SUS functions similarly to the UJT except that it switches at a fixed voltage determined by the breakdown voltage of the built-in avalanche diode, rather than at a fraction (η) of the supply voltage as does the UJT.

Bidirectional Thyristor Switches. Circuit symbols for the most common bidirectional thyristors are shown in Fig. 2–126a. These devices all have current-voltage characteristics similar to that shown in Fig. 2–126b for the triac. As can be seen, these devices can conduct current in either direction and hence are quite valuable in ac switching circuits. The diac is a bidirectional diode thyristor since it has only two terminals, while the triac and SBS are bidirectional triode thyristors. Since the diac has limited application, this discussion will focus on the operating principles of the SBS and the triac. In Section 2–3.7, applications of triacs in power control circuits are discussed in detail.

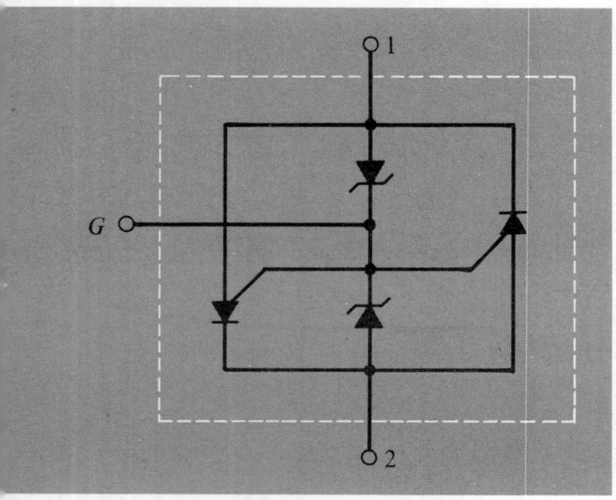

Fig. 2–127 Equivalent circuit of silicon bilateral switch (SBS).

The SBS is very similar to the SUS except for its bidirectional current conduction characteristics. The equivalent circuit of the SBS is shown in Fig. 2–127, and illustrates that the SBS is equivalent to two SUS's connected in reverse-parallel fashion. The control gate is at the anodes of the two internal PUT's, while avalanche diodes are connected between the anodes and the cathodes. The SBS finds its major application in triggering triacs, since it can be switched with voltages of either polarity. The basic SBS trigger circuit is a relaxation oscillator with an ac supply voltage rather than the dc supply voltages used with unidirectional thyristors and UJT's.

The triac has become such an important ac power control device that it will be described here in some detail. The basic pellet structure of the triac is shown in Fig. 2–128a, while a more complete current-voltage characteristic is shown in Fig. 2–128b. The terminals in the triac are normally numbered MT_1 and MT_2 (for main terminals 1 and 2) since anode and cathode terminology is not applicable. Terminal MT_1 is taken as the reference terminal for voltage measurements; current measurements are taken at terminal MT_2 and gate G. Also identified in Fig. 2–128b are two quadrants I and III. As can be seen, conduction with the triac occurs in quadrants I and III. In quadrant I the voltage at MT_2 is positive with respect to MT_1, while the voltage at MT_2 is negative with respect to MT_1 in quadrant III. Also identified in Fig. 2–128b are two important voltages $+V_{BO}$ and $-V_{BO}$, known as breakover voltages. In normal triac use it is desirable to induce conduction by signals applied to the control gate. Thus the ac voltage to be switched should always be less than V_{BO}, since a voltage exceeding the breakover voltage will cause conduction. This relinquishes control over conduction by the gate. Even transient voltages in excess of V_{BO} will cause conduction. The triac remains in its conducting state until the current falls below the holding current, I_H, marked in Fig. 2–128b. One desirable aspect of this feature is that the triac is self-protected against transient voltages exceeding V_{BO}. Hence triacs are not damaged by high voltage transients, although switching the triac into conduction may cause damage to circuits being controlled by the triac.

The triac is an extremely versatile switch and can be triggered by either positive or negative gate currents so long as V_{BO} is not exceeded. Triggering can be accomplished with dc signals, ac signals, or pulses. UJT's, neon lamps, and SBS relaxation oscillators are commonly used. The triggering and conduction characteristics are usually identified by specifying the quadrant in which conduction takes place and the sign of the gate current. Thus the symbol I (+) signifies that the triac is switched in quadrant I by a positive gate current, while I (—) signifies triggering in quadrant I with a negative gate current. The other two possibilities are III (+) and III (—) triggering.

The most sensitive modes of triggering are the I (+) and III (—) modes, although the I (—) mode is only slightly less sensitive. Triggering in the III (+) mode is rather insensitive and often avoided, although specially selected triacs are available for this purpose.

The versatility and simplicity of triac switching have brought about many important ac power control applications, which are discussed in Section 2–3.7. Triacs are commercially available at present with maximum current ratings of up to 40 A and maximum voltages of up to 600 V.

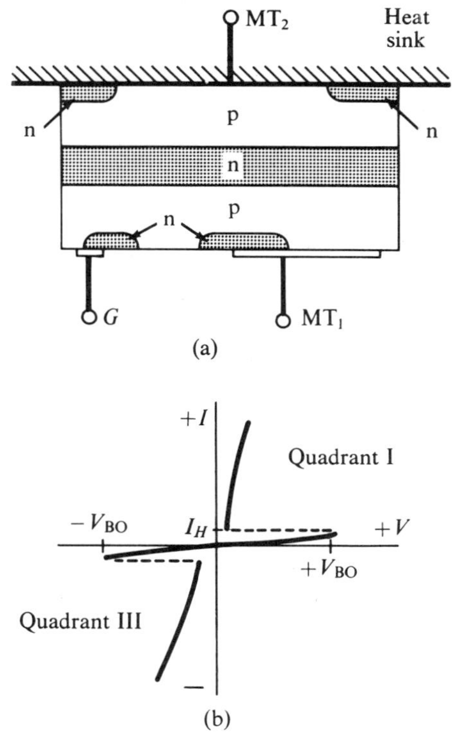

Fig. 2–128 The triac: (a) pellet structure; (b) current-voltage characteristic.

Experiment 2–22 *pnpn Device Characteristics*

A curve tracer is constructed and the current-voltage curves of two common pnpn devices, the SCR and the triac, are obtained. This experiment also demonstrates the relationship between the turn-on time of the thyristor and the gate current.

2-3.4 INFORMATION TRANSMISSION GATES

In an electronics-aided measurement, the information about the quantity being measured is first converted into an electrical quantity. This quantity is an **electrical signal.** In a complete measurement, the signals representing the measurement data may proceed through an electronic signal modifier (such as an amplifier) and then through one or more data domain converters (such as a current-to-voltage converter and a recorder). In instruments of even modest complexity there are a number of signals which represent different characteristics of the system being measured. Examples of related multiple signals include the current through a device and the voltage across it, the voltages from 10 thermocouples measuring a temperature profile, and voltages representing the pH of a solution and its temperature. As the instrument performs the measurements on such signals, certain signal (information) paths must be opened and others closed. Also, many instruments provide several different measurement functions which are achieved by altering the signal paths through the various data modification and transformation elements in the instrument. Circuits used to direct and interrupt the paths of electrical signals are called **information transmission gates.**

The basic elements in information gates are the switching devices described earlier. The application of several switching devices in gating circuits is discussed in this section. While all information transmission gates operate on the same basic switching concepts, there is a significant variation in design and implementation depending on the accuracy of signal transmission required. Two extremes of accuracy requirement are considered in this discussion: analog transmission gates and binary information gates. In the former, any distortion of the electrical signal quantity represents a proportional distortion of the data represented by the signal. Binary information gates are used with signals representing simple two-state information, such as "the temperature is or is not greater than 42°." With such signals, one range of voltage (or current) values represents one signal state (such as "greater than") and another quite different range of voltage values represents the other state (such as "*not* greater than"). With such binary information signals, considerable distortion of signal level can occur with no degradation of the information content of the signal.

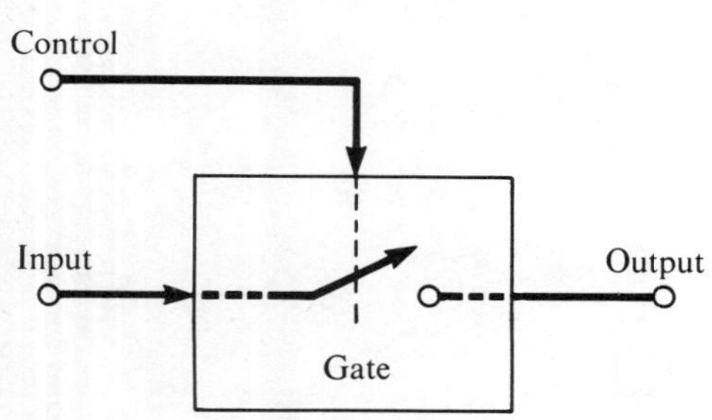

Fig. 2–129 Basic gate.

Analog Transmission Gates

A basic transmission gate is shown in Fig. 2–129. When the gate is **open** or ON, the signal at the gate input appears at the output. When the gate is **closed** or OFF, the output signal is zero. The signal level at the gate control input determines whether the gate is open or closed. An analog gate is designed to transmit from its input to its output an exact reproduction of the

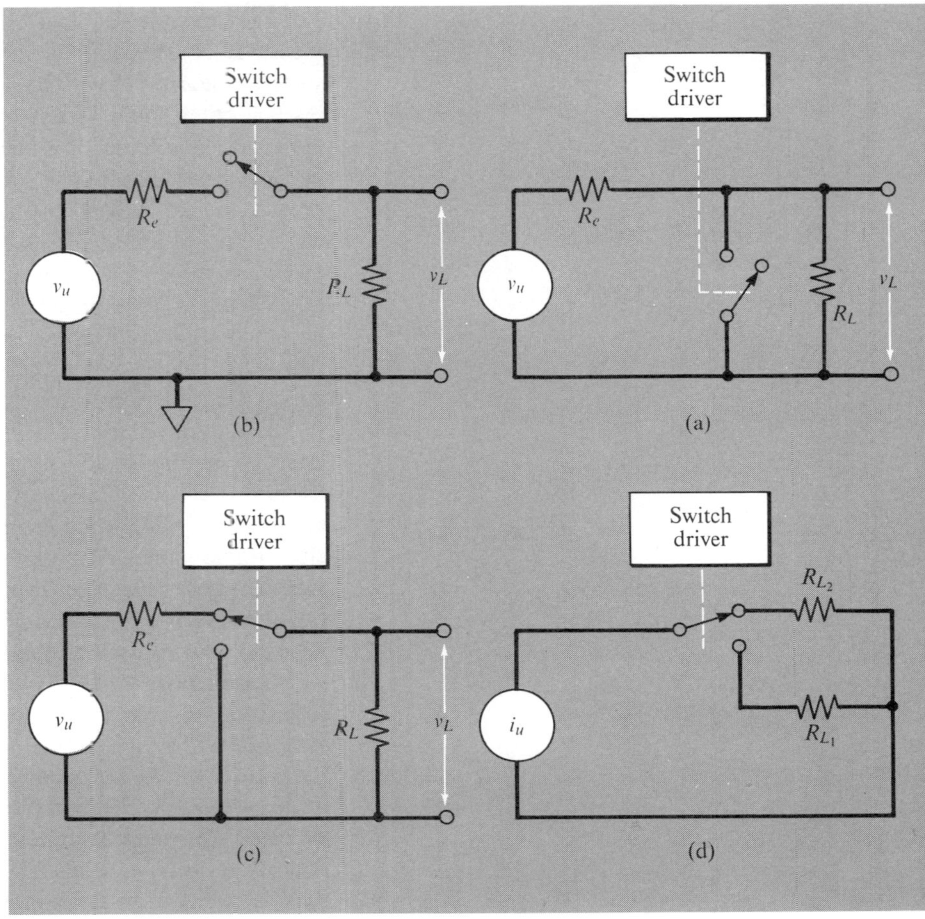

Fig. 2–130 Basic types of analog switches: (a) shunt; (b) series; (c) series-shunt voltage switch; (d) series-shunt current switch.

input waveform during the selected interval when it is open and to have a zero output when it is OFF. Other terms used for such a device are analog switch, linear switch, sampling gate, transmission gate, and linear gate. The simplest analog gate could be a manual switch, but the term gate generally refers to an automatically actuated switch. The first analog gates were made from relays. The modern reed relay form of the mechanical contact gate is still one of the most accurate and simplest analog gates to implement.

The actual switching devices can be arranged in the gate circuit in several different ways. The most common four arrangements are shown in Fig. 2–130 along with the signal sources and loads. The choice among these possibilities depends on the switch characteristics and the gate application requirements. The simple series switch gate is shown in Fig. 2–130b where R_e is the equivalent source resistance. When made from an ideal switch, the output voltage would be $v_u R_L/(R_e + R_L)$ when the gate is open (switch closed) and zero volts when the gate is closed. In order to approach this ideal, it is necessary that the switch's ON resistance, R_{ON}, be very much less than R_L and that its OFF resistance, R_{OFF}, be very much greater than R_L. With the optimum value of R_L it is possible that a transmission accuracy of 0.1% would require a switch with an R_{OFF}/R_{ON} ratio of 10^6 or more. Note that in the closed state, the output terminal is "floating." That is, its only connection to common is through the load. This is an advantage when other signal sources are to be gated to this same load when this gate is closed, but it is a disadvantage if a well-defined zero voltage output is required for the closed-gate state.

The gate circuit of Fig. 2–130a employs a shunt switch. The gate is open when the switch is open, producing the same ideal output voltage as the series switch above. When the switch is closed the switch shorts the gate input and output terminals. To provide a good zero output in the closed-gate state requires that $R_{ON} \ll R_e$. Practical signal sources and switches generally require a resistor in series with the source to increase the effective R_e. At the same time, R_{OFF} must be very much larger than R_L to avoid distorting the signal voltage.

Better switching efficiency with fewer demands on the source and load resistance values are obtained with a combination of series and shunt switches as in Fig. 2–130c. When the gate is open, the circuit acts like the series switch; and when the gate is closed, the source is disconnected (as in the series switch) but the load is shorted (as in the shunt switch). Thus accurate transmission requires that $R_{ON} \ll R_L$ and, for a good zero output, that $R_{OFF} \gg R_{ON}$. Thus the R_{OFF}/R_{ON} ratio requirement for 0.1% accuracy is 1000 times less than for the series switch and the closed output voltage level is well defined at zero. One common application of the series-shunt switch is as a **chopper.** A chopper is a device that is used to alternately transmit and interrupt a signal at regular intervals. The chopping operation is generally performed on a signal so that the input signal level can be accurately compared with a known reference level (often common). For this reason, it is important that the gate circuit used for chopping have a well-defined closed-gate output voltage.

A series-shunt current switch is shown in Fig. 2–130d. This switch alternately directs the signal current i_u through two different loads, R_{L_1} and R_{L_2}. A current switch always has the signal current i_u pass through the switch regardless of the value of the load R_L. The voltage that is switched is a secondary consideration and is equal to $i_u R_L$.

There are other important considerations in classifying analog switches. For instance, it is important whether or not the signal and control paths are isolated and how the switch is controlled. A **galvanically isolated switch** has no currents that flow between signal and control circuits, whereas a **direct-coupled switch** has no galvanic isolation. A **voltage-controlled switch** requires a control voltage to turn it ON, whereas a **current-controlled switch** depends on a control current to turn it ON.

The applications of several solid state switches in analog gating circuits are described below.

Diode Analog Gate Circuits. Even though the diode is a two-terminal device, it is possible to use a diode circuit to gate signals ON and OFF in response to an actuating source. The circuit of Fig. 2–131a shows a

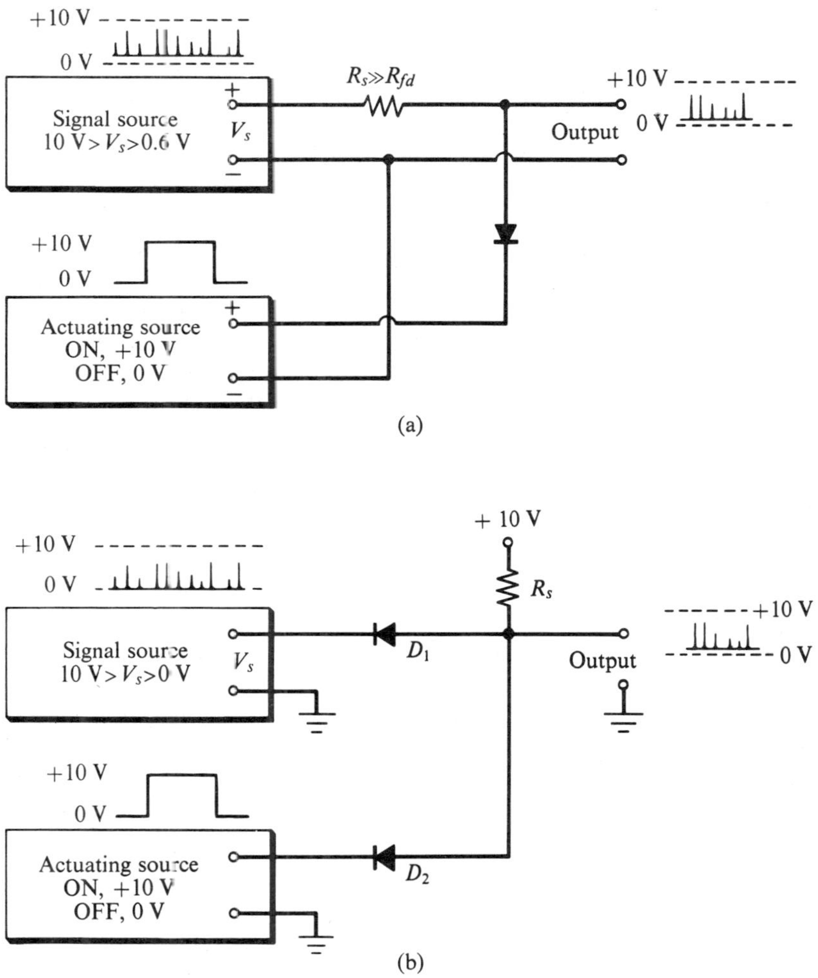

Fig. 2–131 Diode analog gates: (a) shunt switch diode; (b) series-shunt switch.

shunt gate circuit using a diode. When the actuating source output is 0 V, the diode acts as a closed shunt switch shorting the output terminals for any signal source voltage greater than 0.6 V.

When the actuating source output is +10 V, the diode becomes reverse biased and will not affect the transmission of signals of 10 V or less to the output terminals.

The switching circuit of Fig. 2–131a puts a heavy demand on both the signal and actuating sources. In order to maintain no variation in output in the OFF position, the actuating source must be capable of accepting the full signal current at its output without changing the output voltage; i.e., output resistance of the actuating source must be nearly zero in the 0 V state. Another circuit which provides better isolation between the signal and actuating sources uses two diodes as shown in Fig. 2–131b. When the actuating source output is 0 V, diode D_2 conducts the current from the 10 V source through R_s to ground. The output potential is held at 0.6 V, the forward drop across D_2. Positive signals from the signal source give D_1 a reverse bias resulting in no change in current through D_2 or output potential. However, when the actuating source potential is more positive than the signal source, D_1 is conducting and D_2 is reverse biased. The output will now follow V_s and will be equal to V_s plus the forward drop across D_1.

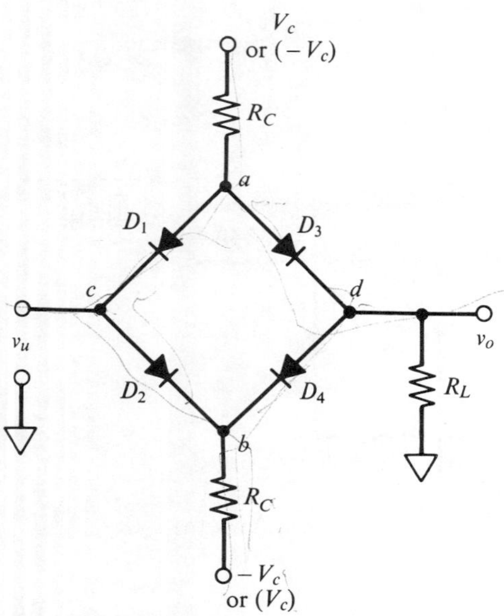

Fig. 2–132 A four-diode analog switch.

The diode analog switches shown in Fig. 2–131 can be improved by adding two more diodes as shown in Fig. 2–132. If the control voltages V_c and $-V_c$ of this circuit are balanced, there will be no voltage offset at the output of the type that resulted from the forward voltage drop across the switching diodes in the series-shunt switch. The required magnitude of the control voltages depends on the amplitude of the input voltage v_u, and is determined by the requirement that the current be in the forward direction in diodes D_1, D_2, D_3, and D_4 when V_c and $-V_c$ are applied as shown in Fig. 2–132.

When the control voltages are suitable all four diodes conduct, and a very low resistance path exists between the input signal v_u and the output load resistor R_L. Therefore the output voltage v_o will faithfully follow the input voltage v_u. It is, of course, important that the load resistor R_L be much larger than the diode resistances and the output resistance of the signal source. It can also be seen in Fig. 2–132 that the voltage drops across the diode pairs D_1-D_3 and D_2-D_4 are in opposition, so that when $v_u = 0$ the output voltage $v_o = 0$.

When the control voltages are reversed to $(-V_c)$ and (V_c), the four diodes are reverse biased and the resistance in series with the output is much larger than the load resistor R_L. Therefore the output voltage is zero for any value of input signal up to the magnitude of the control signal in the reverse biased condition.

Transistor Analog Gate Circuits. The transistor switch shown in Figs. 2–106 and 2–111 can be used in analog gating circuits as shown in Fig. 2–133. In the series switch circuit (Fig. 2–133a), the gate is closed when the transistor base is positive with respect to v_u. The reverse biased base-collector junction makes an open output when the gate is OFF. A negative pulse is used to turn the gate (and the transistor) ON. The output voltage will be 0.1 to 0.2 V less than the input voltage because of $V_{CE(\text{sat})}$ and will follow any input voltage which is more than 0.6 V positive with respect to the base voltage. When negative signal levels are to be switched, an npn transistor is used with a positive-going turn-on signal.

The shunt switch shown in Fig. 2–133b uses an npn transistor to short the output terminals for positive signals when the switch base is driven positive. This switch gives better transmission accuracy for the open gate and a gate-closed output voltage of $V_{CE(\text{sat})}$. The series-shunt combination is shown in Fig. 2–133c.

For precision analog switching BJT's are frequently used in an inverted or common-collector mode as illustrated in the shunt switch gate of Fig. 2–134. It is known that the voltage offset is smaller when a BJT is operated in the inverse mode. Also, the ON resistance can be as low as 2 Ω, and many BJT's can be turned ON and OFF in 10 nsec or less when they are operated in the inverse mode.

The control signal that is applied to the base should be sufficiently more positive than ground potential so that the base-collector junction becomes forward biased when the transistor is to be turned ON. When the transistor is ON, the output voltage v_o is equal to the voltage drop between the collector and emitter, $V_{CE(\text{sat})}$, which can be only 0.2–0.4 mV for base currents greater than 0.5 mA. Therefore, when the transistor is ON the output voltage is close to zero and the gate is closed. When the transistor is OFF the output voltage v_o is equal to the input signal v_u. To turn the transistor OFF and transmit the signal v_u to the output, both the base-collector and base-emitter junctions must be reverse biased.

Field-Effect Transistor Analog Gate Circuits. As might be expected, the field-effect transistor family (JFET and IGFET) make excellent switches for analog gates. They are used in the now-familiar series, shunt, and series-shunt configurations as shown in Fig. 2–130. The FET switch offers high analog transmission accuracy because it does not introduce a junction potential in series with the signal. It also offers the versatility of being

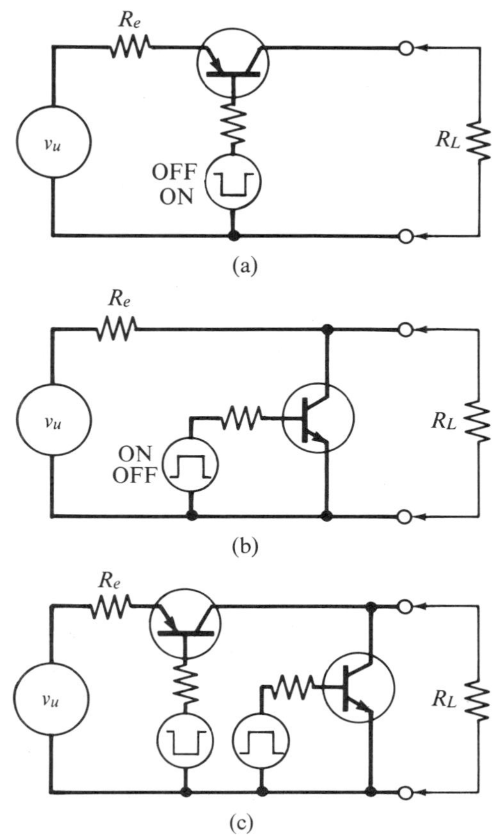

Fig. 2–133 Transistor gate circuits: (a) series; (b) shunt; (c) series-shunt and driver.

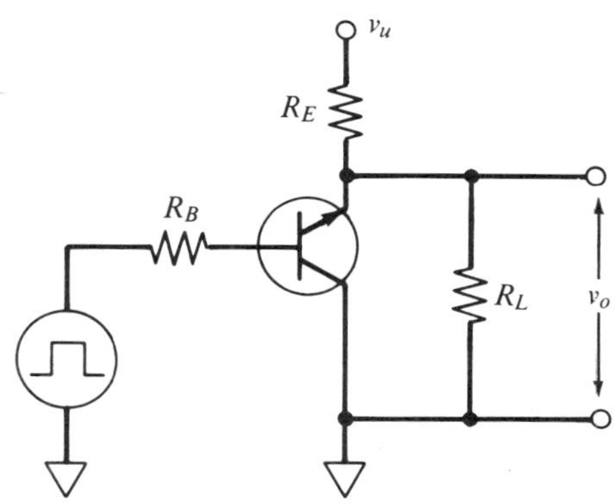

Fig. 2–134 Inverted-mode BJT switch.

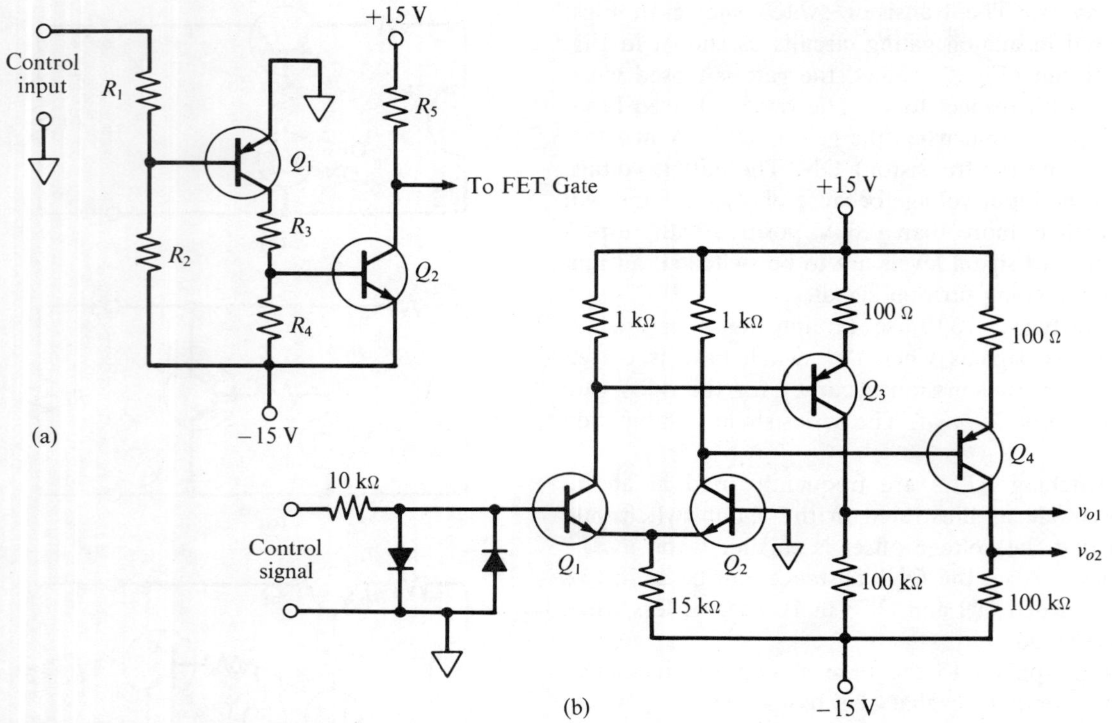

Fig. 2–135 FET switch and driver: (a) Tobey et al., p. 329; (b) Smith, p. 191.

able to conduct current in either direction. Its disadvantages with respect to the BJT switch are higher ON resistance and generally slower switching speed.

The gate voltage of the field-effect transistor switch is provided by a **FET switch driver circuit** which converts an ON-OFF control signal with convenient voltage levels into a signal suitable for turning the FET completely ON and OFF. Two such driver circuits are shown in Fig. 2–135a and b. In Fig. 2–135a, R_1 and R_2 are chosen so that for the more positive value of the control signal (say +4 V), transistor Q_1 will be OFF, and for the more negative value (say +0.5 V), it will be ON. Thus for the more negative control signal value, Q_1 is ON; this causes Q_2 to turn ON and produce ∼−15 V at the FET gate driver output. For the more positive control input level, Q_1 and Q_2 are OFF and the FET gate output voltage is ∼+15 V. This ±15 V gate voltage swing is enough to switch signal levels of ±10 V with most FET devices.

The driver circuit of Fig. 2–135b is convenient for a series-shunt gate circuit or for two series switch gates in an SPST configuration because of the complementary gate outputs provided. Transistors Q_1 and Q_2 have

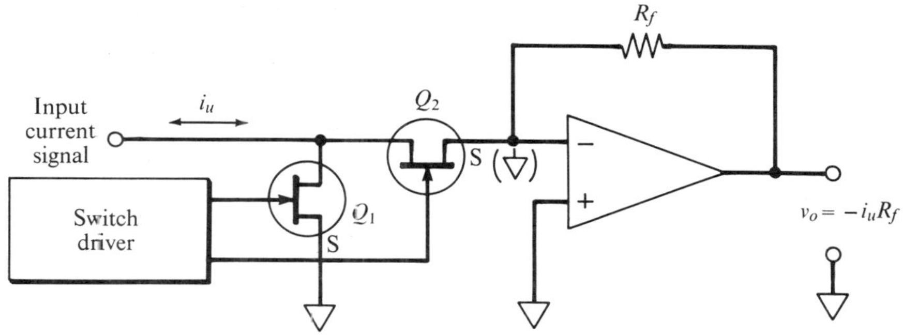

Fig. 2–136 A JFET series-shunt current switch.

the same emitter potential while the base of Q_2 is connected to common. If the control signal is greater than $+0.6$ V, Q_1 is ON; this turns Q_3 ON and produces $\sim+15$ V at the v_{o1} output. At the same time, Q_2 is OFF; therefore Q_4 is OFF and the v_{o2} output is ~-15 V. When the control input is less than -0.6 V, the opposite transistors are conducting and the output voltage levels are reversed. When these "complementary" output voltages are used to drive the gates of two FET switches, the switches will always be in opposite states.

When JFET's are used as current switches, they do not require large amplitude driver voltage changes which reduce the switching speed. For a FET current switch, the drain and source voltages are always zero and the voltage across the switch is zero. Since the driver voltage needs to vary only between zero and the pinch-off voltage (<5 V), the current switch is faster than a voltage switch. A series-shunt current switch is shown in Fig. 2–136 with the input current signal feeding into the summing point of an OA.

When JFET Q_1 is ON, Q_2 is OFF, so that the signal current i_u passes through Q_1 to common. When the driver reverses so that Q_2 is OFF and Q_1 is ON, then the current i_u is connected to the summing point of the OA.

Experiment 2–23 *Analog Switching Circuits*

An analog diode switching circuit is first constructed in this experiment and then tested to determine whether the switch faithfully transmits analog current and voltage signals from input to output when it is ON and effectively disconnects them from the output when it is OFF. Then a JFET switch and pnp transistor driver circuit are constructed and tested for analog transmission. Finally, a commercial analog switch is employed and tested. The experiment is thus designed to show the principles of analog switching and some of the circuits for accomplishing high-accuracy analog transmission.

Binary Information Gates

A binary information signal represents a piece of information for which there are only two possible states. Examples of such information are: the laboratory animal is pressing (not pressing) the button; it is after (before) 8:26 a.m.; the conductivity of the distilled water is less than (equal to or greater than) 1 μmho; or, in general, any question that can be answered yes or no, or any statement that is either true or false. Binary information is encoded electrically by assigning a separate range of signal voltage or current levels to each of the two states, for example, 0 to +1 V for one state and +3 to +5 V for the other. The two signal level ranges are called **logic levels.** The more positive logic level is called HI and the more negative, LO. The logic levels are also sometimes called 1 and 0 for the two numerals in the binary number system, or TRUE and FALSE, though in these latter names there is no implication regarding the relative signal values.

In transmitting a binary information signal, variation of the signal level can occur within the given logic level range without any change in the information content of the signal. Thus binary information gates require a much lower signal amplitude transmission accuracy than do analog gates. However, a signal level which is not within either of the two logic level ranges conveys no information at all (if the information is truly binary) and is to be avoided. Therefore, the circuits of some binary information gates are designed to have only two stable signal levels, which are within the logic level ranges of the system. The discussion below introduces several types of binary information gates, or **logic gates.** For a more thorough discussion of logic gates and applications, the reader is referred to Module 3, *Digital and Analog Data Conversions.*

Diode Logic Gates. It was shown in the previous section that diodes can be used as switches in analog gate circuits. The same basic circuit as that of Fig. 2–131b can also be used to switch logic level signals, as illustrated in Fig. 2–137. This circuit is called an AND logic gate. The reason it is called a "logic gate" can be understood by following the circuit action in Fig. 2–137. The logic levels of the signal source are +4.0 V for a HI logic

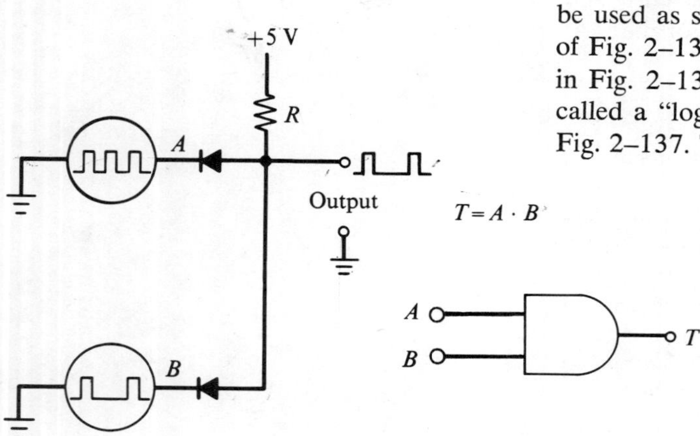

Fig. 2–137 Diode AND gate and symbol.

level and 0 V for a LO logic level. When one input is held at the HI logic level, the output will follow the level changes at the other input; i.e., the gate is open. On the other hand, if one of the inputs is LO, the output will be LO regardless of the input signal; i.e., the gate is closed. For an AND gate all inputs must be HI in order to have a HI level at the output.

Note in Fig. 2–137 that during the first pulse both inputs A and B are at +4.0 V. The diodes are still forward biased between the +5 V supply and the +4 V logic level input signals. The output voltage is thus 4.0 V + V_γ (the voltage drop across the forward biased diode), which is clearly a HI logic level. There is a LO level logic pulse at B during the second pulse at A so that the output is connected through the diode to common; this provides a LO level output. As stated above, a LO level at either input will keep the gate closed.

When diodes are connected as shown in Fig. 2–138, they perform the so-called OR function. That is, a HI level at input A OR B OR C will cause a HI level at the output. During the period when any one input is HI, the logic levels at all other inputs can change without having the changes appear at the output. A common application of the OR gate is for preventing digital signal sources which need to be connected to one point in a circuit from interfering with each other. If input A is held at a HI logic level, then the A-input diode conducts and there is a voltage drop across the output resistor which reverse biases the other diodes so as to isolate the other inputs.

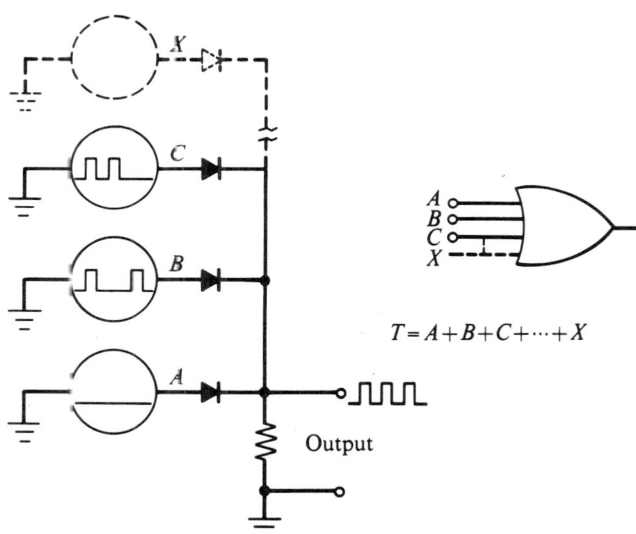

Fig. 2–138 Diode OR gate and symbol.

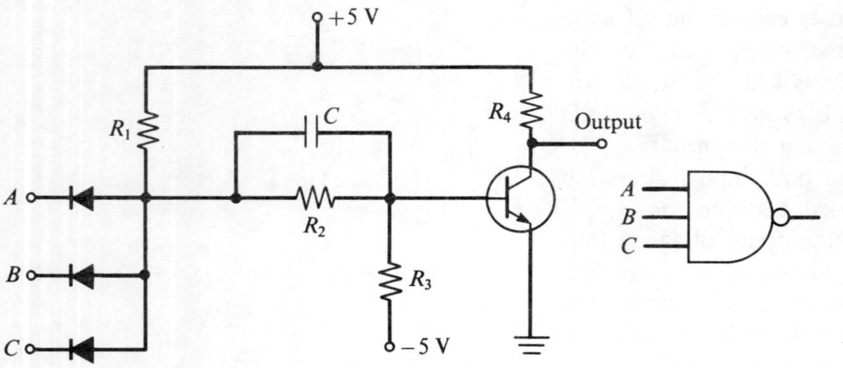

Fig. 2–139 Positive AND gate with inverting amplifier.

Diode-Transistor Logic Gates. A natural development from the diode gates was the addition of a transistor shunt switch to the gate output to reduce the output resistance as shown in Fig. 2–139. This is a diode-transistor logic or DTL gate. The input diodes and R_1 are the AND circuit shown in Fig. 2–137. When the AND output is near 0 V, the transistor is cut off and the output is +5 V, a HI logic level. A HI output from the AND gate is sufficient to saturate the transistor, making the output very near a LO 0 V. The gate function is now the AND function with an inverted (opposite logic level) output. This is called a NOT-AND or NAND gate. The symbol for a NAND gate is shown in Fig. 2–139. When a LO level signal source is connected to one of the inputs, a current of $5/R_1$ A must pass to ground through the signal source without raising the input voltage beyond the upper limit of the LO level. Therefore a signal source for the gate of Fig. 2–139 must have a very low output resistance when its output level is LO. When the output of the NAND gate is LO, the transistor is saturated and conducting heavily. Its output resistance is so low that it can be connected to many gate inputs. The number of gate inputs which can be connected to a gate output without jeopardizing the logic level is called the **fan-out,** an important gate characteristic. Very little current is required of a signal in the HI state, and the amplifier is easily designed to supply enough current at the HI logic level to equal the fan-out limitation set by the LO state conditions.

Integrated Circuits. Many of the currently practiced techniques of semiconductor fabrication involve diffusion of dopants and chemical etching on selected areas of a silicon crystal "chip." These techniques allow for hundreds of transistors and/or diodes to be made on a single chip. Many transistors and diodes can then be interconnected right on the chip to form more complex circuit units called "integrated circuits." Resistors can also be made by careful control of doping area and depth. The doping and etch-

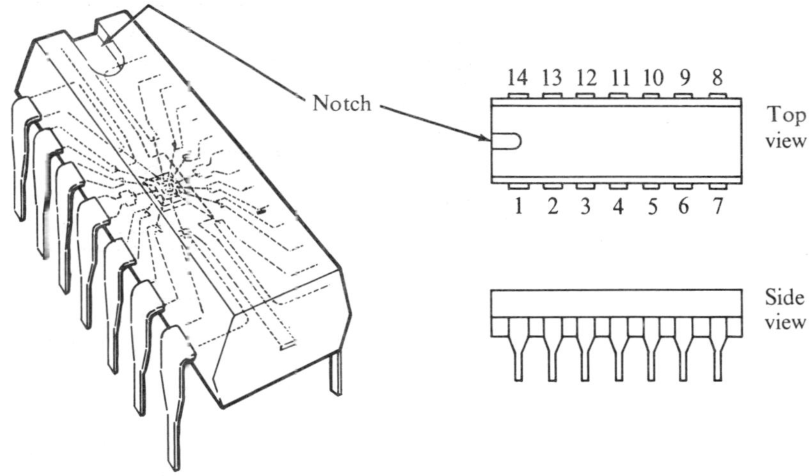

Fig. 2–140 Integrated circuit plastic dual in-line package.

ing operations are carried out first to make the desired array of transistors, diodes, and resistors. Then an oxide insulating layer is put over the chip. This is followed by selective etching through the insulation where the connections to the components are to be made. Next aluminum is deposited on the chip in the form of the required interconnections. The chip is then mounted in a metal, ceramic, or plastic body and connections are welded from the connector pins of the body to aluminum pads on the interconnection pattern where the external connections are to be made. Finally, the body is sealed and the device is given a final check. An example of an integrated circuit is shown in Fig. 2–140.

As semiconductor technology has improved, the number of active devices and components that can be economically put on a single silicon chip has increased greatly. Several complete gates of considerable complexity are now available in a single tiny package, as are many prewired combinations of gates that perform special functions. Indeed, it is the availability of such versatile circuits in convenient and inexpensive form that has sparked a new revolution in scientific instrumentation and many other areas of electronics.

TTL Gates. Available in integrated circuit form are the DTL gates described in the previous section and several other types in many variations. Along with the development of these circuits came the realization that some circuit improvements were possible and economical with IC's which were not practical with discrete components. An example is the TTL gate shown in Fig. 2–141, where Q_1 is a multiple-emitter transistor. Grounding any one or more inputs forward biases transistor Q_1; this puts the collector of Q_1

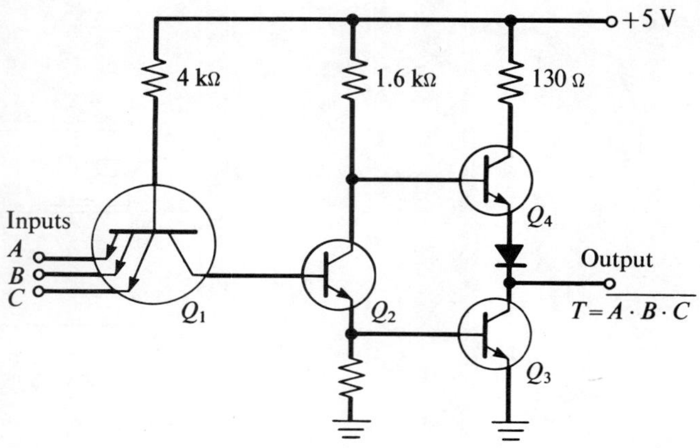

Fig. 2–141 TTL positive NAND gate. (Courtesy of Texas Instruments.)

at a low potential and turns Q_2 OFF. This, in turn, turns Q_4 ON and Q_3 OFF and results in a HI output. If, however, all inputs are HI (or unused) the base-collector junction of Q_1 will conduct, forward biasing the base-emitter junction of Q_2. When Q_2 is ON, Q_3 is ON and Q_4 is OFF, resulting in a LO level at the output. This corresponds to the inverted AND or NAND function.

The circuit is actually very similar to that of the DTL gate, with Q_1 replacing the gate and coupling diodes. Transistors Q_4 and Q_3 are a series-shunt switch providing a larger current capability in the 1 state, and Q_2 provides gain for increased speed. Since the connecting leads inside the integrated circuit (IC) gate are extremely short, lead capacitance and inductance are minimized and very high speed is possible. The first commercial TTL gates had only a 13 nsec propagation delay, but even faster TTL gates are now available. Schottky TTL gates have only about a 3 nsec propagation delay. The fan-out is about 10 and the noise immunity is typically 1.0 V, making the TTL a very convenient general purpose binary gate.

Experiment 2–24 *Binary Information Gates*

It is shown in this experiment that the diode switching circuit of the previous experiment can also be used to transmit binary information, but the considerations for successful operation are different. The diode binary information gate is constructed and tested, and limitations are noted. Then a diode-transistor binary gate is constructed and tested in order to show how a transistor inverter following the diode gate can improve performance. Then an integrated circuit binary information gate is used. The input and output levels are determined for a TTL NAND gate and advantages and limitations over the diode and diode-transistor gates are noted.

2–3.5 PRINCIPLES AND APPLICATIONS OF LIGHT-ACTIVATED SWITCHES

In recent years there have been numerous new devices introduced which are light activated. These devices convert radiant energy into electrical current or voltage or a variable resistance. In Module 1 and Section 2–1 of this module light-sensitive devices were discussed as transducers for measurements of light intensity. They also have important uses as switches for power control and for electrical isolation. In this section the basic principles of light-activated semiconductors are first discussed with emphasis on photodiodes, phototransistors, and light-activated thyristors, such as the light-activated SCR (LASCR). Then the most commonly used radiation sources, such as tungsten lamps, neon lamps, and light-emitting semiconductors, are briefly discussed. This section concludes with a presentation of source-detector combinations for achieving electrical isolation and for accomplishing several important switching functions.

Light-Activated Switches

A wide variety of light-sensitive devices is available. The cadmium sulfide, cadmium selenide, and cadmium telluride photocells convert radiant energy into a variable resistance. These devices were discussed in Section 2–1 of this module. Other devices such as photovoltaic cells, which were described in Module 1, produce voltage and current from incident radiant energy. This discussion will focus on the newer semiconductor junction light-sensitive devices, which include light-activated diodes, transistors, and thyristors.

The Photodiode. As explained in Section 2–3.1, the current through a reverse biased pn junction is due to the energetic creation of electron-hole pairs in the depletion region at the junction. In addition to thermal generation of electron-hole pairs, electromagnetic radiation of the proper energy (wavelength), when incident on the depletion region, can be absorbed by the material and create electron-hole pairs as illustrated in Fig. 2–142a. Provided the rate of charge carrier production by absorbed electromagnetic radiation is very much larger than the rate of thermal generation, the reverse bias current is directly proportional to the photon flux (light intensity). The characteristic curves shown in Fig. 2–142b demonstrate this linear relation for a wide range of reverse bias voltages.

If the photodiode is operated with enough reverse bias to prevent the recombination of charge carriers in the depletion layer, the photocurrent is given by

$$-I_p = b_\lambda Q_e P_k,$$

where b_λ is the quantum efficiency, Q_e is the electron charge, and P_k is the incident photon flux (photons/sec) on the junction.

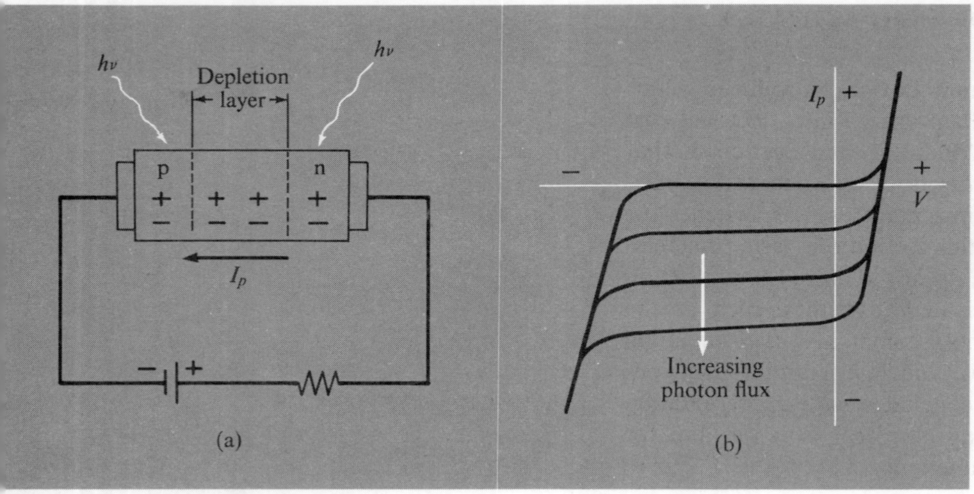

Fig. 2–142 The photodiode: (a) pictorial showing electron-hole pair production; (b) current-voltage characteristics for different photon fluxes.

The photodiode can provide currents in the μA range and is extremely fast-responding. Transit times of charge carriers across the junction are in the subnanosecond range; frequency response is therefore in the gigahertz region.

Phototransistors. The ordinary phototransistor is similar to the photodiode except that internal gain is provided. A simplified pictorial of the phototransistor is shown in Fig. 2–143a, along with an equivalent circuit and the common circuit symbol. In the equivalent circuit (Fig. 2–143b) the collector-base junction, which serves as the photodiode, is separated from the transistor for ease in understanding its operation.

Light striking the photodiode gives rise to a photocurrent I_p which supplies the base current of the transistor. The base current is amplified by the dc current gain, h_{FE}, of the transistor so that the emitter current I_E is given by

$$I_E = I_p(h_{FE} + 1). \tag{2–127}$$

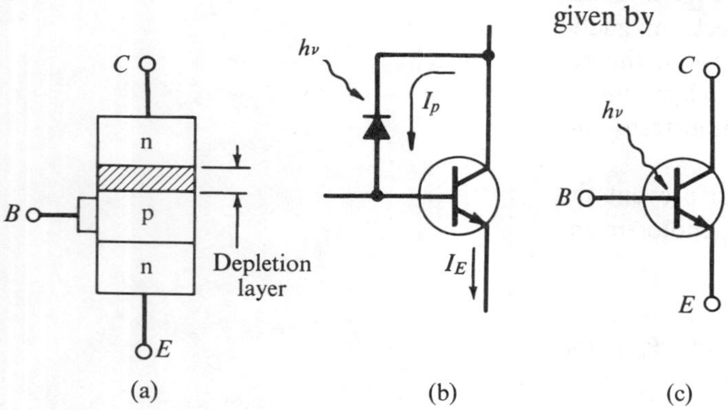

Fig. 2–143 The phototransistor: (a) pictorial; (b) equivalent circuit; (c) circuit symbol.

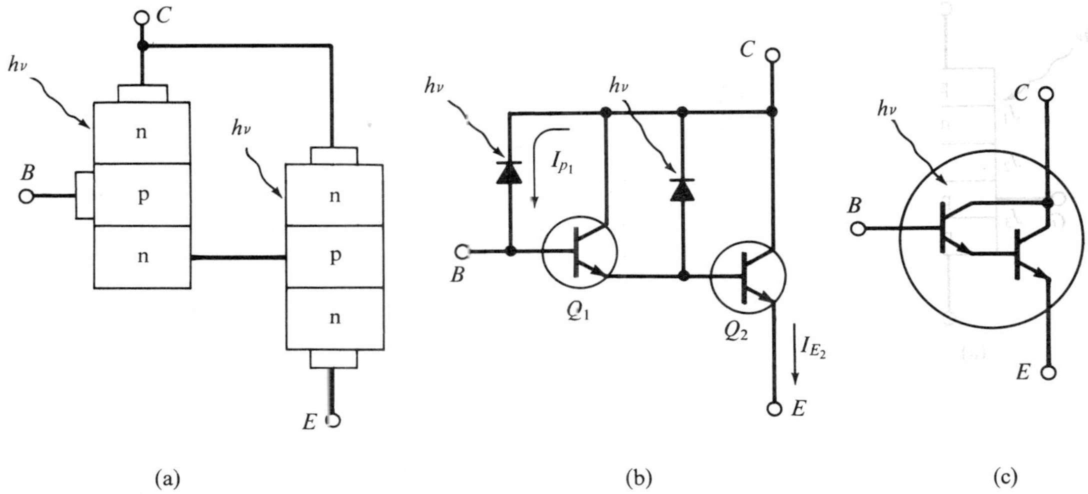

Fig. 2–144 Photodarlington amplifier: (a) pictorial representation; (b) equivalent circuit; (c) circuit symbol.

Typical phototransistors can amplify the photodiode current by 10–100 times. Switching speeds of phototransistors must be carefully considered for some applications. Typical delay times are 1–2 μsec and rise times are of the order of 4–8 μsec.

A newer type of phototransistor called a photodarlington amplifier is illustrated in Fig. 2–144. The equivalent circuit (Fig. 2–144b) illustrates that the photodarlington consists of two coupled phototransistors. The emitter current from Q_2, I_{E_2}, is approximately given by

$$I_{E_2} \approx I_{p_1} h_{FE1} h_{FE2}. \tag{2–128}$$

The fact that the current is the product of two dc current gains accounts for the high sensitivity of the photodarlington amplifier. Photodarlingtons can provide output currents of the order of hundreds of milliamperes. However, their switching speeds are somewhat slower than those of normal phototransistors.

The Light-Activated SCR (LASCR). The LASCR is a relatively recent device, whose basic operating principles can be understood with the aid of Fig. 2–145. If forward voltage is applied from the anode to the cathode of the LASCR, junctions J_1 and J_3 are forward biased. Conduction does not occur, however, because J_2 is reverse biased in the absence of light. When light strikes the silicon, free electron-hole pairs are created as in the photodiode. These pairs created near the depletion layer can move across J_2. If sufficient light is present, the current in the reverse biased diode becomes large enough to increase the net current gain of the two transistors above unity and the SCR goes into conduction. Thus the basic firing cri-

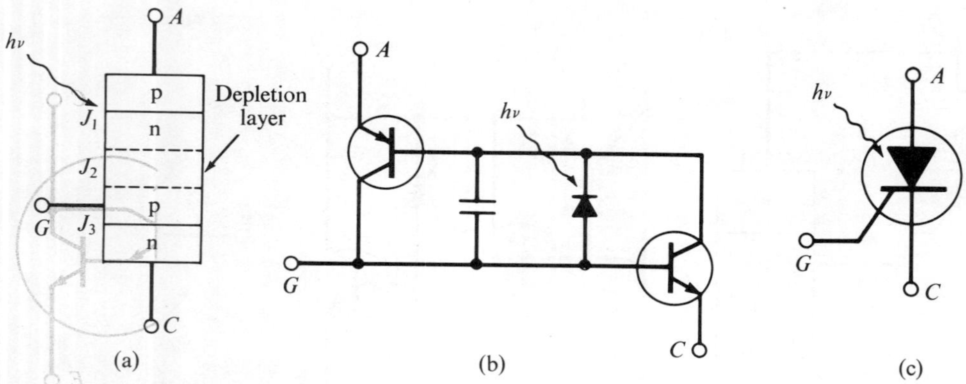

Fig. 2–145 The light-activated SCR (LASCR): (a) pictorial representation; (b) two-transistor equivalent; (c) circuit symbol.

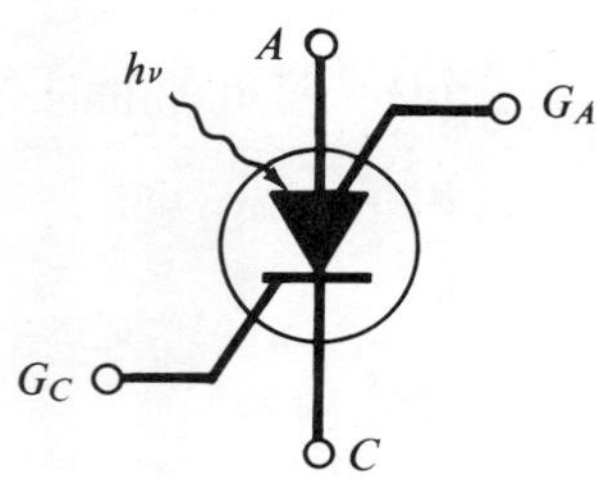

Fig. 2–146 Circuit symbol of light-activated silicon-controlled switch (LASCS).

terion is the same as for a normal SCR with the exception that there is an additional light-generated current which can cause conduction. Like normal SCR's, the LASCR can also be switched by the injection of current into the gate. Hence the LASCR is a very versatile switching device.

LASCR's are available which can switch amperes of current and block several hundred volts in the reverse direction.

The Light-Activated Silicon Controlled Switch (LASCS). Perhaps the most versatile of all pnpn devices is the LASCS, which has both anode and cathode control gates in addition to being sensitive to light. The circuit symbol for the LASCS is shown in Fig. 2–146. The same techniques used in switching the SCR and LASCR can be used with the LASCS. However, the presence of an anode gate allows turn-on to be accomplished with negative pulses at the anode, and turn-off with positive pulses.

Light Sources

A variety of light-emitting devices are used in conjunction with light-activated switches to provide numerous important control and isolation applications. The emitted radiation from the light source as "seen" by the photodetector depends on many variables, including the relative sensitivity of each as a function of wavelength, the geometry, use of light filters and light pipes, and others. It is important that the source have a long stable life, and where many sources are to be used the cost is important. Where distance between the source and light-activated device must be great, the intensity and focusing are especially important. Incandescent lamps, neon bulbs, electroluminescent devices, light-emitting injection diodes, ruby and gas lasers, and others have been used.

Incandescent Lamp. The tungsten filament lamp is inexpensive and has a wide spectral output which is similar to blackbody radiation as shown in

Fig. 2–147. However, the rise time of the output light is poor (typically 100 msec), and tungsten bulbs are susceptible to vibration breakage and burnout. They are used primarily with systems that have a low pulse rate.

Neon Lamp. The neon lamp is very inexpensive, has long life, and the required input power is much less than for the incandescent bulb. Response time is in the microsecond range. Most of the light output is limited to two orange spectral lines, the maximum available output is quite limited, and a high voltage supply is needed. This limits the flexibility of the optical link, and a detector must be chosen that is sensitive to the orange lines.

Semiconductor Sources. Light-emitting diodes (LED's) and junction lasers are dependent on semiconductor pn junctions which emit photons as a result of hole-electron recombinations in the structure. In silicon and germanium diodes the energy from recombinations is delivered to the crystal primarily as heat, but for junction diodes made from gallium arsenide (GaAs), gallium phosphide (GaP), and gallium arsenide phosphide (GaAsP) the recombination energy is converted mostly into emitted photons. This light is either incoherent and concentrated in a narrow bandwidth, as illustrated in Fig. 2–147 for a GaAsP LED, or it can be coherent and concentrated primarily in a narrow line. When emitting coherent light it is called a **junction laser.**

These diodes have very fast response times in the nanosecond range and have good dynamic range and linearity. Present diode devices have their light output typically in the infrared region (about 900 nm), the red region (650 nm), or the green region (570 nm) so that a detector must be used that is sensitive in this region.

The circuit symbol for a LED is illustrated in Fig. 2–148, along with a plot of relative intensity vs. current input for a GaAsP LED. For a LED it is customary to define the quantum efficiency or electrical-to-optical conversion efficiency as the ratio of the number of photons out to the num-

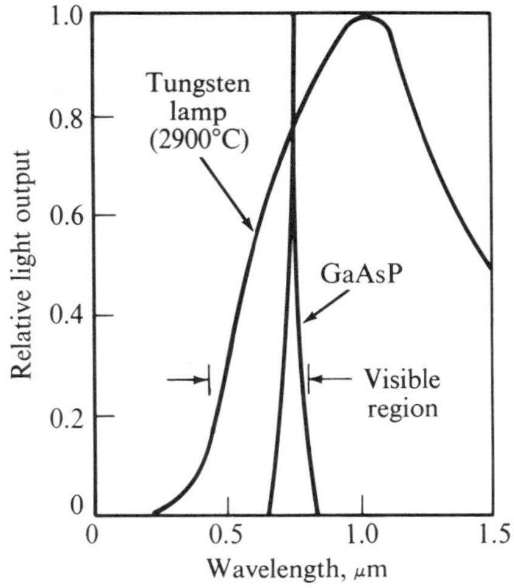

Fig. 2–147 Relative output of sources vs. wavelength.

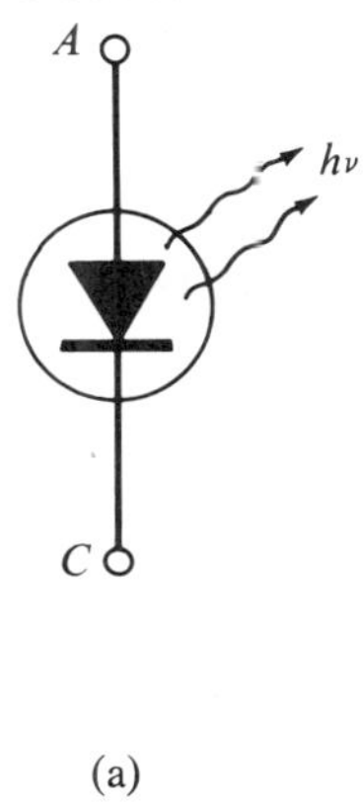

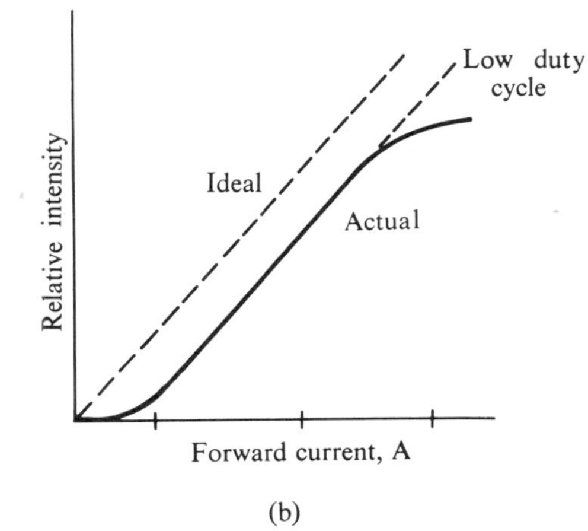

Fig. 2–148 Light-emitting diodes: (a) circuit symbol; (b) power output vs. forward current for a GaAsP LED.

ber of electrons in. LED's are often used in a pulsed mode to increase the output intensity, since the LED can withstand higher peak currents under low duty cycle pulsing conditions.

Optical Coupling for Electrical Isolation

The combination of a light source with a photosensitive device inside an opaque enclosure can perform electronic functions that are difficult by other means. The information from the light source and associated devices can be transferred to the photodetector without any electrical coupling. Light is the coupling link. Because the light source terminals and photodetector terminals are isolated, the pair forms an almost ideal four-terminal network in which there is essentially no interaction between input and output circuits (although the 1–2 pF capacitance typical between terminals of some pairs can sometimes be significant).

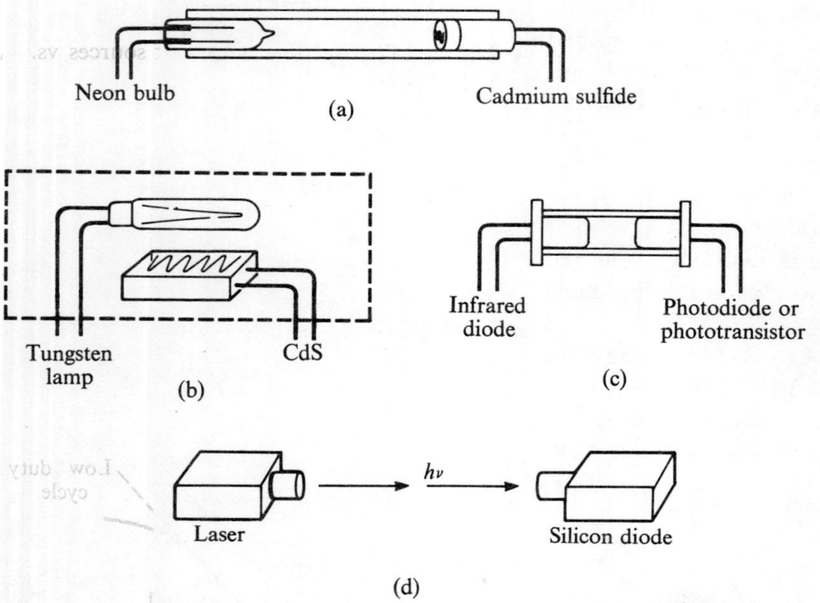

Fig. 2–149 Source-detector pairs for opto-electronic devices.

Some typical optical links matching source detector pairs are illustrated in Fig. 2–149. In recent years completely encapsulated optical couplers have become available commercially. Figure 2–150a shows an optical coupler made from a LED and a phototransistor. When input current is applied to the LED, radiation is produced and transduced by the phototransistor back into an electrical current. Complete isolation of the input and output circuits is accomplished by optical coupling and no feedback can occur. Figure 2–150b shows a typical plot of output current vs. input current for the optical coupler of Fig. 2–150a.

Optical coupling has several other important advantages. It eliminates ground loops and isolates noise sources. Switching can be accomplished on a very short time scale. Also, optical coupling can be used to shift voltage levels as in binary information transmission.

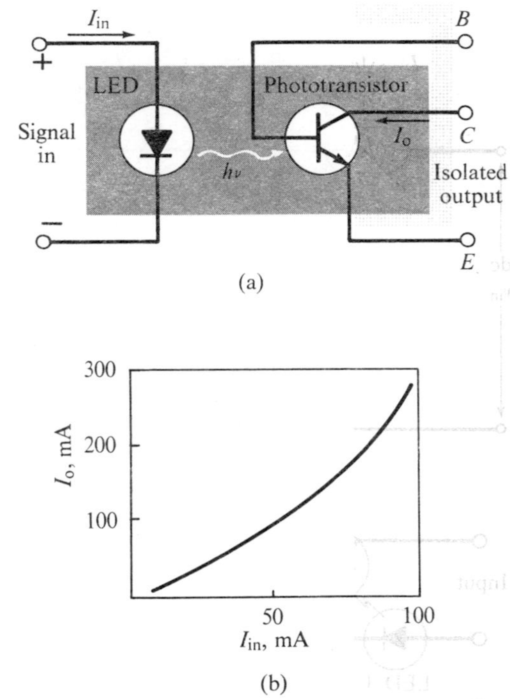

Fig. 2–150 Optical coupler: (a) source-detector pair; (b) output current vs. input current.

Applications of Optoelectronic Devices

The applications of semiconductor optoelectronic devices could be classified into two groups: one group in which the device acts as a continuously variable resistor (rheostat), and another group that depends only on the low ON resistance and high OFF resistance for switching applications. A few examples are presented to illustrate some of the many uses of source-detector pairs.

The use of the rheostat characteristics of a CdS detector together with a tungsten source is illustrated in Fig. 2–151 for an illumination control. A portion of the light radiation is made to fall on the CdS cell, which is in series with resistor R_{ai}. The voltage drop across R_{ai} is fed to an amplifier which provides an output signal to a series power element. The effective resistance R_s of the series element depends on the input to the amplifier, which depends on the light incident on R_c. The amplifier and power series element are connected in such a way that an increase in lamp intensity for any reason will cause a decrease in R_c. This causes an increase in the amplifier input voltage $v_{ai} = V_r R_{ai}/(R_{ai} + R_c)$, which causes the resistance R_s of the series element to increase, thus decreasing the lamp current. The resulting optoelectronic feedback, therefore, brings the light illumination down toward its original value, and tends to hold illumination constant, as "seen" by the CdS detector.

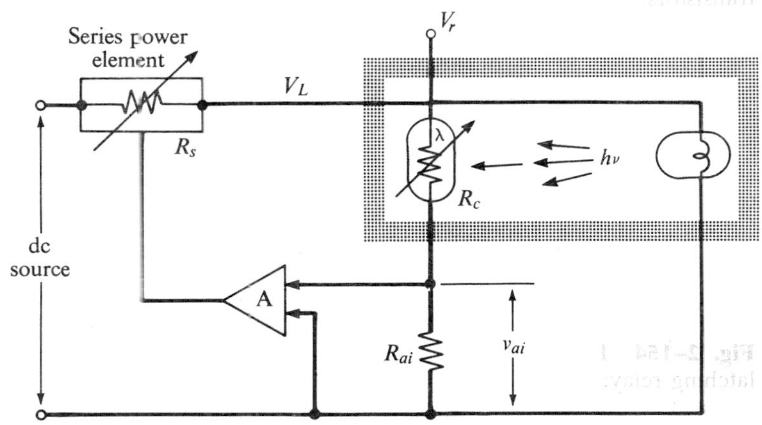

Fig. 2–151 Illumination control using a tungsten source CdS detector.

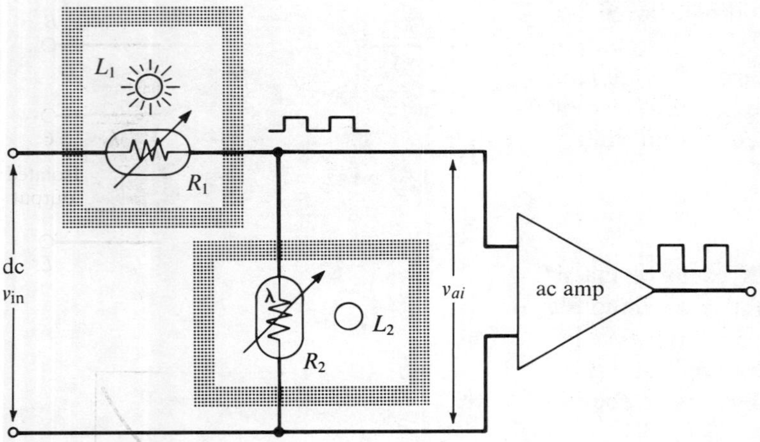

Fig. 2–152 Neon source–cadmium selenide chopper.

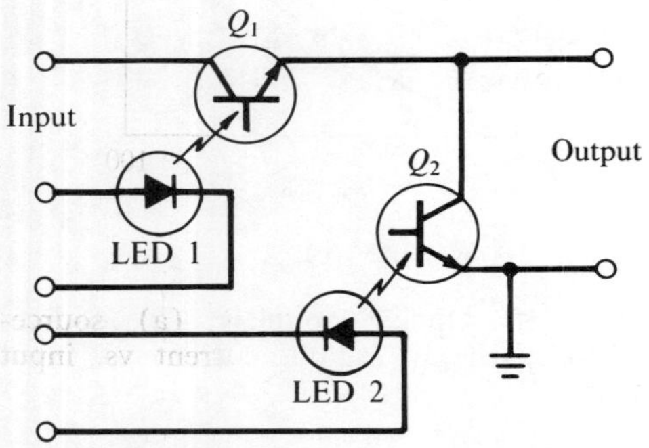

Fig. 2–153 A series-shunt chopper circuit using light-emitting diodes (LED's) and phototransistors.

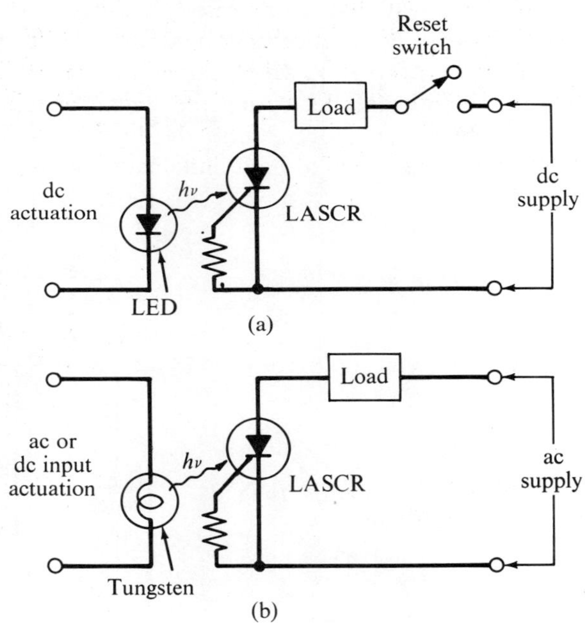

Fig. 2–154 Light-actuated relays: (a) dc latching relay; (b) ac nonlatching relay.

A widely used switch application of the neon source CdSe detector is a chopper (i.e., a modulator to convert dc to ac voltage). This is illustrated in Fig. 2–152. Lights L_1 and L_2 are alternately turned ON and OFF so that when L_1 is ON, L_2 is OFF. Therefore, when R_1 is LO (e.g., 1 kΩ), R_2 is HI (e.g., 1 MΩ) and amplifier input $v_{ai} = v_{in}R_2/(R_1 + R_2) \approx v_{in}$; and when R_1 is HI, R_2 is LO, so that $v_{ai} \approx 0$. The amplifier input voltage is therefore a chopped signal proportional in amplitude to the input dc signal v_{in}. This chopped signal can now be advantageously amplified or manipulated as an ac signal. A similar series-shunt chopper using LED's and phototransistors is shown in Fig. 2–153.

A semiconductor analog of an electromechanical relay can be simply accomplished with a LED and a LASCR as illustrated in Fig. 2–154a and b. In addition to offering complete isolation, the solid state relay has the advantages of microsecond response, no contact bounce, and very small size. The dc latching relay shown in Fig. 2–154a requires a dc supply. When the lamp is off, the LASCR is nonconducting and no current flows through the load. When the lamp is turned on by a control signal, the LASCR conducts and current passes through the load. The LASCR remains in conduction even after the lamp is turned off until it is reset by removing the supply voltage. A nonlatching relay is illustrated in Fig. 2–154b. It uses an ac supply for the LASCR. The LASCR resets during the negative half-cycle of the ac supply voltage. Current will flow through the load only when the lamp is energized during the positive half-cycle of the LASCR supply voltage.

Only a few of the important applications of optoelectronic switches have been discussed in this section. Other applications include the reading, converting, or translation of known information from punched cards or paper tape. Major applications are being found in communication systems and other areas beyond the scope of this module.

Experiment 2–25 *Optoelectronic Switching Circuits*

Several different light sources and detectors are characterized in this experiment for applications in optoelectronic switching. Experience is gained in wiring a BJT driver circuit for operating a tungsten lamp, and neon lamps and light-emitting diodes (LED's) are also characterized. The photoconductive cell and phototransistor are used as detectors. A photoconductive cell optoelectronic chopper is connected as a modulator for converting a dc signal to an ac signal. A series chopper, a shunt chopper, and a series-shunt chopper are studied.

2–3.6 PRECISE SWITCHING FOR SIGNAL LEVEL CONTROL AND WAVEFORM GENERATION

Switches which are actuated when a preset signal level is exceeded are very widely used in electronic instrumentation. Switch devices which are signal level actuated include the breakdown diode, the unijunction transistor, the gas-discharge lamp, and any switching device controlled by the output of a comparator. Level-actuated switches can be used to keep signals within safe or desired limits when the switch actuating levels are set at the boundaries of the desired range. Signal limiting circuits of this type, called clipping circuits, using a variety of switch types, are described in this section.

Another very interesting and useful application of level-actuated switches is in the increasingly popular function generator. The basis of most generators which produce a variety of waveforms, including triangular and square waves, is integrating circuits controlled by level-actuated switches. A discussion of representative techniques and circuits used in the generation of sawtooth, triangular, square, sine, and pulse waveforms is presented in this section.

Level-actuated switches are also used to convert an analog-encoded signal (such as the voltage level from a strain gauge bridge) to a binary information signal (such as the information that the strain on gauge A is greater, or less, than the strain on gauge B). Many applications of this type appear in Module 3, *Digital and Analog Data Conversions*.

Clipping Circuits

Clipping or limiting circuits are used to prevent voltages from exceeding desired positive or negative maximum values, to cut off either positive or negative pulses, and to isolate or eliminate various sections of waveforms. This is accomplished by a variation of the analog transmission gate, which is open when the signal is within the set limits and which provides the limit voltage at the output when the input exceeds the limit value. Biased pn junction diodes and breakdown diodes are the most commonly used switching devices in clipping circuits.

The pn junction diode can be thought of as a voltage-actuated switch with the actuation voltage threshold at 0.6 V. The clipping circuits which result when the diode is used in the series and shunt switching positions are shown in Fig. 2–155a and b respectively. In the series circuit, the diode is OFF (gate closed) except when v_{in} is more positive than +0.6 V. Signal voltages on the positive side of the threshold are followed at the output, while voltages less than 0.6 V are disconnected from the load by the reverse biased diode. Thus the negative input voltages are "clipped off." The shunt diode in Fig. 2–155b becomes conducting for v_{in} less than −0.6 V, thus cutting the output voltage off at −0.6 V. In each case, if the diode

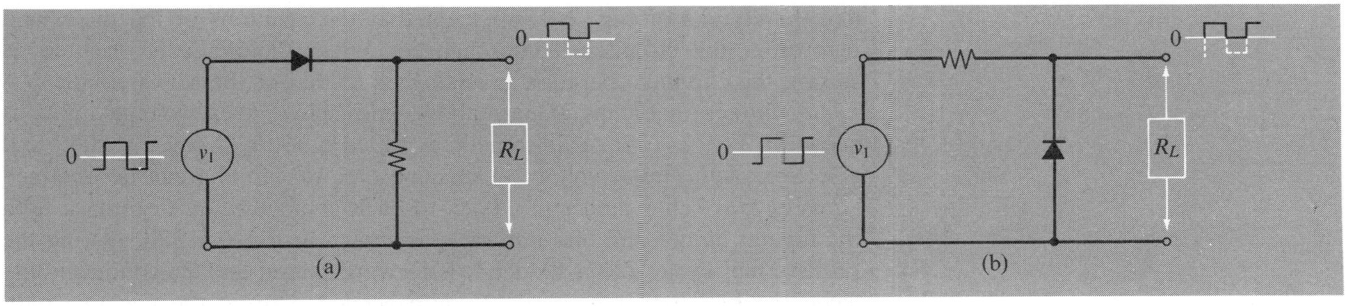

Fig. 2–155 Limiters: (a) series; (b) shunt.

were reversed, the negative signal voltages would be transmitted and the positive values cut off.

It is often desirable to set the clipping threshold at some voltage other than 0 V. This is accomplished with pn junction diodes by placing in series with the diode a bias voltage source which is equal to the desired threshold voltage. Examples for a series diode limiter are shown in Fig. 2–156. In the circuit of Fig. 2–156a, the diode will conduct only when v_{in} is greater than the bias voltage, V. When the diode is not conducting, the output voltage is V V. Thus the circuit clips off all input voltage excursions on the negative side of V V. This will be true for any value or polarity of bias voltage V. The effect of reversing the diode is shown in Fig. 2–156b. In this case, $v_o = V$ when v_{in} is greater than V and the diode is OFF. However, when v_{in} is less than V, the diode conducts and $v_o = v_{in}$. Thus voltages greater than V are clipped off. Shunt clipping circuits using biased diodes are just

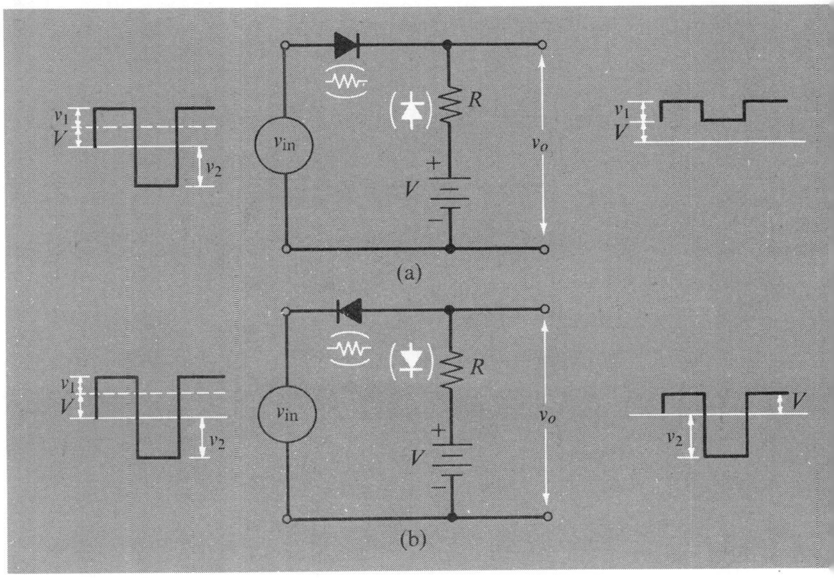

Fig. 2–156 Clipping above and below a fixed voltage level: (a) clipping below; (b) clipping above (component positions for equivalent shunt circuits shown in parentheses).

like the circuits of Fig. 2–156 except that the positions of the diode and resistor in the circuit are interchanged. When the diode is oriented as shown, the clipping operation is analogous to that of the series circuit.

A limiter that clips signals above and below fixed voltage limits is shown in Fig. 2–157a. It is sometimes called a **slicer,** because it slices out a section of the input voltage. The same type of clipping can be obtained by using two Zener diodes in a back-to-back configuration, to replace both the regular diodes and bias batteries, as shown in Fig. 2–157b. During the positive half-cycle, Zener diode Z_1 is forward biased and Z_2 conducts only when the breakdown potential V_z is exceeded. During the negative half-cycle Z_2 is forward biased and clipping occurs at the Z_1 breakdown voltage. Perfectly symmetrical clipping using the circuit of Fig. 2–157b would require the breakdown potentials of Z_1 and Z_2 to be perfectly matched. This requirement is avoided by using a single Zener diode in the diode bridge circuit shown in Fig. 2–157c. On the positive half-cycle, D_1 and D_3 are

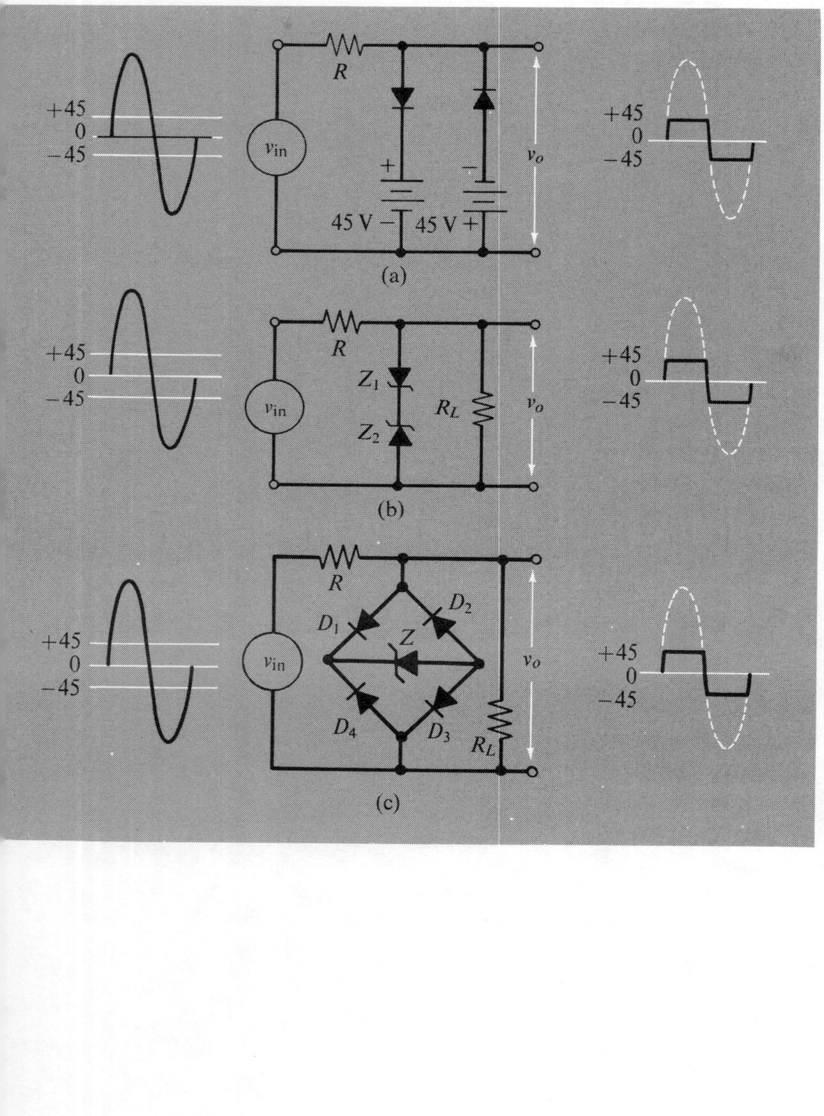

Fig. 2–157 Slicer circuits: (a) use of diodes and bias batteries; (b) use of Zener diodes; (c) use of single Zener diode in diode bridge.

forward biased, placing the input voltage v_{in} as a reverse bias across the Zener diode. When $v_{in} > V_z + 2V_f$, the diode bridge circuit acts as a shunt switch clipping the output voltage at that value. On the negative half-cycle, D_2 and D_4 are forward biased and v_{in} again appears as a reverse bias on the Zener diode.

The four-diode analog switch of Fig. 2–132 (repeated in Fig. 2–158) is also used as a limiting circuit. The output voltage, v_o, will follow v_{in} as long as all the diodes are forward biased. Assume that there is a maximum positive voltage for point A in the circuit, $V_{A(\max)}$. If v_{in} exceeds $V_{A(\max)}$, D_1 becomes reverse biased. Since D_4 and D_2 are still ON, $V_B = v_{in}$ and $v_o = V_{A(\max)}$, making D_3 reverse biased. Now we can see that v_o is the $+V_C$ supply voltage divided by R_1 and R_L. Therefore,

$$v_{o(+\max)} = +V_C R_L/(R_1 + R_L).$$

By similar arguments

$$v_{o(-\max)} = -V_C R_L/(R_2 + R_L),$$

which results in a clipping of the input voltage at these values. The clipping points are easily adjusted by varying the supply voltages or the series resistors. It is easy to see how the "window" of voltages between clipping levels decreases as the supply voltages decrease and how it disappears completely when the supply voltage polarities are reversed.

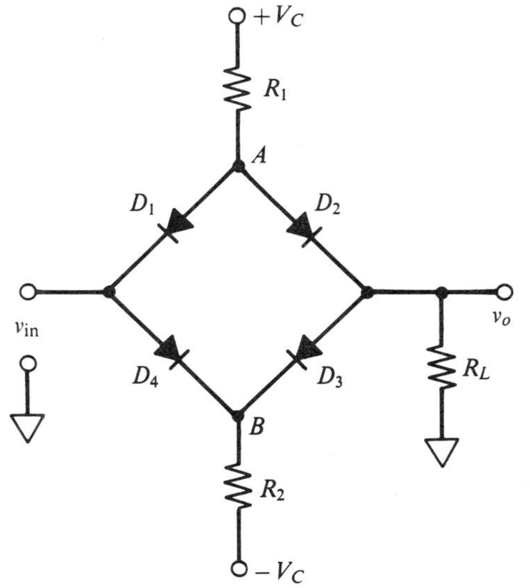

Fig. 2–158 Diode bridge limiter.

Clipping Circuits with Operational Amplifiers. Clipping circuits are frequently used in conjunction with operational amplifiers to keep the input signals within safe bounds, to keep the amplifier output from limiting, or to provide a precision limiting function. When the ideal output voltage of an operational amplifier is greater than its maximum practical value (for instance, when an OA with a $v_{o(\max)}$ of +12 V is connected as a gain-of-100 amplifier to an input of 0.2 V), the amplifier limits. Some types of OA's recover from a limit state very slowly in comparison to their normal response. Thus, when an OA is used for an application in which it is likely to limit, it is desirable to provide a circuit that will automatically switch to a lower gain as the amplifier output voltage approaches its maximum value. This is accomplished by putting a shunt limiting circuit in the negative feedback path of the OA circuit as shown in Fig 2–159. Since the output voltage must equal the voltage across the feedback element when the amplifier is in control, the output voltage cannot exceed the turn-on

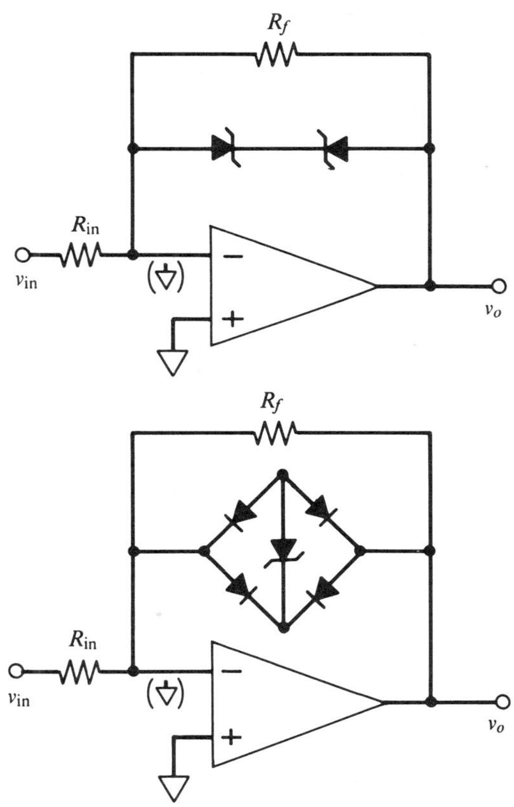

Fig. 2–159 Fast recovery limiting of OA output voltage.

threshold voltage for the shunt switches connected from the output to the inverting input. These OA output limiting circuits may be used with any OA circuit with a virtual ground at the inverting input. When used with the inverting amplifier circuit shown in Fig. 2–159, the resulting circuit is called a **bounded** amplifier. Such an amplifier has its normal gain (R_f/R_{in}) between the limits and breaks cleanly to zero gain at the set output voltage limits.

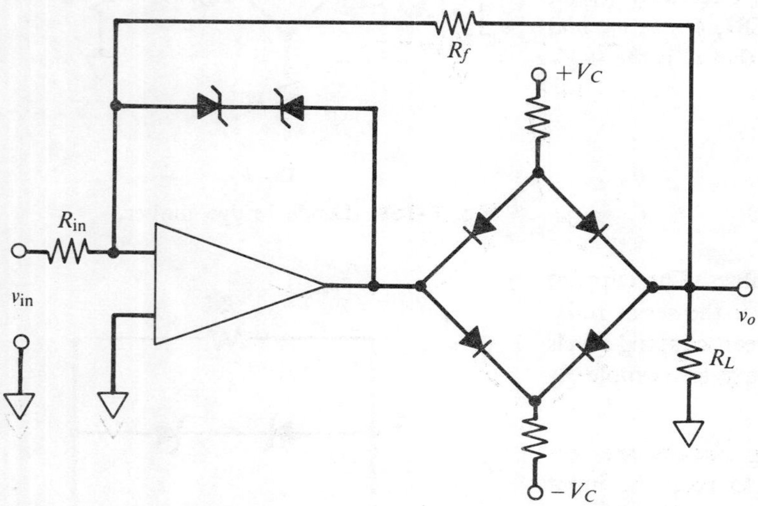

Fig. 2–160 Precision bounded amplifier.

A precision limiting circuit is shown in Fig. 2–160. This circuit uses the diode bridge limiter of Fig. 2–158 in the feedback circuit. Since the fed-back voltage is taken after the diode bridge, any offset or loss in the bridge is compensated for by the amplifier output. The voltage limits are the same as those for Fig. 2–158 except that R_f must be considered to be in parallel with R_L. When the output voltage is at a limit, the normal feedback path is open and the OA itself limits and goes out of control. This is prevented by the Zener diodes, for which the breakdown voltage should be between the desired output voltage boundary and the maximum OA output voltage.

Clamping Circuit

The clamping circuit is usually designed to "clamp" the top or bottom of a waveform to a fixed dc level, which may be zero, while preserving its shape and amplitude. It is sometimes called a "dc restorer," because it can restore the dc level that may have been shifted in amplifier or coupling circuits. A simple dc restorer is illustrated in Fig. 2–161 together with the positive and negative applied voltages and the initial waveforms during dc restoration.

The input voltage is a 500 Hz square wave with amplitude varying from +10 to −10 V, or 20 V peak to peak. Assume that the capacitor is uncharged and that the square wave is suddenly applied as shown by the input waveform in Fig. 2–161. The +10 V is applied across the diode so that it conducts. The effective resistance is now the diode ON resistance in parallel with R_L. Therefore the time constant for charging the capacitor is short. Assuming that the period between alternations of the square wave is equal to one charging time constant, the capacitor will charge to 63% of the applied voltage, or about 6.3 V, before the input voltage drops to −10 V. The instant the voltage drops, the diode stops conducting and the effective resistance is R_L. Now the time constant for the discharge of the capacitor is much longer and it does not discharge significantly during the negative half-cycle.

On the next positive half-cycle the diode does not conduct until the voltage has exceeded 6.3 V, because this is the voltage retained on the capacitor from the previous cycle, and it opposes the applied input. The net voltage applied across the diode at the start of the second cycle is 10 − 6.3 = 3.7 V. During the second positive half-cycle the capacitor will charge some more and will add 2.3 V (63% of 3.7 V) to the previous 6.3 V, to give a total of 8.6 V. Again during the negative half-cycle the discharge of the capacitor is insignificant, so that the 8.6 V on the capacitor opposes the positive swing of the third half-cycle. The capacitor continues to charge to 63% of the remaining voltage until the charge is essentially 10 V.

After the capacitor is charged to 10 V, it is apparent that, during each positive half-cycle of the input, the output will be 0 V. During each negative half-cycle of the input, the output will be −10 − 10 = −20 V. In effect the entire input waveform has shifted downward at the output so that the top is on the zero axis. The bottom of the waveform could be clamped to the zero axis instead by simply reversing the diode. In the case of narrow pulses it is necessary to clamp to the base, not the peak, to provide a sufficient portion of the cycle for charging.

A dc restorer can be biased to clamp the bottom or top of the waveform at some preselected voltage by placing a voltage source in series with the diode as in the shunt limiting circuit.

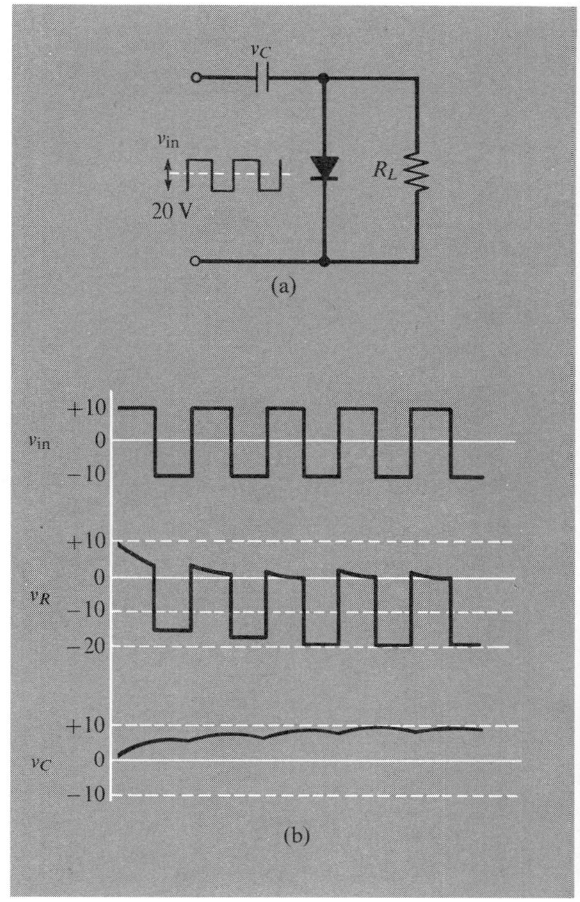

Fig. 2–161 Direct current restorer or clamping circuit: (a) clamping circuit; (b) waveforms.

Experiment 2–26 *Diode Clipping and Clamping Circuits*

Several diode clipper circuits are connected and tested so as to illustrate how they can slice off sections of a waveform. A diode clamp is also connected, and its ability to restore the dc component in an input square wave after it has been coupled through an *RC* circuit is investigated.

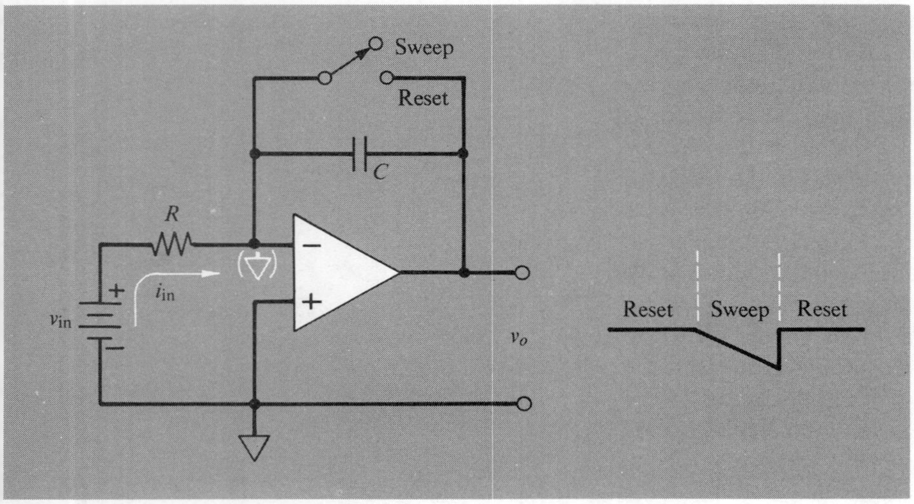

Fig. 2–162 OA integrator connected as a simple sweep generator.

Linear Sweep Generators

A signal that changes at a constant rate is generally obtained by integrating a constant input signal. The circuit most often used for this purpose is the operational amplifier integrator introduced in Section 2–2.3 and shown in Fig. 2–162. The constant voltage v_{in} supplies a current $i_{in} = v_{in}/R$ to the summing point of the OA. The amplifier output voltage v_o is always such that the summing point S is maintained at nearly 0 V (virtual ground) while the current i_{in} passes through the feedback circuit (switch or C). When the switch is closed (reset position), there is no resistance between point S and the output, so they must be at the same potential and $v_o = 0$ V. When the switch is opened (sweep position), the current i_{in} causes a continuous accumulation of charge on C. The charge q at any instant is $i_{in}t$, where t is the time in seconds since the sweep began. The voltage across the capacitor $v_C = q/C = i_{in}t/C$. The voltage $v_o = v_C$ (since S is at 0 V), so

$$v_o = -v_{in}t/RC \tag{2–129}$$

and the sweep rate dv_o/dt in volts per second is the value of v_o after $t = 1$ sec, or

$$\frac{dv_o}{dt} = \frac{-v_{in}}{RC}. \tag{2–130}$$

For example, if $v_{in} = +5$ V, $R_{in} = 100$ kΩ, and $C = 0.1$ μF, the sweep rate would be $-5/(10^5 \times 10^{-7}) = -5 \times 10^2$ or -500 V/sec. If the OA has a maximum output voltage of ± 10 V, the sweep will be linear for

$10/500 = 1/50$ sec before limiting at -10 V. When the switch is closed, the capacitor discharges through the switch and v_o returns quickly to zero. This simple sweep circuit is useful for many laboratory applications for sweep times in the range of microseconds to minutes.

To produce a repetitive, periodic sweep signal, it is necessary to provide an automatic periodic reset for the circuit of Fig. 2–162. This can be done in two ways. One is to use the output of an oscillator to close the reset switch momentarily at regular intervals. With this method, the values of R and C affect the sweep rate and the maximum amplitude, but not the frequency, because that is controlled by the oscillator. The other method is to use a level-actuated switch to momentarily close the reset switch when the desired maximum sweep voltage is reached. This technique is illustrated in Fig. 2–163. A generalized comparator-operated switch is shown. As the sweep output, v_o, increases toward the comparison voltage, v_c, the comparator output is positive. A fraction of the comparator output provides v_c. When this level is reached, the comparator output becomes negative, turning the switch ON and charging v_c to 0 V. Thus the switch will remain closed until the capacitor is discharged, and then it will open again. In the level-control method of automatic resetting, the R- and C-values affect both the sweep rate and the frequency, but not the amplitude, because that is controlled by the comparator.

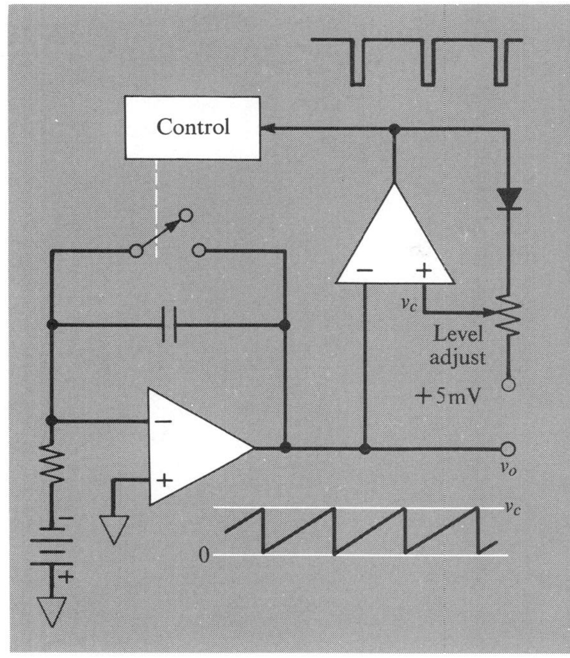

Fig. 2–163 Sweep generator with automatic level switch reset.

A very simple sort of sweep generator/oscillator which operates on this same principle uses a neon bulb as a level-controlled switch to discharge the capacitor. The circuit is shown in Fig. 2–164. When the capacitor is charged through R to the starting voltage of the neon lamp, the lamp conducts, discharging C to the maintenance voltage v_m. The resistor R is too large to allow the minimum current required for discharge in the lamp; so it stops conducting and the charge on the capacitor starts to increase once again. The sweep is not perfectly linear because the charging rate is not constant. A flash of light is produced during each discharge. Another very similar circuit is the unijunction relaxation oscillator of Fig. 2–119, described in Section 2–3.3.

Triangular and Square Wave Generators

Another way to discharge the capacitor is to reverse the integrator input voltage polarity so that the current is reversed. This will cause the capacitor to discharge at the same rate it was charged and produce a triangular wave

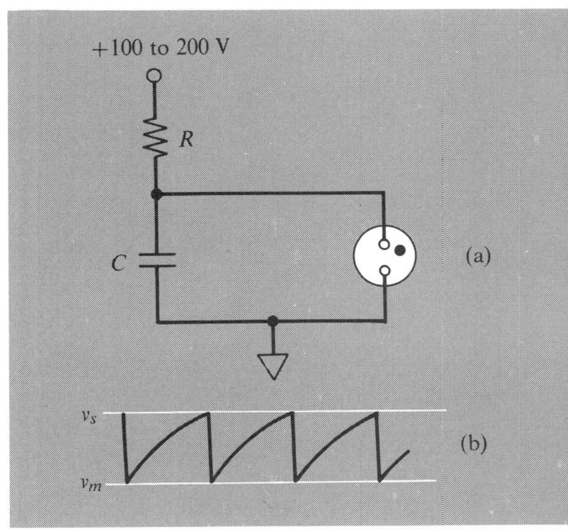

Fig. 2–164 Neon bulb sweep generator/oscillator: (a) circuit; (b) waveform.

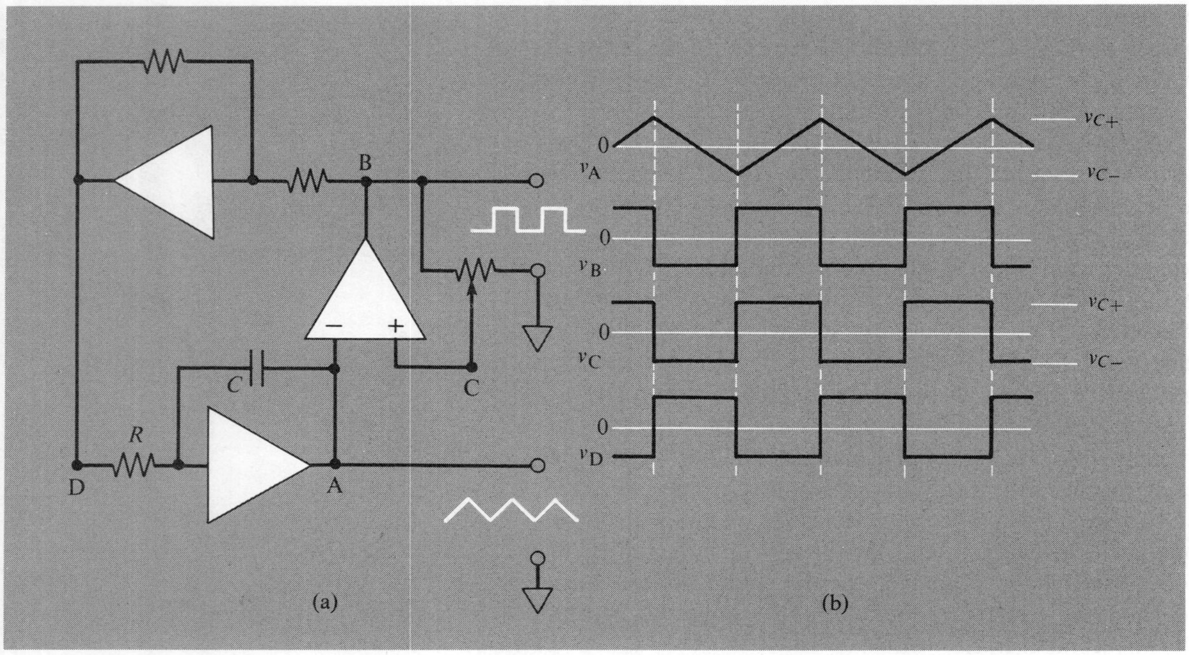

Fig. 2–165 Triangular waveform generator: (a) circuit; (b) waveforms.

at the output. If the triangular waveform is to be repeated, the input voltage polarity should be reversed again when the capacitor has been discharged to its starting voltage. Thus an automatic reversal of the input voltage is required at each limit of the sweep voltage. This can be accomplished with two comparators, one with the threshold set at the positive limit and one at the negative limit. Another way is to change the threshold setting to the other limit each time the threshold is crossed. This technique is shown in Fig. 2–165. At the start of the waveforms shown, the input voltage D is negative, causing the integrator output voltage A to sweep in the positive direction. Because the threshold voltage C is more positive than A, the comparator output is positive. A fraction of the comparator output voltage B provides the threshold voltage C. When the integrator output A just exceeds the threshold voltage *C*, the comparator output B, and consequently the threshold voltage C, become negative. This change in state at B is often used to operate a polarity reversing switch for the voltage at D. In the circuit shown, the B voltage is simply inverted by the third operational amplifier to provide a voltage at D of the proper polarity. As the waveforms show, the signal at A is a triangular wave and the signal at B is a square wave. It is this basic triangular waveform generator which is used in the popular laboratory **function generator,** which has triangular, square, and sometimes sine wave outputs. The value of *R* is varied to change the frequency (and sweep rate) over one or two orders of magnitude. Different values of C are switched in to cover the entire frequency range.

The basic generator concept described above can be adapted to provide several additional waveforms and capabilities. If the integrator input voltage is supplied by an external source, the frequency of oscillation will be proportional to that voltage. Thus the function generator becomes a **voltage controlled oscillator** (VCO). Of course, the comparator output must operate a reversing switch on the input voltage. If the positive and negative integrator input voltages are not equal, the capacitor will charge and discharge at different rates and produce asymmetrical waveforms. The result would be a sawtooth instead of a triangle and pulses instead of a square wave. This is often done intentionally to produce those waveforms.

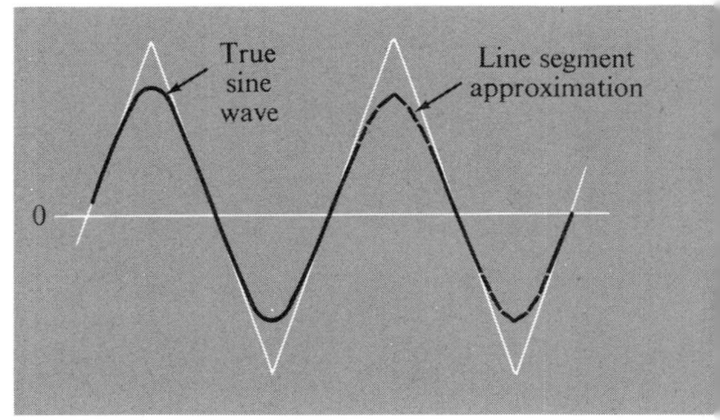

Fig. 2–166 Synthesizing a sine wave from a triangular wave by line segment approximation.

Sine Wave Synthesis. In the function generator, the basic waveform is the triangular wave and the sine wave must somehow be derived from it. This is usually accomplished by a diode wave-shaping circuit. The triangular waveform and the desired sine wave are shown in Fig 2–166. In order to obtain the sine wave from the triangular wave, an attenuator is needed which will attenuate the larger input voltages much more than the smaller ones.

A nonlinear voltage divider to serve this purpose can be made with diodes and resistors as shown in Fig. 2–167. The single diode in Fig. 2–167a is connected as a shunt clipping switch with a resistor in series. Therefore, for $v_{in} < v_b$, the output signal is the unattenuated input signal.

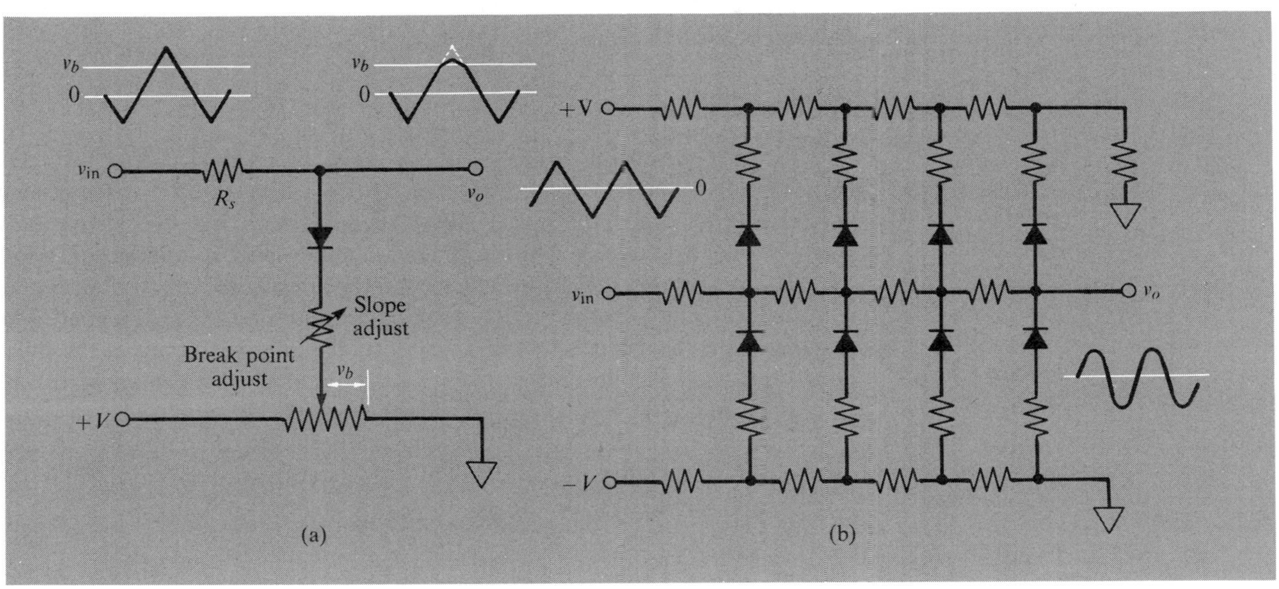

Fig. 2–167 Diode wave shaping: (a) single break; (b) multiple break.

When v_{in} is greater than v_b, the diode will conduct a current proportional to $v_{in} - v_b$. This current causes an *IR* drop in R_s proportional to $v_{in} - v_b$. Thus the slope of the output signal changes at v_b. The point of the slope change is set by the **breakpoint,** v_b, adjustment. The amount of slope change depends on the value of the slope-adjust resistor. Additional diodes and resistors can be added to provide additional breakpoints and slope changes until the desired waveform is approximated by a series of line segments. The line segment approximation is shown in Fig. 2–166, and a multiple-break circuit (with four positive and four negative breakpoints) is shown in Fig. 2–167.

The sine wave output produced by diode wave shaping in a good function generator is generally of high quality. The line segments are not observable with an oscilloscope. The function generator approach is the best way to produce low-frequency (below 20 Hz) sine waves, since they are difficult to produce with quality by standard harmonic oscillator circuits.

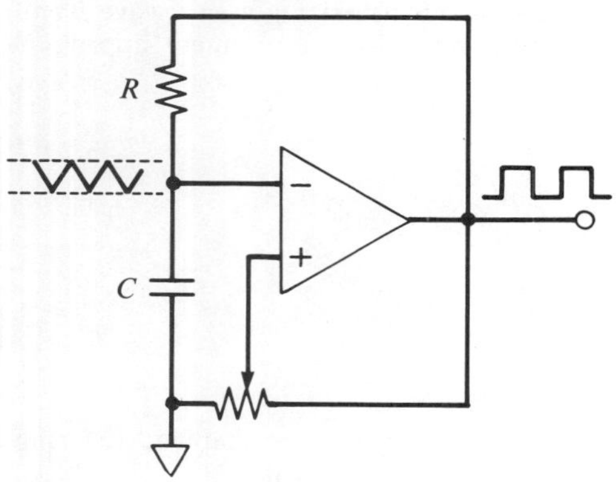

Fig. 2–168 OA square wave generator.

Square Wave Generator. The function generator of Fig. 2–165 produced a square wave simultaneously from the level switching comparator output. If the linearity of the integrating circuit is not important, that is, if the square wave is the desired output waveform, the circuit can be simplified considerably. A simple passive *RC* integrating circuit can take the place of the active OA integrator and, since the *RC* circuit does not invert as the OA integrator does, the inverting amplifier is also unnecessary. The resulting circuit is shown in Fig. 2–168. A circuit of this type, which essentially snaps from one state to another but has no stable state, is sometimes called an **astable multivibrator.**

Pulse Generators

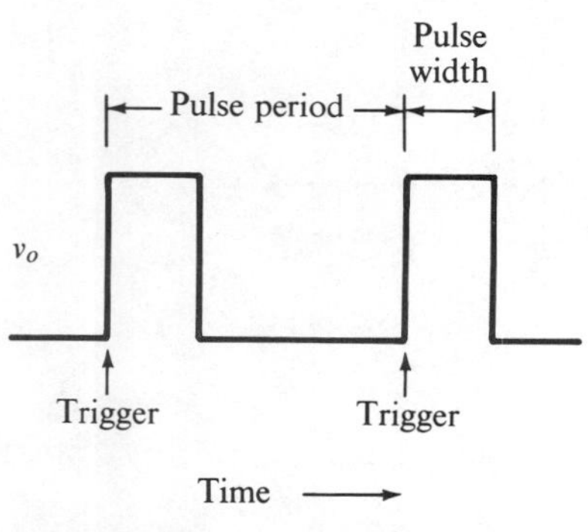

Fig. 2–169 Output signal from a pulse generator.

A pulse is a waveform in which the voltage or current changes quickly from its normal steady value and then returns to the normal value after a period of time called the **pulse width.** Such a circuit has two states, only one of which is stable, and is thus called a **monostable multivibrator.** Most laboratory pulse generators produce "square" pulses of the type shown in Fig. 2–169. The pulse is initiated by some sort of "trigger" signal. The pulse width is determined by the time constant of an *RC* circuit in the pulse generator. The astable circuit of Fig. 2–168 can be made monostable by putting a diode shunt clipping switch across the capacitor. The resulting circuit is shown in Fig. 2–170. Because of the diode, the capacitor cannot charge in the positive direction. Thus the stable state occurs when the output is positive and the threshold voltage at the + input of the OA is positive. A trigger circuit is added which can convert a sharp negative-going transition of a trigger signal into a negative pulse of sufficient magnitude to make the threshold voltage negative. When the trigger pulse occurs, the output goes negative, holding the threshold negative and charging the capacitor in the negative direction. When the capacitor voltage reaches the threshold, the output returns to its stable positive state. The pulse width depends on, and is approximately equal to, the *RC* time constant.

Monostable multivibrators are available in convenient integrated circuit form, which makes the assembly of such circuits from discrete components increasingly rare. If two monostable circuits can be connected to trigger each other, the resulting circuit is astable with a total period of oscillation equal to the sum of the two monostable pulse widths. This type of astable circuit has the advantage of providing independent control of the width of the positive and negative portions of the resulting waveform.

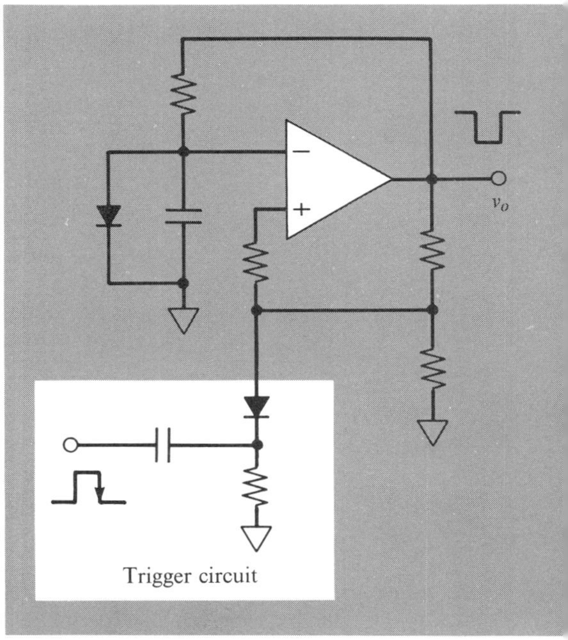

Fig. 2–170 Pulse generator.

Experiment 2–27 *Waveform Generators*

This experiment illustrates that a general purpose function generator can be constructed from a basic operational amplifier linear sweep generator. The function generator produces variable-frequency sine, square, and triangular waveforms. A linear sweep generator with automatic reset is first constructed and tested. Then the circuit is modified to produce the various other waveforms. Some characteristics and limitations of the function generator are then studied.

Experiment 2–28 *Multivibrators*

In this experiment various multivibrator circuits are constructed and tested. The basic astable, monostable, and bistable multivibrators are constructed using operational amplifier circuits. The limitations and advantages of these circuits are noted.

2–3.7 POWER CONTROL WITH SOLID STATE SWITCHES

The solid state devices discussed in previous sections find many important applications for power switching in addition to the low-level signal switching applications discussed in Section 2–3.4. Power transistors are often used for ON-OFF switching of dc power to various loads. Since the switching considerations are identical to those discussed in Section 2–3.3, BJT power switches will not be considered in this section. Instead this section will be devoted to the use of SCR's and triacs in power control applications. Power switching can be classified into two categories: in the first, **static switching**, the controller is used to turn the power applied to the load completely ON and OFF. In the second category of power switching, often called

phase control, the ac supply is switched to the load for a precisely controlled fraction of each cycle so as to control the magnitude of the average power applied to the load.

This section discusses some of the basic considerations in static and phase control and presents several important applications. The reader is referred to the GE *SCR Manual** for a more exhaustive discussion.

Static Switching with Thyristors

The SCR and the triac are important controllers for a variety of static switching applications. Switching of both ac and dc can be accomplished, although with the latter the SCR must be turned off by one of the methods discussed in Section 2–3.3. The uses of SCR's and triacs in static ac switching, static dc switching, protective circuits and time delay relay circuits are considered below.

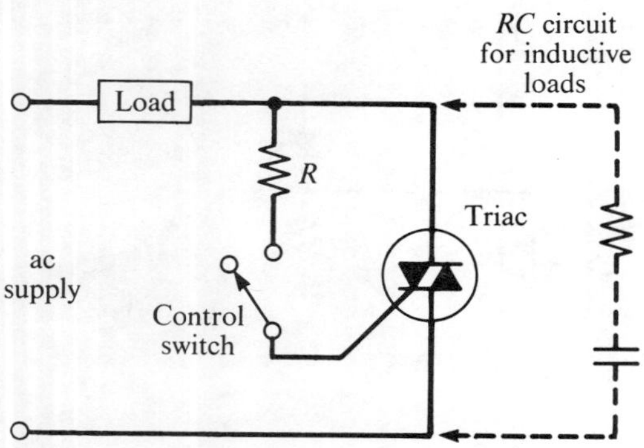

Fig. 2–171 Basic ac static switch.

AC Switching Circuits. High-speed switching of ac power to a load can be readily accomplished with a triac circuit similar to that shown in Fig. 2–171. A control device such as a reed switch or light-activated switch is used to trigger the triac into conduction and thus to apply the ac voltage to the load. Resistor *R* is used to limit the gate control current to the triac to a safe value. The triac has many advantages over electromechanical relays for simple static switching applications. Contact bounce is completely eliminated with triacs and contact wear is no longer a factor. Hence the lifetime of such a solid state switch is considerably greater than that of an ordinary relay. The amount of power that can be switched in this manner is determined by the triac ratings, which should be carefully considered in power switching. If loads containing inductance are to be switched, the *RC* network shown in dotted lines in Fig. 2–171 should be included to protect the triac from voltage transients (recall that a rapidly changing current through an inductor produces a large counter emf). A similar but more complex ac switching circuit can be constructed with two SCR's connected in inverse-parallel fashion. This latter circuit should be used if the ac supply has a frequency greater than about 400 Hz, since triacs are normally limited to lower frequencies.

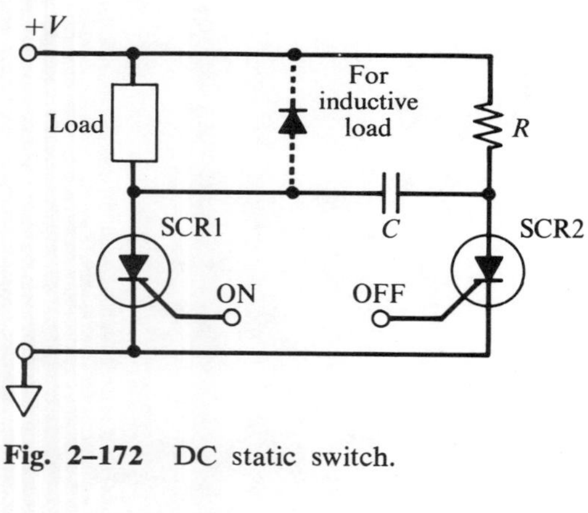

Fig. 2–172 DC static switch.

DC Switching Circuit. A static SCR switching circuit for control of dc through a load is shown in Fig. 2–172. The circuit is basically a bistable multivibrator or flip-flop. A turn-on control signal applied to the gate of SCR1 triggers the SCR into conduction and applies the dc voltage to the load. The application of the supply voltage begins to charge capacitor *C* through resistor *R*. When the turn-off signal is applied to the gate of SCR2,

* *SCR Manual,* 1972, fifth edition, General Electric Company, Syracuse, N.Y.

SCR2 conducts (acts like a short circuit) and capacitor C is essentially connected across SCR1. Since the right-hand plate of C was charged positively with respect to the left-hand plate, this reverse biases SCR1 momentarily and turns it OFF. The current through the load then decreases exponentially to zero.

SCR Protection Circuits. Because SCR's are capable of rapid power switching, they are often used for protection against transient voltages, overvoltages, and current shorts.

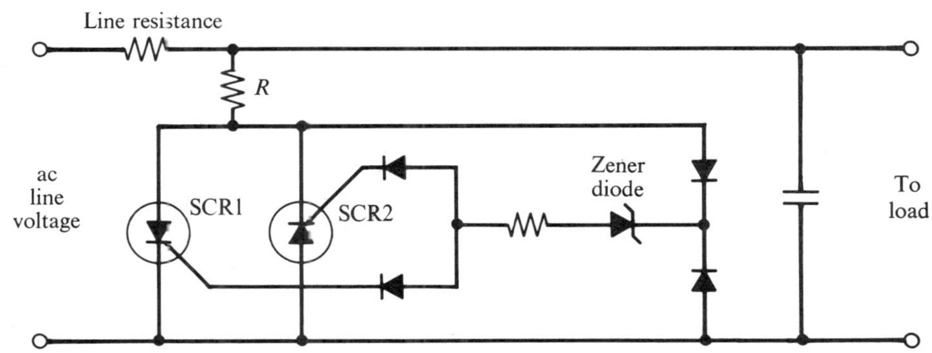

Fig. 2–173 Transient line voltage protection circuit with SCR's.

Figure 2–173 shows an SCR circuit that is useful in protecting loads from transient line voltage surges. Any transient surge which exceeds the Zener voltage of the Zener diode causes either SCR1 or SCR2 to trigger. Which SCR triggers depends on the line voltage polarity at the instant of the transient. Resistor R is a current limiting resistor which protects the SCR. The loading effect of the SCR conduction causes the transient voltage to be dropped across the resistance of the ac line rather than across the load. When the line voltage reverses polarity, the SCR which was conducting turns OFF.

An interesting circuit used in power supply protection is the so-called "crowbar" circuit shown in Fig. 2–174. This circuit provides protection against both line voltage surges and possible short circuit loads. Two UJT's are used to trigger the SCR's. UJT1 provides a trigger pulse if overvoltages are present, while UJT2 triggers the SCR if load currents exceed the maximum value. Potentiometer P_1 sets the maximum desired value of the dc supply. If the supply should exceed the value set, the peak-point voltage of UJT1 is exceeded; this supplies a trigger pulse to the SCR. The entire dc supply voltage is then applied to the circuit breaker, which opens the main supply line. Short circuit protection is provided by UJT2. The voltage across resistor R_1 is proportional to the load current. When the

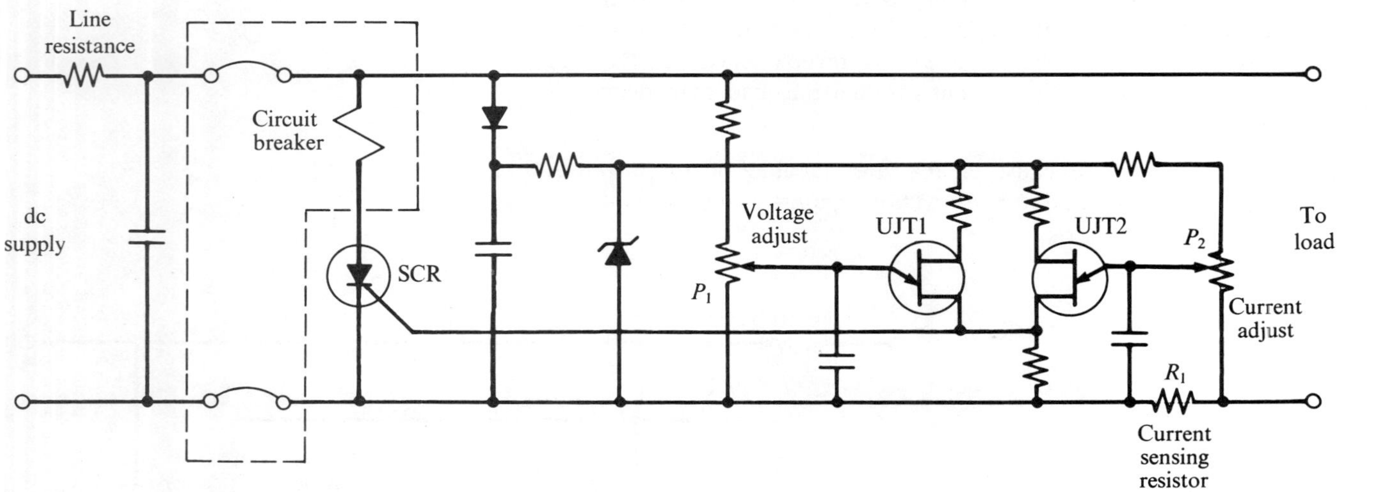

Fig. 2–174 Crowbar overvoltage and overcurrent protection circuit.

voltage across R_1 exceeds the maximum desired value set by potentiometer P_2, UJT2 triggers the SCR into conduction. As before, this shuts down the dc supply by operating the circuit breaker. Many supplies are protected against only overvoltage or overcurrent instead of both. Hence, for only-overvoltage protection UJT2 and the circuitry associated with it are often left out. Likewise for only-short-circuit protection UJT1 and its associated circuitry are deleted.

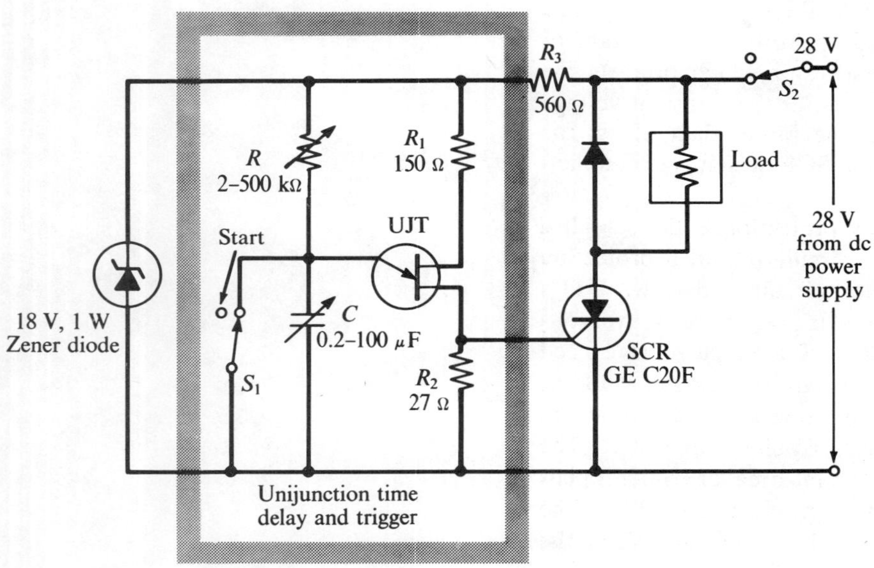

Fig. 2–175 Precision solid state time delay circuit.

Time-Delay Relay Circuits. The SCR is often combined with UJT, neon lamp, or PUT relaxation oscillators to provide precision time delays. A typical application of the unijunction transistor is shown in Fig. 2–175. The UJT is used to provide a precise time-delayed trigger pulse to turn the SCR power control circuit ON. The emitter-base 1 voltage for the UJT is supplied by the voltage across the capacitor C. At the start, C is shorted by switch S_1. When S_1 opens, the capacitor C charges through R from the stable 18 V Zener-regulated supply. The voltage across C is determined by the RC time constant and Zener voltage source. When the capacitor charges to the UJT peak-point voltage, V_p, the UJT fires. This provides a pulse across R_2 which triggers the SCR. The full supply voltage (less the small voltage drop across the SCR) is then applied across the load terminals.

The voltage across the emitter-to-base 1 of the UJT drops to less than 2 V when the SCR triggers, so that a low voltage is maintained on capacitor C, but this varies as a function of time.

The upper limit on the time delay depends on the specified accuracy, the V_p of the UJT, the leakage current of capacitor C, and the temperature. The upper limit for R is determined by the requirement that the current to the UJT emitter be large enough for it to fire (larger than I_p).

To reset the circuit for another timing cycle, the dc supply must be removed from the SCR by opening switch S_2 and the capacitor C should be shorted by closing S_1.

Phase Control Switching Circuits

In phase control switching, as opposed to the static switching discussed in the previous section, the ac supply is switched to the load for a precisely controlled fraction of each cycle. This is the basic way in which power is controlled in lamps, motors, heaters, etc. The thyristor is a basic component in this mode of switching. Control of the average power applied to a load is obtained by triggering the thyristor at a specific point in time (phase) on the ac waveform. The thyristor then conducts for the remainder of the half-cycle.

Phase control can take many forms. Only one half-cycle of the ac wave might be used (half-wave control), or both half-cycles might be used (full-wave control), depending on the application. Although SCR's are used in many phase control applications, the triac is a more versatile controller since it conducts in both directions and is self-protecting against transients.

SCR and Triac Power Controllers. An example of a circuit that can be used for control of power in a load is shown in Fig. 2–176. The section of the circuit enclosed in the block acts as a switch to provide current through the load only during a discrete selected fraction of each half-cycle. The RC circuit determines the portion of each half-cycle required to charge

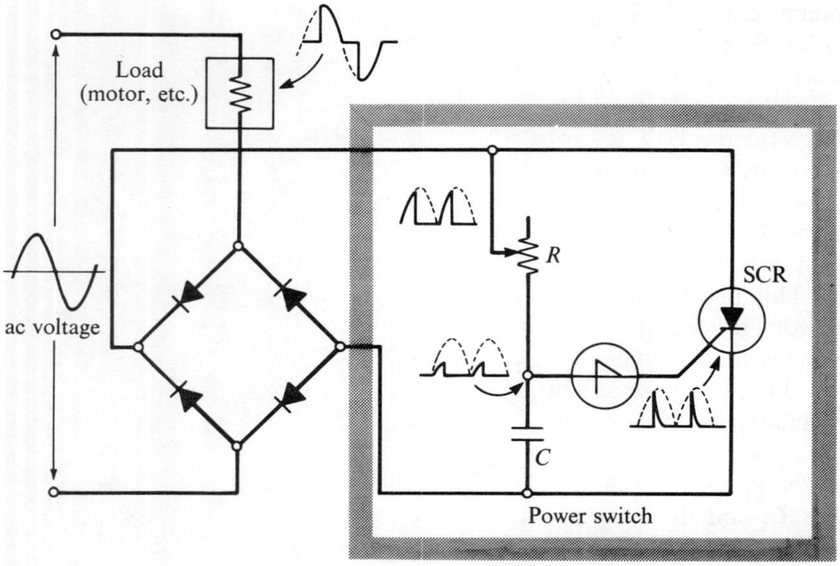

Fig. 2–176 Use of SCR to control power in a load.

the capacitor C to a sufficient voltage for the four-layer diode to turn ON, which provides the necessary gate signal for the SCR to turn ON. Because the power delivered to the load in each half-cycle depends on the portion of each half-cycle required to charge the capacitor to the firing potential of the diode and the SCR, it is easy to change the power to the load by varying resistor R or capacitor C. The voltage waveforms shown in the various portions of the circuit of Fig. 2–176 as related to the input supply voltage illustrate the circuit action. Note that because the effective resistance of the ON SCR is very small, the circuit current is essentially determined by the load resistance. This circuit uses full-wave power control.

A similar circuit designed to operate a 60 Hz motor with full-wave phase control using a triac is shown in Fig. 2–177. The diac is used to trigger the triac and resistor R_1 adjusts the trigger point (phase).

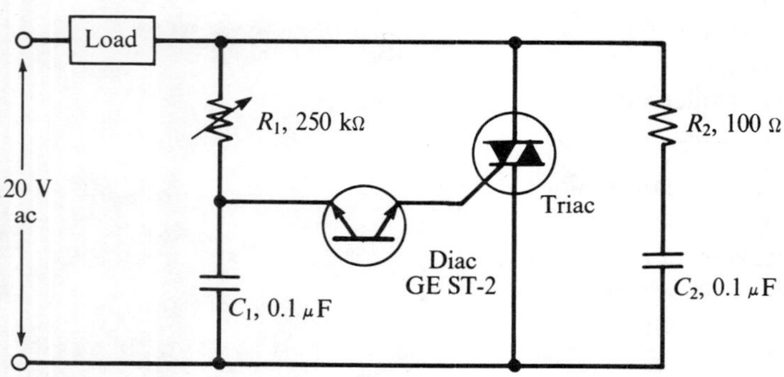

Fig. 2–177 Full-wave phase controller for motors.

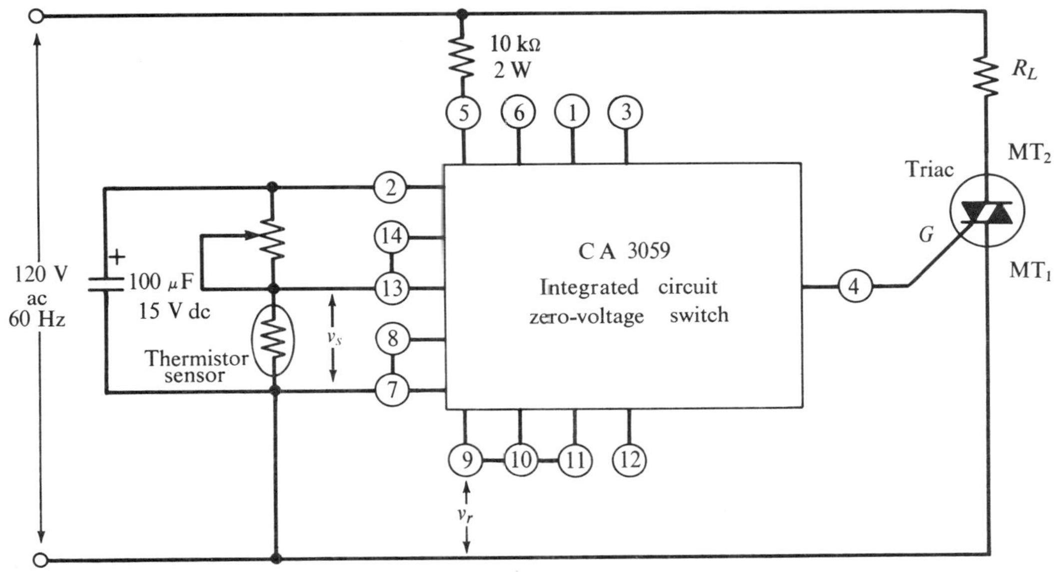

Fig. 2–178 Zero voltage switched triac circuit.

Zero Voltage Switching

Whenever power is switched ON and OFF very rapidly as with solid state switches, high-frequency electromagnetic interference (EMI) is generated. Although thyristors eliminate interference problems caused by rapid current interruptions in inductive loads, special circuits are required to eliminate EMI during turn-on. Experimentally it has been shown that the minimum amount of EMI is generated in ac power switching if the thyristor is turned ON when the ac waveform crosses zero voltage. Likewise the minimum EMI is created when the switch turns OFF the instant the current through it drops to zero.

A variety of circuits have been proposed for sensing the ac waveform zero crossing and providing a trigger pulse for the SCR or triac. However, recently zero voltage switches have become available in integrated circuit packages and their use in power switching is increasing at a rapid rate. A triac on-off temperature controller which uses an IC zero voltage switch is shown in Fig. 2–178. The integrated circuit is a combination zero voltage threshold sensor and a trigger circuit. Hence the triac is turned ON very near the zero crossing by a trigger pulse from the IC to the triac gate whenever the voltage v_s across the thermistor sensor exceeds the reference voltage v_r. Such a zero voltage switched triac circuit almost completely eliminates generated EMI. In the past few years encapsulated modules have become commercially available which include a zero voltage switch and triac. These solid state ac relays are rugged and simple to use.

Experiment 2–29 *Control of Power in a Load with Solid State Switches*

Several circuits employing solid state power switches are constructed. This experiment illustrates how SCR's and triacs can be used to accomplish power supply protection, and several modes of light source control. A circuit is studied as an overvoltage protector. Then a lamp output is controlled by a triac circuit. A circuit is constructed to turn the lamp off after a precise amount of radiation has been received by the detector, and to control the power applied to the lamp in order to keep the intensity constant.

PROBLEMS

1. In the nonideal switch circuit of Fig. 2–70, the switch has an open resistance $R_{so} = 900\ \text{k}\Omega$, a closed resistance $R_{sc} = 80\ \Omega$, and an equivalent capacitance $C = 400$ pF. It is desired to switch a 5 V source V_s to a 50 kΩ load R_L. (a) What are the voltage drops across the switch and load for both states of the switch? (b) What are the two time constants, τ_o on opening the switch and τ_c on closing the switch?
2. A silicon pn junction is described by Eq. (2–95) with $I_i = 10^{-10}$ A at 300°K. Calculate and plot the *I-V* curve of the diode for voltages of 0.30, 0.40, 0.50, and 0.60 V.
3. The diode switching circuit of Fig. 2–82 was graphically analyzed and the diode found to have a forward static resistance $R_{fs} = 65\ \Omega$, and a reverse static resistance $R_{rs} = 5\ \text{M}\Omega$ at the operating points. Calculate the voltage across the load for the two values of the voltage source.
4. Find the theoretical forward dynamic resistance R_{fd} of a diode with a forward current of 0.8 mA.
5. In the transistor switching circuit of Fig. 2–106, $h_{FE} = 100$, $V_{CE(\text{sat})} = 0.2$ V, and $V_{BE(\text{ON})} = 0.6$ V. The input signal varies between 0 and +1 V, $R_B = 4\ \text{k}\Omega$, $R_L = 1\ \text{k}\Omega$, and $V_{CC} = 5$ V. (a) When the input signal is +1 V, is the transistor in saturation? (b) Calculate the collector current, I_C, when the transistor is ON, and the ON resistance of the transistor.
6. The transistor switch circuit in Fig. P–6 is used to control current through a relay coil. The relay coil has a resistance of 200 Ω and requires 20 mA to operate. (a) Does the transistor have to be in saturation to energize the relay coil? Explain. (b) If a +4 V signal at the control input is to turn the relay ON, what current will be drawn from the source ($V_{BE(\text{ON})} = 0.6$ V)? (c) What is the minimum value of h_{FE} the transistor can have and still energize the relay coil?

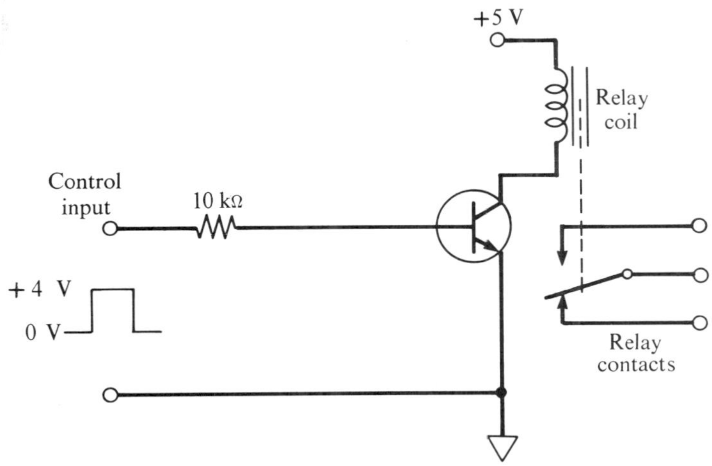

Fig. P–6

7. A switching transistor is characterized by a collector-emitter cutoff current ($\beta_N I_{CO}$) of 20 μA at $V_{CE} = 15$ V and $V_{CE(\text{sat})} = 0.4$ V at $I_C = 50$ mA and $I_B = 500$ μA. For a switching application with $V_{CC} = 15$ V and $R_L = 1$ kΩ, estimate the **ON** and **OFF** resistances of the transistor

8. A unijunction transistor (UJT) has a peak-point voltage of 15 V at a value of $V_{BB} = 19$ V. If $V_\gamma = 0.6$ V, what is the intrinsic standoff ratio of the UJT?

9. (a) For stabilization of the peak-point voltage against temperature variations, an external resistor R_2 can be added in series with B_2 of the UJT circuit of Fig. 2–116. If $R_2 \ll R_{BB}$, show that the peak-point voltage can be approximated by

$$V_p \approx V_\gamma + \eta V_B - \frac{\eta V_B R_2}{R_{BB}},$$

where V_B is the total supply voltage between R_2 and B_1. (b) Show what value R_2 is necessary to cancel the temperature dependence of V_γ, which is the major contributor to temperature variations in V_p.

10. The series-type analog switch of Fig. 2–130b is constructed using a FET switch. The **ON** and **OFF** resistances of the switch are 50 Ω and 100 kΩ respectively. If the source v_u has an output of 5 V and an equivalent resistance R_e of 100 Ω, and $R_L = 1$ kΩ, calculate v_L for the two states of the switch.

11. Consider the four-diode analog gate of Fig. 2–132. The voltages V_c and $-V_c$ are determined by the condition that the current in each of the four diodes must be in the forward direction. If the diode forward resistances

are negligible, show that the minimum value of V_c required is

$$(V_c)_{\min} = v_u\left(2 + \frac{R_c}{R_L}\right).$$

12. (a) A reverse biased photodiode has a quantum efficiency of 0.3 at the wavelength of an incident light beam of photon flux 10^{12} photons/sec. What is the photocurrent I_p? (b) A phototransistor has a dc current gain of 50. If it is operated under the same conditions as the photodiode in part (a), what is the emitter current I_E?

13. (a) In the diode series limiter of Fig. 2–155a, v_{in} is a square wave which has values of -5 V and $+5$ V. The diode has an ON static resistance of 50 Ω and an OFF static resistance of 500 kΩ at the operating points. If the load resistance is 5 kΩ, what are the output voltages for the two states of the source? (b) Repeat the calculation for the shunt limiter of Fig. 2–155b and the same conditions as in part (a).

14. Design diode clipping circuits which will limit the output voltage v_o to the following limits regardless of the range of input voltages: (a) $0 < v_o < +5$ V, (b) $-10\text{ V} < v_o < +10$ V, (c) $+50\text{ V} < v_o < +75$ V.

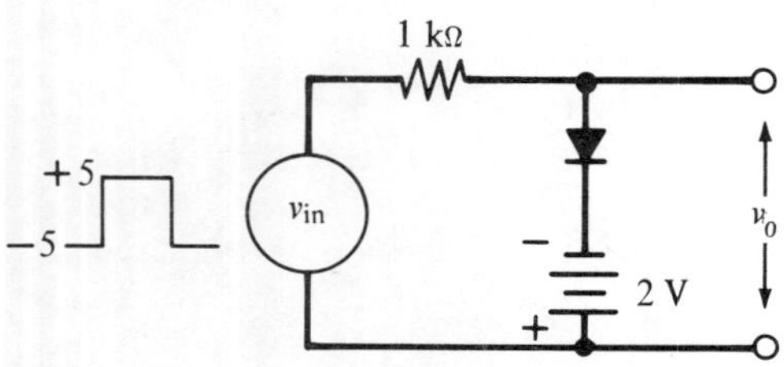

Fig. P–15

15. The shunt limiter of Fig. P–15 was analyzed graphically, and R_{fs} was found to be 50 Ω, while $R_{rs} = 1$ MΩ. Calculate v_o for the two values of v_{in}.

16. The OA linear sweep generator of Fig. 2–162 has $v_{\text{in}} = +1.5$ V, $R = 250$ kΩ, and $C = 0.5$ μF. (a) What is the sweep rate of the generator? (b) If the maximum output voltage of the OA is ±12 V, for how many seconds will the sweep remain linear, before limiting?

Experiment 2–14 *Characteristic Curves and Load Line for an Ideal Switch*

Equipment: ADD
OA card
breadboard connection system
oscilloscope

Experience is gained with a curve tracer for obtaining characteristic curves of an ideal switch. The load in a switch circuit is varied to show the variation of the load line and to gain experience with graphical techniques which are also applicable in studying diode and transistor characteristics in subsequent experiments.

Before beginning this experiment read Equipment Appendix 5, which describes the complete breadboard connection system. *Answer the following question.*

What voltages are available on the power patch card when it is attached to the digital power module?

E2–14.1 Construction of Curve Tracer

The characteristic curves of a device can be readily displayed on an oscilloscope by using a system that is connected as shown in Fig. E2–21. The output voltage of a sweep generator is applied across a voltage divider made up of a resistance R_L in series with the device under test. The voltage drop across R_L is a measure of the current i through the device and is connected to the vertical input of an X-Y oscilloscope, and the voltage applied across the device is connected to the horizontal input. If the level and amplitude controls of the generator are adjusted to provide voltages over the applicable range of the device, a current-voltage (I-V) characteristic curve is repeatedly plotted on the scope face for every sweep. Note that for the circuit of Fig. E2–21 to operate, the oscilloscope common and the sawtooth generator common must not be connected together.

The magnitude of the output ramp voltage from the signal generator of the analog digital designer (ADD) is not sufficient to cover the operating ranges of several devices to be studied; therefore it is amplified with an OA circuit as illustrated by the diagrams in Fig. E2–22.

The operational amplifier is wired as an inverting summing amplifier. The output voltage v_o is the inverted sum of the level and sawtooth input voltages (v_{level} and v_{sawtooth}, respectively) according to the relationship

$$-v_o = v_{\text{sawtooth}} \cdot \frac{100\ \text{k}\Omega}{R_{\text{in}}} + v_{\text{level}},$$

where R_{in} is the resistance of the amplitude control. However, the output voltage of the operational amplifier is limited to about $\pm$10–13 V. Therefore, if the sawtooth amplitude and the level control are set so that the sum exceeds the amplifier limit, the output waveform will appear truncated at the limit value. The amplitude and level controls can be adjusted to provide sweep outputs of any desired span in the range $+10$ V to -10 V. A diagram showing how to make connections to the OA circuit card after inserting it into the ADD is shown in Fig. E2–22b. Connect the OA sweep circuit as shown in Fig. E2–22b.

E2–14.2 Use of Curve Tracer

Make the following adjustments while observing the output of the circuit of Fig. E2–22 on the oscilloscope.

Set the generator at 100 Hz and the amplitude control at about the midpoint. Adjust the level control to be just short of the positive limit. Adjust the amplitude control for maximum excursion with only slight limiting at the negative extreme. *Check* to see that there is no chassis–common jumper (possibly through the three-wire line cord) on the power card of the ADD for this experiment if the scope is grounded.

Record below the sawtooth generator output waveforms.

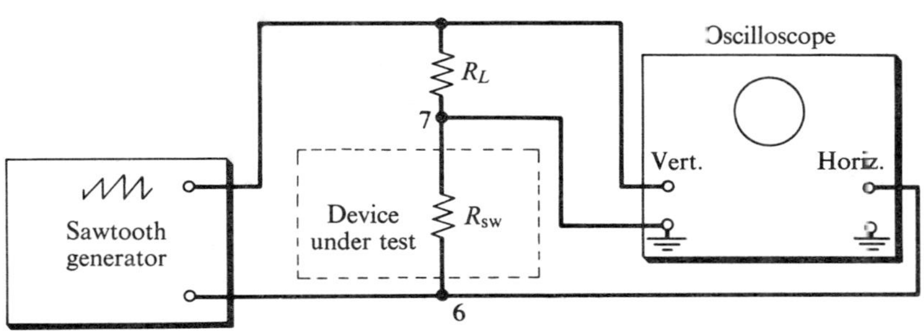

Fig. E2–21 Block diagram of curve tracer for obtaining characteristic curves of devices.

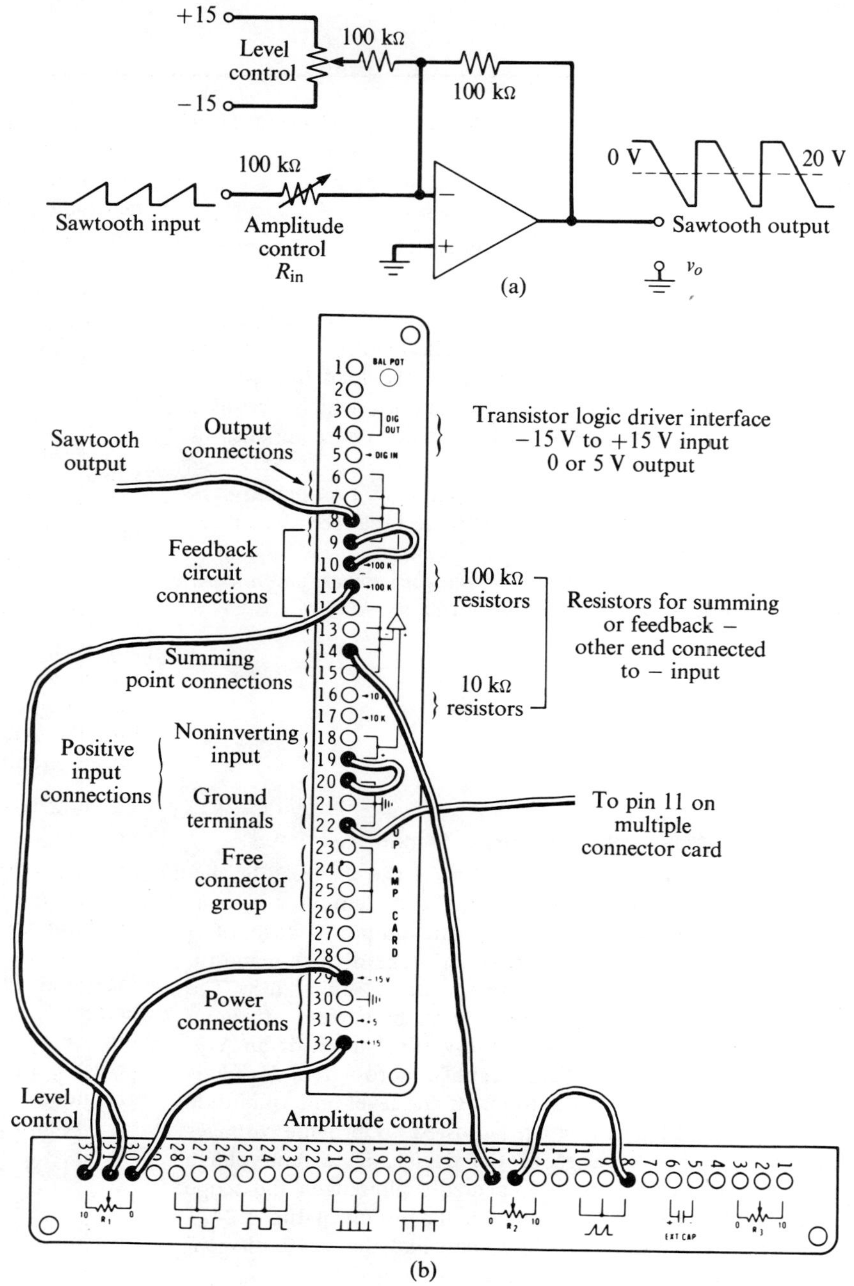

Fig. E2–22 Operational amplifier circuit for curve tracer sweeps: (a) schematic diagram; (b) OA card connections.

The output of the sawtooth generator of Fig. E2–22 will now be connected to various devices (loads) in order to obtain their characteristic current-voltage curves. A simple resistor will first be used, and then an ideal switch will be tested.

E2–14.3 Resistor Current-Voltage (I-V) Curves and Load Lines

Connect the sawtooth output of the OA generator of Fig. E2–22 to the series circuit of R_L and the device under test, R_{sw}, and make the oscilloscope connections, as shown in Fig. E2–23. Use a 10 kΩ resistor for R_L and a 100 kΩ potentiometer (pot) for R_{sw}. *Observe the current-voltage curves for R_{sw} in the middle of its range and at its two extremes. Record the waveform below.*

While varying R_{sw} rapidly, note that the first- and third-quadrant load lines can be observed. *Record the load lines.*

E2–14.4 "Ideal" Switch

Replace the pot R_{sw} in Fig. E2–23 with a wire jumper contact which can perform as an ideal switch. Connect and disconnect the jumper and note the *I-V* characteristic for each state. *Record the characteristic curves and the operating points for maximum output voltage from the sweep generator.*

Keep the curve tracer connected for use in Experiment 2–16.

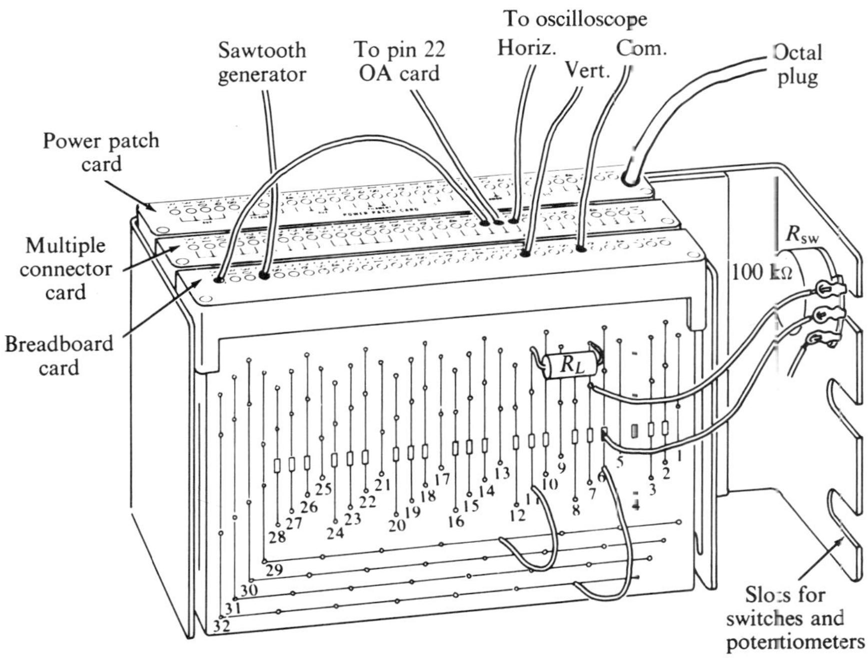

Fig. E2–23 Pictorial of breadboard connection system for circuit of Fig. E2–21.

Experiment 2–15 *Relay Characteristics*

Equipment: ADD
relay card
oscilloscope
VTVM or FET VM

The operate, release, and contact bounce times can be critical for suitable operation of some electronic circuits. It is the purpose of this experiment to observe these times and determine their significance in practical circuits. Also, the pull-in and drop-out currents are determined for a relay.

E2–15.1 The Relay Card

Examine the relays mounted on the printed-circuit relay card. Note the type of relay and the type of contacts. Follow the three contacts from one of the relays to the connectors

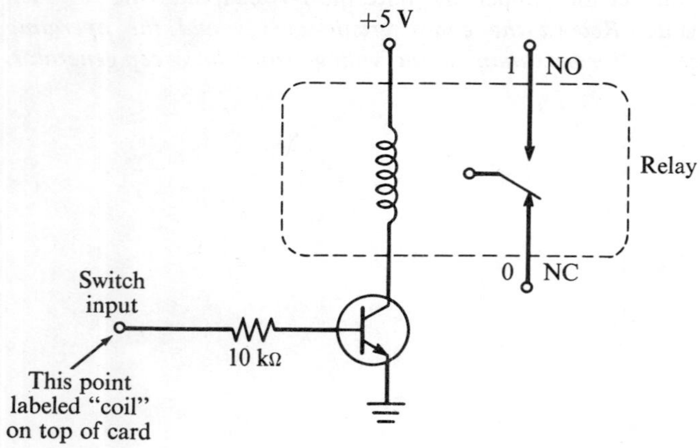

Fig. E2–24 Transistor drive circuit for relay.

accessible at the top of the card, and note how they are labeled. Note that the coil of the relay is not connected directly to the top terminal labeled "coil." The coil is driven by a transistor switch circuit as shown in Fig. E2–24. Such transistor switching circuits will be studied in later experiments.

A +5 V signal applied to the input switches the transistor from OFF to ON to provide a current exceeding the "pull-in" current for operating the relay.

Why is it useful to have a transistor drive circuit for the relay?

E2–15.2 "Pull-in" and "Drop-out" Currents

With the +5 V supply disconnected, measure the resistance of the relay coil by connecting the VTVM or FET VM on a suitable ohms scale between the collector of the transistor driver and the +5 V terminal, as illustrated in Fig. E2–25.

Report the measured coil resistance.

R_{coil} = ______________________

Now connect the +5 V supply to the relay card, and a variable voltage supply to the input resistor, as shown in Fig. E2–25. Also keep the VTVM connected across the relay coil, but switch to the +5 V dc range.

Vary the input voltage gradually until the relay just pulls in and *record the voltage across the relay coil.* Now decrease the input voltage until the relay armature just drops out and *record the voltage across the relay coil.*

v_{coil} on pull-in = ______________________

v_{coil} on drop-out = ______________________

Using Ohm's Law and the measured voltages and coil resistance, *calculate the relay pull-in and drop-out currents. Show your calculations below.*

Pull-in current = ______________________

Drop-out current = ______________________

E2–15.3 Measurement of Operate Time

The pictorial diagram in Fig. E2-26 shows one possible procedure for connecting the relay at the top of the card into a circuit utilizing the "patch" capability of the relay card, multiple connector card, and timing module contacts. Note that the OFF-ON characteristics of the square wave generator are used for the switch *S*. That is, 0 and +5 V are

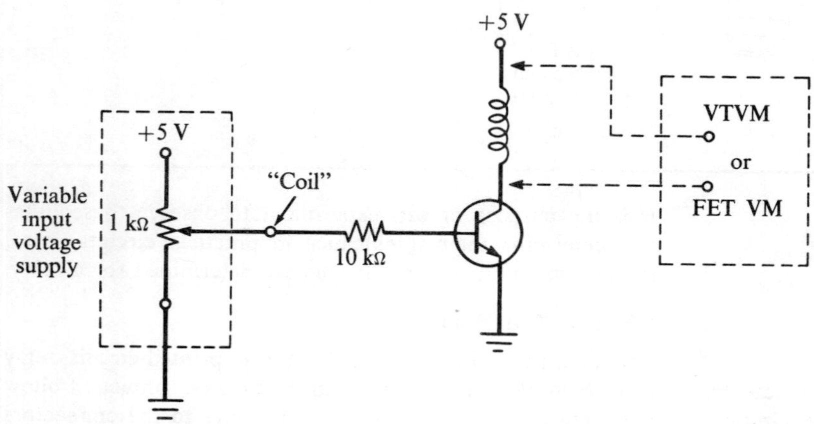

Fig. E2–25 Circuit for measuring pull-in and drop-out currents.

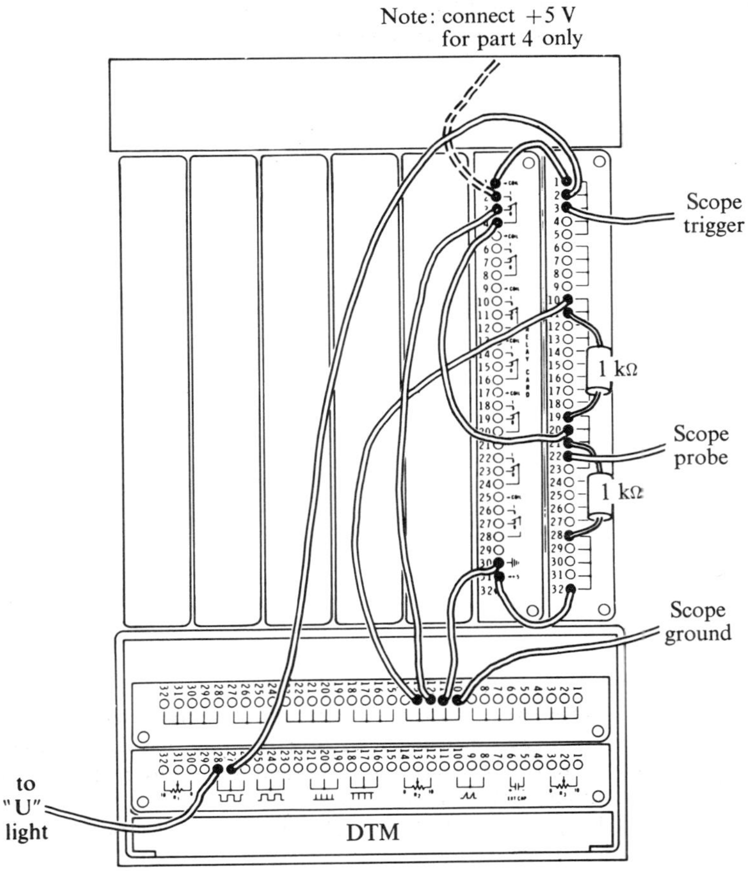

Fig. E2–26 Pictorial of relay circuit for measuring operate time.

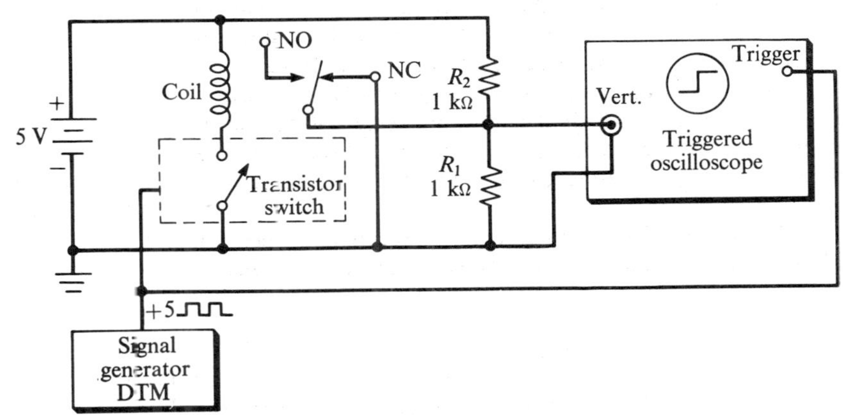

Fig. E2–27 Schematic of relay circuit for measuring operate time.

alternately applied to the relay coil. The repetition rate of OFF-ON operation can be varied by the square wave generator period controls on the front panel of the digital timing module. This switching method will enable the triggered scope to be adjusted more easily and the trace to be observed continuously.

Connect a circuit similar to the circuits shown in Figs. E2–26 and E2–27 and also connect the NO relay contact to common. *Measure the operate time required to break the normally closed contacts. Vary the period and observe whether there is any change in operate time. Estimate the limits of operation.*

Suggested scope settings at the start: (1) vertical gain, 2 V/cm; (2) sweep, 2 msec/cm.

Vary the scope settings to obtain the most accurate readings of the measured times. *Report all results and explain below.*

E2–15.4 Operate, Transfer, and Bounce Times

After measuring the operate time as illustrated in Fig. E2–27, measure the operate, transfer, and bounce times as follows.

Connect the NC contact to common as shown in Fig. E2–27. However, connect the NO contact of the relay to +5 V instead of to the common. This should give the waveforms on the dual-beam scope, as shown in Fig. E2–28.

Note that as the square wave frequency is increased, a point is reached at which the movable relay contact never touches the NO contact before returning. Also, note the effect of the higher frequencies on the operate, transfer, and bounce times.

Measure both the transfer and bounce times. Vary the OFF-ON *period and observe whether there is any change in times. Record the waveforms observed in the space below and report all measurements.*

Note that, if the function of the circuit requires the closing of the normally open contacts, then the **operate time** is the sum of the operate time and the transfer time, but should not include bounce time.

Why not include bounce time in the operate time?

Vary the switching period. Does the period change the measured times?

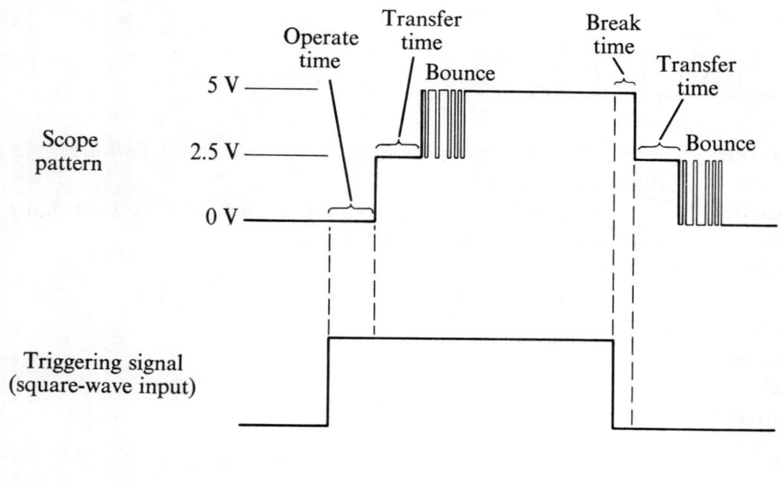

Fig. E2–28 Scope waveform showing operate, transfer, and bounce times.

Experiment 2–16 *Characteristic Curves and Resistance of a Diode*

Equipment: ADD
OA card
breadboard connection system
oscilloscope
silicon signal diode
VOM or VTVM

The characteristic curve for a silicon diode is obtained with an oscilloscopic curve tracer, and the significant diode parameters are measured from the curve. The values obtained from the forward biased diode characteristic curve are used to plot log I vs. V so as to determine the logarithmic relationship over a few decades. The dynamic resistance of the diode is also determined and compared with the static resistance at a current of 1 mA. The problems of using an ohmmeter to measure diode resistance are determined.

E2–16.1 Diode Characteristic Curves

Insert a 1 kΩ resistor as R_L in Fig. E2–23 and replace R_{sw} with a silicon signal diode, with the cathode (banded end) connected to the ADD common and the horizontal input of the scope.

The usual orientation for I-V curves may require the use of inverting inputs on the horizontal amplifier, but if these are not available the forward bias characteristics will appear in the second quadrant as a result of the voltage sweep polarity of the sawtooth generator.

Draw the observed characteristic curve in the space below.

E2–16.2 Forward Bias Characteristics of a Diode

To obtain an accurate measure of the logarithmic relationship between the current and the voltage it is necessary to acquire expanded-scale data. Change the oscilloscope horizontal sensitivity to 0.1 V/div. Use several different vertical sensitivities and obtain 5 current-voltage points over a two-decade (100-fold) range of currents on the I-V curve.

Plot log I vs. V *and check the linearity of the curve. Show the plot below.*

Calculate the slope of the log I vs. V *curve and compare the experimental value with the theoretical slope* [$\log I_f = \log I_r + (39/2.303)V$].

Experimental slope ________ Theoretical slope ________

Determine the forward voltage drop across the diode at 1 mA *current and the dynamic resistance R_{fd} at this current. Compare the measured value of R_{fd} to the theoretical value* [$26\eta/I$ (mA)] *and to the empirical value* [$45/I$ (mA)].

Forward voltage drop = ______________

Experimental R_{fd}	Theoretical R_{fd}

Explain below why the readings are greatly different for each meter range.

E2–16.3 Measurements with an Ohmmeter

Measure the forward and reverse resistances of the above diode using the ohmmeter function of a VTVM or VOM. *Use several resistance ranges and record the measured resistance values.*

Meter range	R_f	R_r

Leave the curve tracer connected for subsequent experiments.

Experiment 2–17 *Reverse Breakdown and Zener Diodes*

Equipment: ADD
OA card
breadboard connection system
oscilloscope
6.8 V Zener diode
DVM

In this experiment the reverse characteristic curve for a Zener diode is observed and the significant parameters of the Zener diode are measured from the curve.

E2–17.1 Reverse Characteristic Curve

Obtain a 6.8 V Zener diode and insert it in the circuit of Fig. E2–21 in place of R_{sw} (or the silicon signal diode of the previous experiment). Adjust the sawtooth generator so that the reverse characteristic of the Zener diode is observed. *Record the reverse characteristic curve below.*

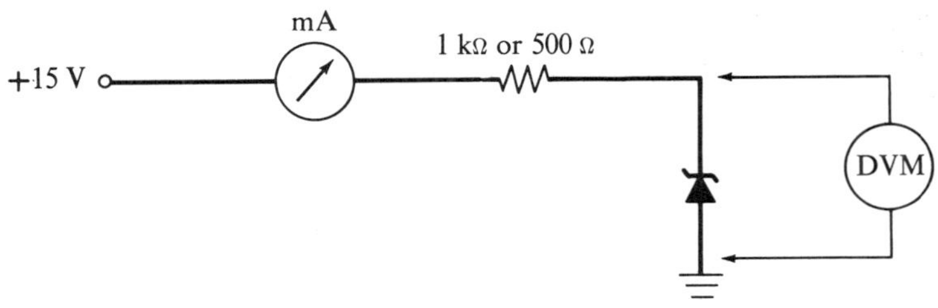

Fig. E2–29 Measurement of dynamic resistance of Zener connected as a shunt regulator.

Current	Voltage drop	Dynamic resistance

Leave the curve tracer connected for later experiments.

E2–17.2 Dynamic Resistance of a Zener Diode

Connect the circuit shown in Fig. E2–29, and measure the reverse voltage drop across the Zener diode for currents of 10 and 20 mA. Use a DVM to measure the voltages and a VOM to measure the exact currents.

From the voltages and currents measured, calculate the dynamic resistance of the diode for 10 *and* 20 mA *reverse current. Average the two values.*

Average dynamic resistance = ____________

Experiment 2–18 *Field-Effect Transistor (FET) Characteristics*

Equipment: ADD
OA card
oscilloscope
MPF 105 or 2N5459 FET
VRS

The characteristic curves for a FET are obtained from a curve tracer and the load line and ON-OFF characteristics are measured.

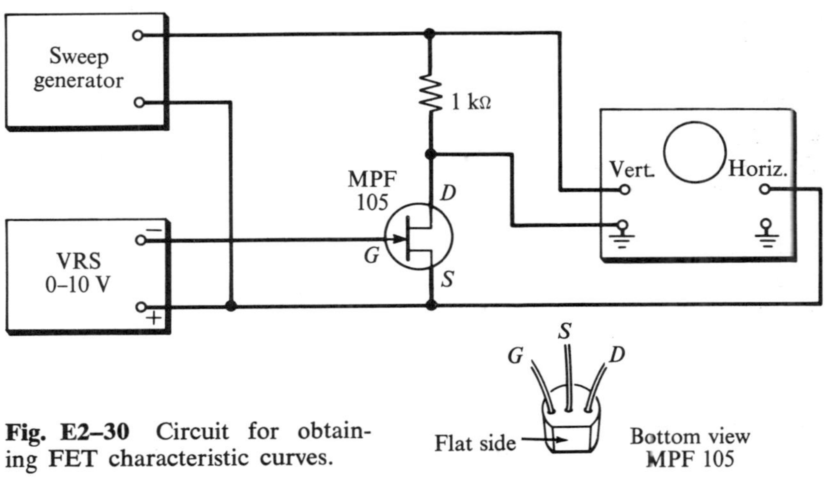

Fig. E2–30 Circuit for obtaining FET characteristic curves.

Connect the sweep generator output of the curve tracer of Fig. E2–22 to the circuit of Fig. E2–30. Vary the VRS voltage from 0 to 10 V and obtain the characteristic curves and load line. *Photograph the curves and attach the photograph below.*

Connect the 1 kΩ drain load resistor to +5 V rather than to the sweep generator output. Vary the VRS voltage until the FET is in saturation and cutoff. *Determine the* ON-OFF *characteristics and report them below.*

Keep the curve tracer connected for subsequent experiments.

Experiment 2–19 *Transistor Characteristics*

Equipment: ADD
OA card
VRS
oscilloscope
npn transistor

The collector characteristic curves for an npn transistor that is connected in a common-emitter circuit are obtained by means of an oscilloscopic curve tracer. If a scope camera is available, the curves should be photographed. The recorded curves are to be used to evaluate several dc parameters. The values of $V_{CE(\text{sat})}$ and $V_{CE(\text{OFF})}$ are measured and compared with the expected values. The dc beta β_N or h_{FE} is determined from the characteristic curves and compared with typical values for the specific type of transistor.

E2–19.1 Connection of Transistors

Transistors can be wired on the breadboard card in several ways. Their leads can be inserted directly into the connections on the side of the breadboard card, generally into a set of three in-line connectors. When this technique is used, the transistor leads should be bent to conform to the connector spacing *before* insertion. Three small holes can be drilled near the front edge of the desk-top chassis for this purpose. The transistor leads should be inserted into these holes before insertion into the breadboard connectors. An alternative approach is to solder No. 22 hookup wire to the lugs of a transistor socket. A high-quality "test-type" socket is recommended. Then insert the other end of the hookup wire into the desired connectors.

Keep in mind that the transistor connections are generally shown as **base** diagrams (the leads as you would see them looking up toward the bottom of the transistor). However, when used on the breadboard card, they are viewed from the top.

Check to see that your transistor connections are not reversed left for right. Improper connections to a transistor can destroy it.

E2–19.2 Characteristic Curves

Connect the circuit shown in Fig. E2–31 using a 2N3393 npn transistor or its equivalent. Set the horizontal and vertical sensitivities each at 10 V/div, and the voltage reference source (VRS) output at 5 V. Use the same sweep generator as in Fig. E2–22.

Adjust the amplitude of the sweep generator so that only the forward characteristic section and a short portion of the reverse current characteristic are observed, as shown in Fig. E2–32. Increase the scope sensitivity to 1 V/div on each axis and locate the zero points for convenient observation of the forward characteristics. Vary the VRS output from 0 to 10 V to obtain (photograph) a set of transistor collector characteristic curves. *Attach the photograph below.*

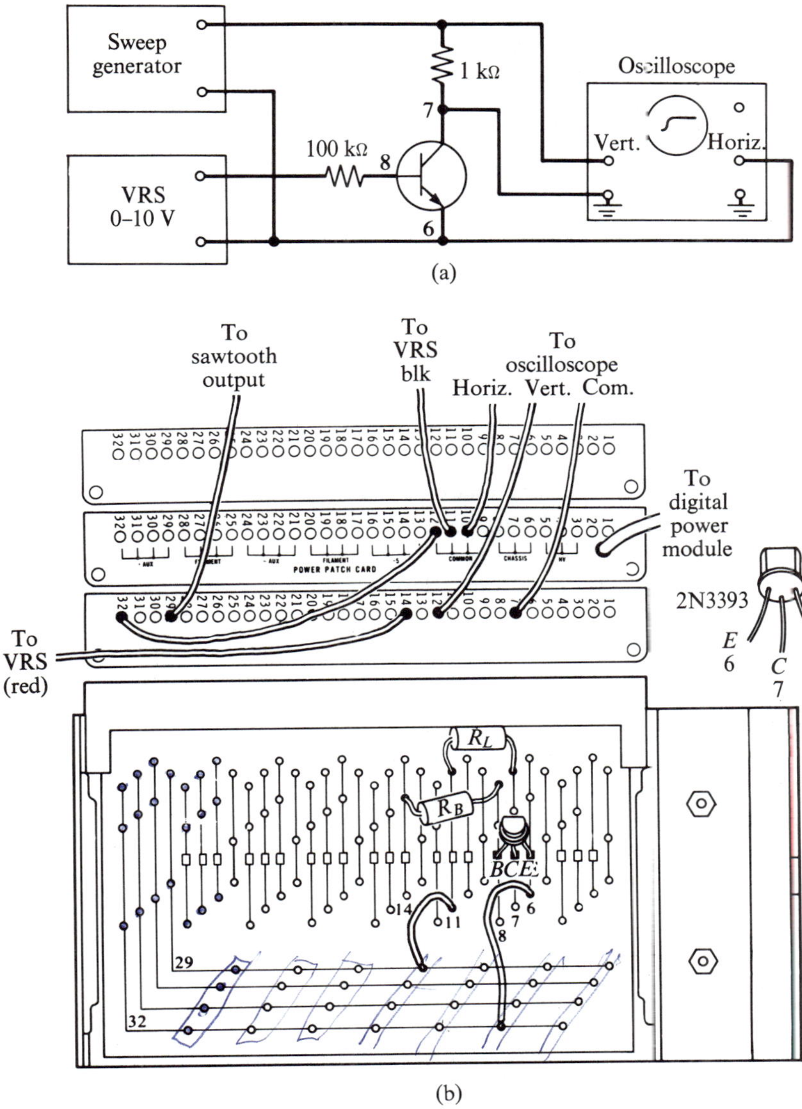

Fig. E2–31 Circuit for obtaining transistor characteristic curves: (a) schematic; (b) pictorial using breadboard connection system.

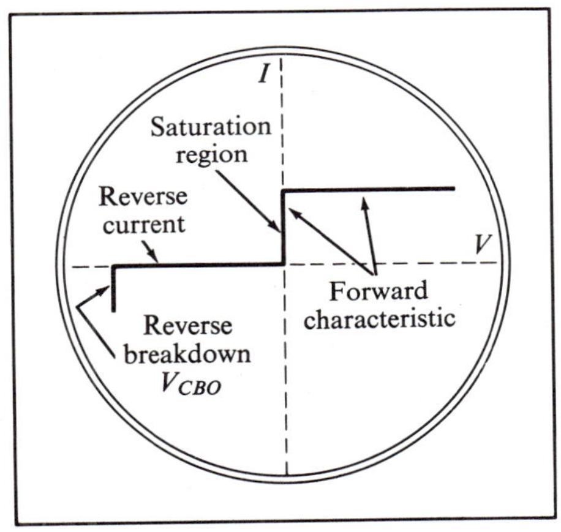

Fig. E2–32 Transistor characteristic curve.

E2–19.3 The Collector-Emitter Voltage for a Transistor Switch ON (in Saturation) and OFF

With the VRS set at 9 V, adjust the sweep generator dc level so that the maximum collector current is 6 mA. Vary the base current from −5 to +100 μA with the VRS and observe the cutoff and saturation regions and the load line. Using the momentary ZERO button, switch the VRS output between 0 and 10 V to observe the ON and OFF states of the transistor switch.

Record the waveforms below.

Connect the 1 kΩ collector load resistor of Fig. E2–31 to +5 V rather than to the sweep generator output. Switch the base current from 0 to 100 μA to cause the transistor to switch from saturation to cutoff states as shown by the position of the dot on the oscilloscope.

When the transistor is ON, *measure* $V_{CE(\text{sat})}$ *and calculate the equivalent* ON *resistance. Also measure* $V_{CE(\text{OFF})}$. *Compare the measured values of* $V_{CE(\text{sat})}$ and $V_{CE(\text{OFF})}$ *with the expected values.*

Measure $V_{CE(\text{sat})}$ with the DVM.

$V_{CE(\text{sat})}$ measured = ____________________

$V_{CE(\text{OFF})}$ measured = ____________________

$V_{CE(\text{sat})}$ expected = ____________________

$V_{CE(\text{OFF})}$ expected = ____________________

E2–19.4 The Direct Current Gain

From the characteristic curves obtained in E2–19.2, *measure* h_{FE} *and compare the measured values with typical values for the 2N3393 npn transistor or other type of transistor that you might be testing.*

If you have several transistors of the same type, note that the h_{FE} values vary considerably from unit to unit.

h_{FE} measured = ____________________

From the measured h_{FE} *calculate what the base resistance should be to ensure saturation if* V_B *and* V_{CC} *are both* +5 V *and* $R_L = 1$ kΩ.

R_B = ____________________

Experiment 2–20 *The Basic Bipolar Junction Transistor (BJT) Switch*

Equipment: ADD
transistor
oscilloscope
VRS

The basic BJT switch circuit is connected and tested so that its important characteristics can be observed. The inverting action of the transistor is first observed. If a high-speed oscilloscope is available, the switching times are measured. Then the transistor switch is connected to a load so as to control the current in an external load. The non-saturating transistor switch circuit is also connected and tested and its characteristics are compared with those of the saturated switch.

E2–20.1 The Transistor Inverter Circuit

Connect the switching circuit of Fig. E2–33. Set the generator frequency at about 10 kHz and measure the amplitude of the square wave output. Use a fixed 110 kΩ base resistor as shown in Fig. E2–33b.

Observe the shape of the output and compare it with the input signal on a dual-trace or dual-beam scope. Note the signal inversion. Explain the rounding of the transitions between the ON and OFF states.

Record the input and output waveforms, one above the other in the space below. Label the axes. Remember that at saturation the voltage across the transistor is low, while at cutoff V_{CE} is high. *Identify these regions on the waveforms.*

E2–20.2 Transistor Switching Times

Set the square wave frequency in the switching circuit of Fig. E2–33 to the highest frequency available by switching the DTM to the external capacitor position and placing a 100 pF capacitor in the EXT. CAP. terminals on the DTM. Observe the transitions of the switch from ON to OFF and OFF to ON. *Record the waveforms below and measure the rise and fall times.* The rise time is the time required for the output to fall from $0.9v_o$ to $0.1v_o$, while the fall time is the time required for the output to increase from $0.1v_o$ to $0.9v_o$.

Rise time = ____________________

Fall time = ____________________

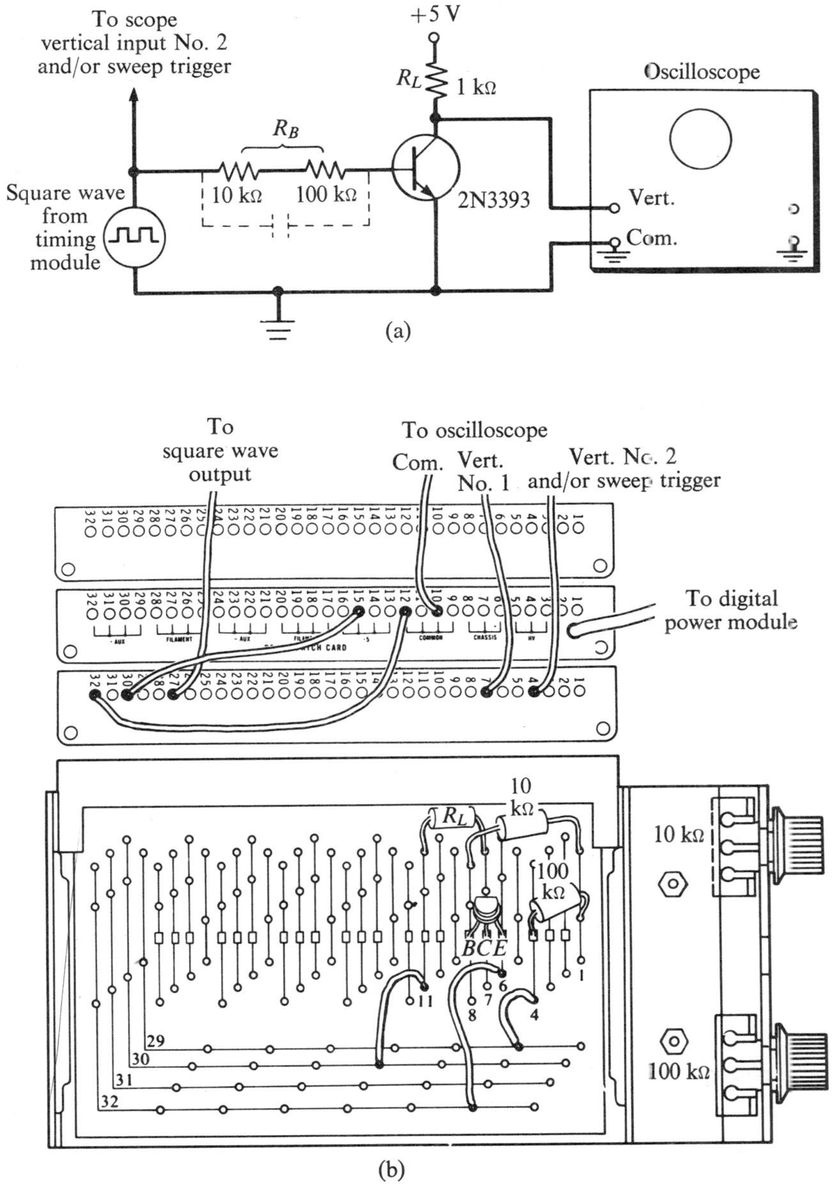

Fig. E2–33 Transistor switching circuit: (a) schematic; (b) pictorial.

Now place a 56 pF "speed-up" capacitor in parallel with the base resistor R_B and note the reduced transition times. *Measure the rise and fall times as before.*

Rise time = ______________________________

Fall time = ______________________________

E2–20.3 Connection of Loads to Transistor Switching Circuits

A transistor switching circuit similar to the light-driver circuits in the BIM is connected in order to control the current through a lamp. Connect the circuit shown in Fig. E2–34 on the breadboard card. The +170 V is available at the top of the power patch card. Turn the lamp ON and OFF

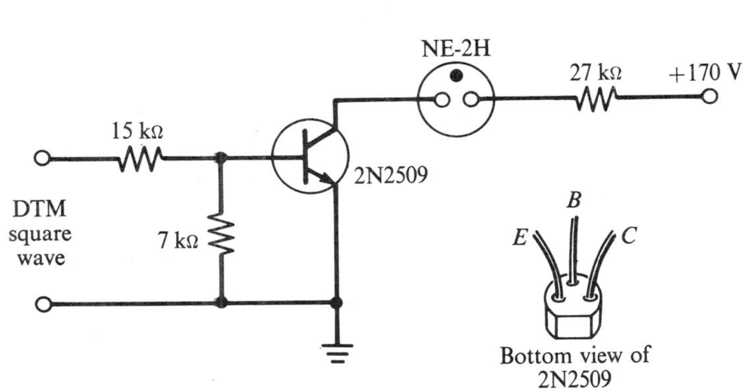

Fig. E2–34 Transistor lamp driver.

by applying a 1 Hz square wave from the DTM. Observe that the state of the lamp follows the input signal as the frequency is changed and that only the cathode is surrounded by a glow. Be careful to connect the transistor correctly because it can be destroyed by the high voltage. Note that the BIM contains 10 circuits identical to that of Fig. E2–34 for driving the 10 indicator lamps.

Disconnect the DTM from the switch input and connect the VRS. *Determine the base current (note: not the VRS output current) required to turn the lamp* ON *by varying the VRS voltage, and inserting a VOM.*

Minimum base current = ____________________

E2–20.4 Nonsaturating Switching Circuit

Connect the nonsaturating switch of Fig. E2–35 on the broadboard card. Apply a 100 kHz, 5 V square wave from the DTM to the switch input. *Compare the storage times with and without the diode and explain the advantages of avoiding saturation.*

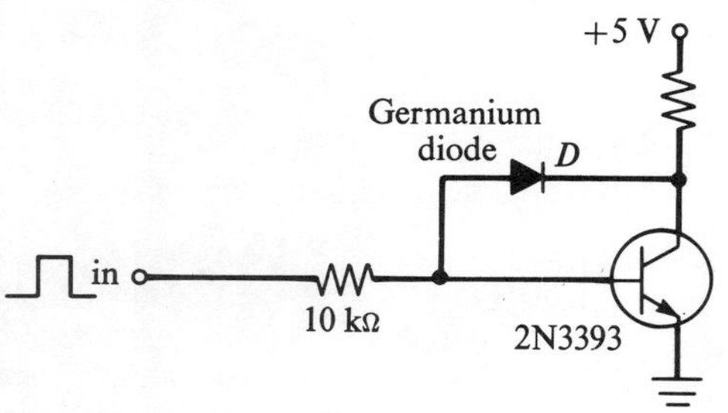

Fig. E2–35 Nonsaturating switch.

Experiment 2–21 *Unijunction Transistor (UJT) Trigger Circuit*

Equipment: ADD
relay card
transistor power supply (TPS) or low voltage power supply (LVPS)
2N4894 unijunction transistor
oscilloscope
capacitance substitution box
resistance substitution box

A unijunction transistor is combined with an *RC* network to provide a simple time delay and trigger circuit. The range of useful time delay is investigated and the circuit signals are observed.

Before beginning this experiment read Equipment Appendix 6, which describes the transistorized power supply (TPS) and the low voltage power supply (LVPS). *Answer the following questions.*

What range of voltages is available from the TPS?

What range of voltages is available from the LVPS?

What are the current limitations of the TPS and LVPS?

Connect the unijunction time delay and trigger circuit of Fig. E2–36. Use a relay on the relay card controlled by a BIM switch for switch S_1. Use the capacitance and resistance substitution boxes for C and R. Set C at 0.22 μF and R at 1 MΩ. Observe the output across R_2 on an oscilloscope as S_1 is opened. *Measure the time between the opening of switch S_1 and the observation of a pulse across R_2.* Change the *RC* time constant to several different values and *repeat the time delay measurements.* Observe also the output across the timing capacitor.

R, Ω	C, μF	RC, sec	Time delay, sec

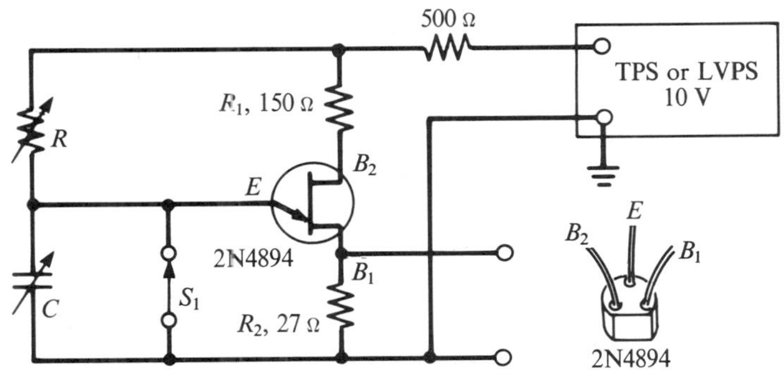

Fig. E2–36 Unijunction transistor time delay and trigger circuit.

Determine the useful range of time delay with the UJT circuit.

Range ______________________________

Experiment 2–22 *pnpn Device Characteristics*

Equipment: high voltage power supply (HVPS) or universal power supply (UPS)
oscilloscope
ADD
OA card
VRS
25 Ω, 2 W resistor
1 kΩ, 1% resistor
TIC47 SCR
40526 triac

A curve tracer is constructed and the current-voltage curves of two common pnpn devices, the SCR and the triac, are obtained. This experiment also demonstrates the relationship between the turn-on time of the thyristor and the gate current.

Before beginning this experiment, read Equipment Appendix 7 which describes the high voltage power supply (HVPS) and the universal power supply (UPS). *Answer the following questions.*

What range of voltages can be obtained from the HVPS?

What range can be obtained from the UPS?

E2–22.1 Current-Voltage Characteristics of an SCR

Assemble the circuit of Fig. E2–37a.

Note: The circuit common is floating and is *not* identical with ground. The true ground is the oscilloscope ground, and precautions must be taken to keep the common and ground separate. Be certain that the VRS −terminal is not jumped to ground. Also check the ADD common to be sure it has not been jumped to the chassis.

The oscilloscope displays the current through the SCR on the vertical channel and the voltage across the SCR on the horizontal channel. The VRS is used to vary the gate current, i_g.

What is the function of the OA follower?

Vary i_g from 0 to 10 mA (0 to 10 V on the VRS) and observe the *i-v* curve. *Determine the following parameters:*

Reverse blocking resistance (resistance in third quadrant) = ______________________________

Forward blocking resistance (resistance in first quadrant before SCR conducts) = ______________________________

Forward conducting resistance, corrected for external circuit = ______________________________

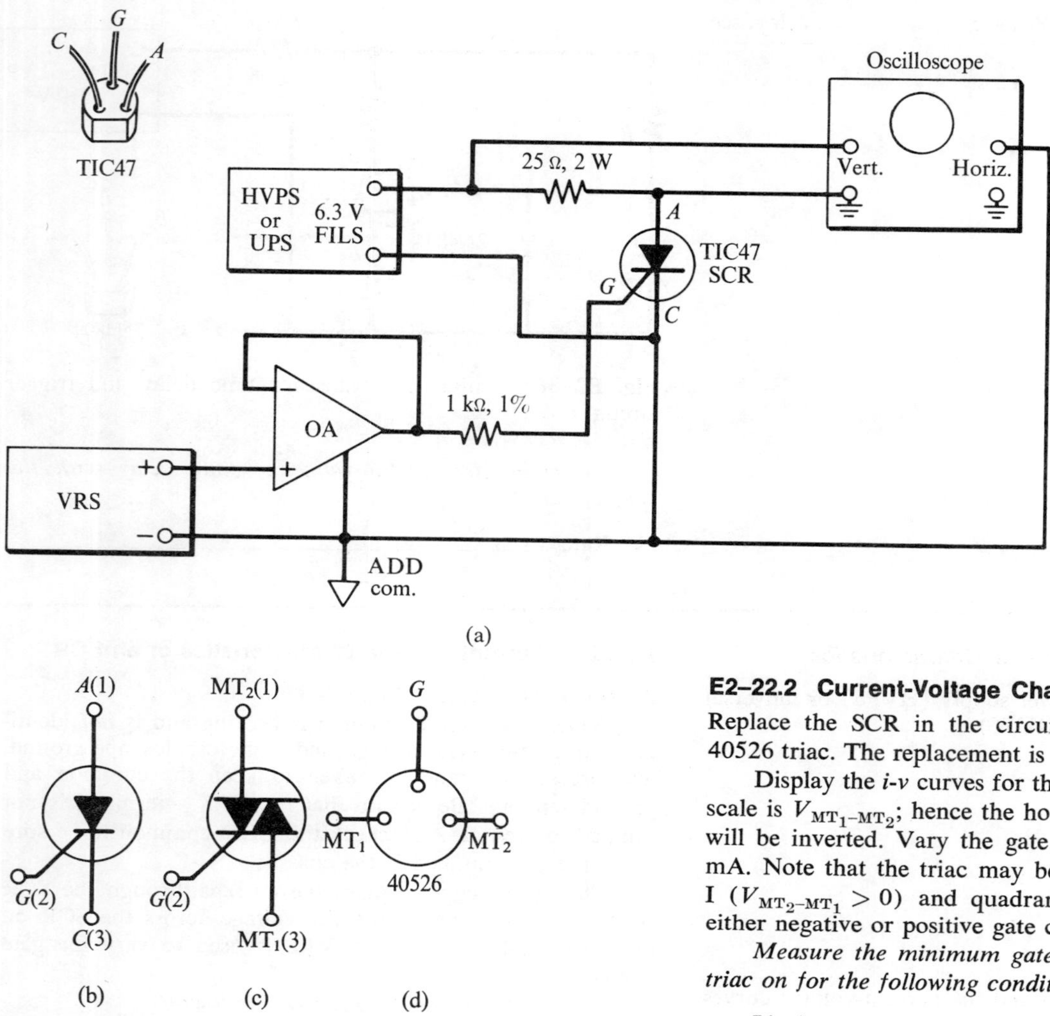

Fig. E2–37 Characteristic curves of pnpn devices: (a) curve tracer; (b) TIC47 SCR connections; (c) 40526 triac equivalent connections; (d) base diagram of 40526 triac.

Minimum gate current needed to switch from OFF to ON = ______________________

Minimum holding current = ______________________

Adjust the gate current (VRS) until the SCR just begins to conduct. Increase the oscilloscope's vertical sensitivity until the breakover voltage V_{BO} can be observed. Note that V_{BO} decreases at constant gate current.

Explain below.

E2–22.2 Current-Voltage Characteristics of a Triac

Replace the SCR in the circuit of Fig. E2–37a with the 40526 triac. The replacement is shown in Fig. E2–37c and d.

Display the *i-v* curves for the triac. Note that the voltage scale is $V_{MT_1\text{-}MT_2}$; hence the horizontal axis of the *i-v* curve will be inverted. Vary the gate current from -10 to $+10$ mA. Note that the triac may be triggered in both quadrant I ($V_{MT_2\text{-}MT_1} > 0$) and quadrant III ($V_{MT_2\text{-}MT_1} < 0$) with either negative or positive gate current.

Measure the minimum gate current needed to turn the triac on for the following conditions:

I(+) ______________________

I(−) ______________________

III(+) ______________________

III(−) ______________________

The remaining measurements are all made with the gate current positive. Operate the oscilloscope in the dual-trace mode, triggered off the line. *Sketch the current and true voltage waveforms (remember that the voltage is inverted by the oscilloscope) and note the gate current for the following:*

Triac ¼ ON (conduction phase angle = 90°)

i_g = ______________________

Triac ½ ON (conduction phase angle = 180°)
i_g = ______

Triac ¾ ON (conduction phase angle = 270°)
i_g = ______

Triac ON (conduction phase angle = 360°)
i_g = ______

Increase the oscilloscope sweep rate and vertical sensitivity so that the current-time relationship at the zero crossing can be seen. Note that the triac (and semiconductors in general) does not turn on immediately. *Measure the slope of the current-time curve at the minimum gate current and maximum gate current* (10 mA).

$$\frac{di}{dt}(i_{g\,\text{min}}) = \rule{6cm}{0.4pt}$$

$$\frac{di}{dt}(i_g = 10\text{ mA}) = \rule{6cm}{0.4pt}$$

Since electromagnetic interference (EMI) is proportional to di/dt, and triacs are routinely employed to switch hundreds of amperes, minimization of di/dt is important.

Optional: Propose a circuit which will minimize di/dt and can be constructed from components and instruments previously examined. *Do not build the circuit.*

Experiment 2–23 *Analog Switching Circuits*

Equipment: ADD
transistorized or low voltage power supply (TPS or LVPS)
VRS
VOM or DVM
two silicon signal diodes
MPF 105 or 2N5459 JFET
2N3702 pnp transistor
EU-900-JA analog switch card

An analog diode switching circuit is first constructed in this experiment and then tested to determine whether the switch faithfully transmits analog current and voltage signals from input to output when it is ON and effectively disconnects them from the output when it is OFF. Then a JFET switch and pnp transistor driver circuit are constructed and tested for analog transmission. Finally, a commercial analog switch is employed and tested. The experiment is thus designed to show the principles of analog switching and some of the circuits for accomplishing high-accuracy analog transmission.

Before beginning this experiment read Equipment Appendix 8 which describes the analog switch card. *Answer the question below.*

What are the maximum analog voltages which can be applied to the FET switches on the analog switch card?

E2–23.1 Analog Diode Switching Circuit

Connect the two-diode analog switching circuit of Fig. E2–38. Before attaching the TPS or LVPS to the circuit, adjust the variable output voltage to +10.0 V. Use switch A on the BIM as the actuating source. Attach a DVM at the output.

With switch A at the 0 V ("0" position), measure the output voltage while varying the analog input voltage (VRS) between 0 and +5 V.

Does the analog input voltage influence the output voltage? ________________

Now change the actuating source (switch A) to +5 V ("1" position) and measure the output of the analog switching circuit while varying the VRS from 0 to +5 V. *Record the results.*

Analog input voltage (VRS)	Output voltage

Explain below how the analog switch works and why the output voltage is slightly higher than the analog input voltage.

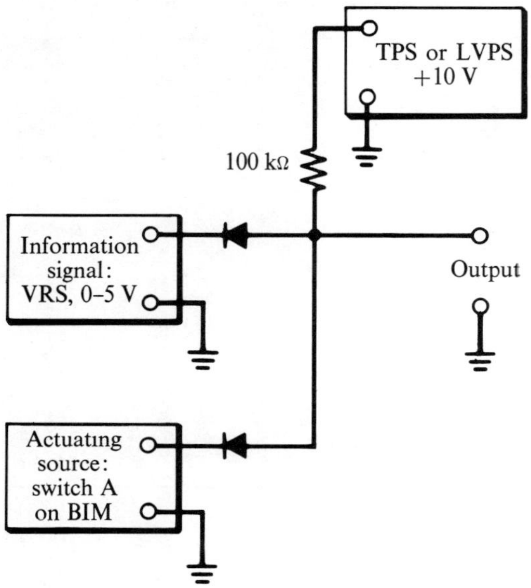

Fig. E2–38 Analog diode switch.

E2–23.2 JFET Analog Switch

Connect the FET switch and pnp transistor driver circuit of Fig. E2–39. Set the switch control input to 0 V and attach the VRS set at 1 V to the analog input. Note that the FET switch is closed (ON) so that the input voltage is transmitted to the output. Do not exceed 5 V at the analog input. Apply various VRS voltages from 0 to 5 V at the analog input and *measure the input and output voltages accurately. Record these voltages below.*

Analog input voltage	Output voltage

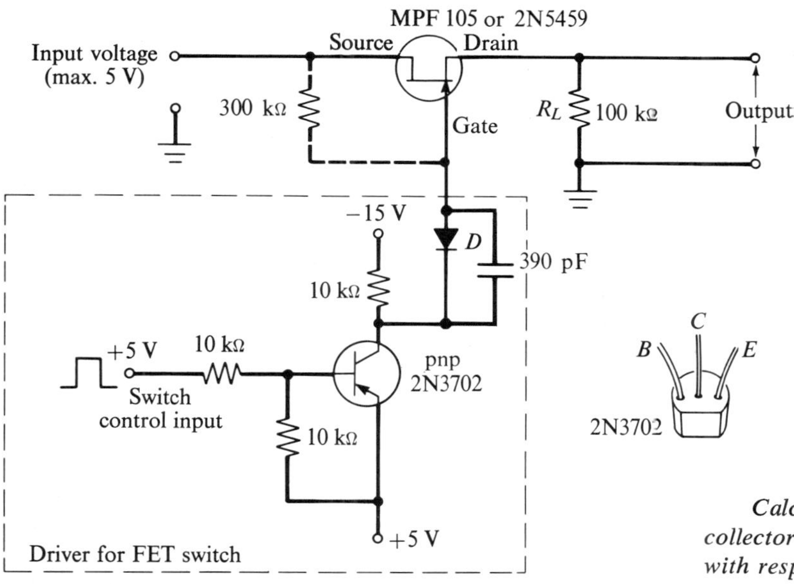

Fig. E2–39 n-channel JFET switch.

Calculate the effective **OFF** *resistance. Measure the collector voltage of the pnp transistor and note its polarity with respect to common.*

Consider the FET switch to be a simple resistance when **ON** and *determine the effective* **ON** *resistance* R_{ON}. *Explain how you measured* R_{ON}.

R_{ON} = ______________________

Now apply +5 V to the control input and *measure the output while varying the VRS analog input voltage.*

Analog input voltage	Output voltage

E2–23.3 Analog Switch Card

Insert the analog switch card in the ADD. Note that each switch contains a driver circuit similar to that in Fig. E2–39. Six of the switches require 0 V to close the switch, while two switches require +2.4 to +5 V for switch closure. Attach the VRS as the analog input and a BIM switch output as the switch control signal. Apply various input voltages and *measure the input and output voltages of the switch with the switch control signal in each state.*

Switch control	Analog input voltage	Output voltage

Connect the square wave output of the DTM to the switch control input and switch the input signal at rates up to 10 kHz or more. *Record the output waveforms for various analog input voltages.*

Experiment 2–24 *Binary Information Gates*

Equipment: ADD
VRS
LVPS or TPS
VOM or DVM
2N3393 npn transistor
two silicon signal diodes
EU-800-JC NAND gate card
resistance substitution box

It is shown in this experiment that the same diode switching circuit of the previous experiment can also be used to transmit binary information, but the considerations for successful operation are different. The diode binary information gate is constructed and tested, and limitations are noted. Then a diode-transistor binary gate is constructed and tested in order to show how a transistor inverter following the diode gate can improve performance. Then an integrated circuit binary information gate is used. The input and output levels are determined for a TTL NAND gate, and advantages and limitations over the diode and diode-transistor gates are noted.

Before beginning this experiment, read Equipment Appendix 9, which describes the NAND gate card. *Answer the questions below.*

How many two-input NAND gates are on the card?

How many four-input NAND gates are on the card?

An unused input is the same as a __________ *signal.*

E2–24.1 Diode Binary Information Gate

The basic diode switching circuit of Fig. E2–38 can also be used as a binary information gate. If the information to be transmitted is not continuous but limited to two levels, high (+5 V) and low ($\approx$0 V), the circuit of Fig. E2–38 is called an AND gate because the output is high (+5 V) only when both the information signal (VRS) and the actuating source are in the high state (+5 V).

Construct the diode AND gate of Fig. E2–38. Adjust the VRS to 0 V and measure the output voltage for both positions of the actuating source (switch) A. Change the VRS to +5 V and measure the output voltage for both positions of switch A. Consider any voltage less than 1 V to be a low state and any voltage greater than 2 V to be a high state, and *fill in the table below for all four possible combinations of information signal and actuating source.*

Information signal	Actuating source	Output
Low	Low	
Low	High	
High	Low	
High	High	

If only binary (high-low) information is to be transmitted, *discuss below why the switching considerations are different from those for switching a continuous analog input signal.*

E2–24.2 Diode-Transistor Binary Information Gate

A breadboard layout for a three-input diode-transistor (DTL) binary gate is shown in Fig. E2–40. *Before wiring the circuit, draw the circuit diagram below.*

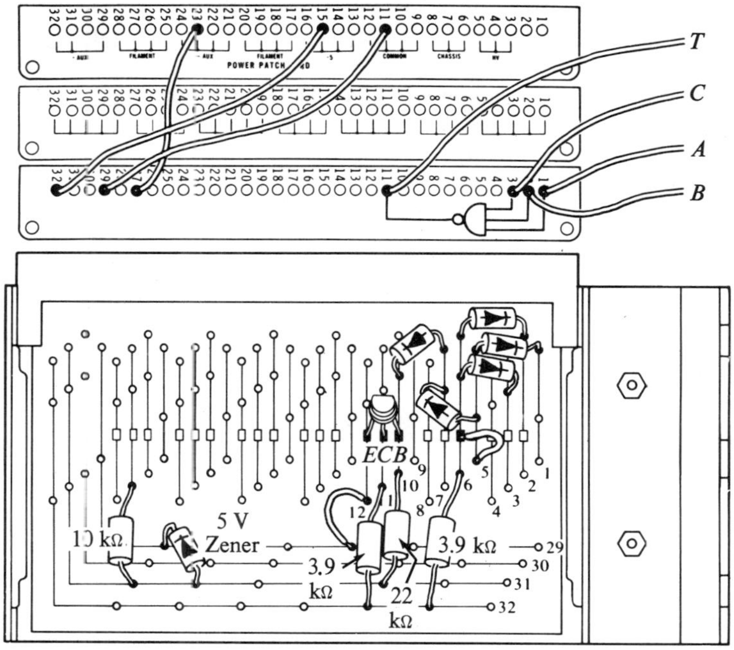

Fig. E2–40 Pictorial of DTL NAND gate.

Construct the circuit of Fig. E2–40. Connect the gate output T to an indicator light on the BIM. *Determine whether the output is high* (1) *or low* (0) *for each of the eight possible combinations of BIM switches A, B and C. Fill in the table.*

Switch positions			Output
A	*B*	*C*	*T*
0	0	0	
0	0	1	
0	1	0	
0	1	1	
1	0	0	
1	0	1	
1	1	0	
1	1	1	

Explain why this circuit is called a NAND (NOT AND) gate.

Disconnect the A and B switches from the inputs and connect a resistance substitution box (RSB) between one input and ground. Remove the indicator light from the output and connect a VOM or DVM. Increase the RSB resistance and note the output voltage.

Does the output high level change as the input resistance is increased?

In what direction?

What are the limitations of this DTL gate in terms of driving other binary information gates?

E2–24.3 Integrated Circuit NAND Gate

Insert the EU-800-JC NAND gate card in the ADD. Connect a variable voltage source (VRS) to one of the two-input NAND gate inputs. CAUTION: ALWAYS STAY IN THE RANGE OF 0 TO +5 V. *Measure the output voltage as a function of input voltage. Use the DVM or other high input resistance meter to measure both input and output voltages.* This is necessary because the gate can load the voltage source.

Input voltage	Output voltage

With the VRS set at 0 V, insert a current meter between the VRS + output and the NAND gate input. *Measure the input current for a 0 V input and note its direction.*

Why is this gate called a current sinking gate?

Now apply BIM switches to the two-input NAND gate inputs. Connect the output to a BIM light. *Determine the output state for the four possible input switch positions and verify its NAND operation.*

Switch positions		
A	*B*	Output
0	0	
0	1	
1	0	
1	1	

Experiment 2–25 *Optoelectronic Switching Circuits*

Equipment: ADD
OA card
oscilloscope
CM47 tungsten lamp
light-emitting diode
phototransistor
2N2509 npn transistor
NSL 364 photoconductive cell

Several different light sources and detectors are characterized in this experiment for applications in optoelectronic switching. Experience is gained in wiring a BJT driver circuit for operating a tungsten lamp, and neon lamps and light-emitting diodes (LED's) are also characterized. The photoconductive cell and phototransistor are used as detectors. A photoconductive cell optoelectronic chopper is connected as a modulator for converting a dc signal to an ac signal. A series chopper, a shunt chopper, and a series-shunt chopper are studied.

E2–25.1 Source Characteristics

Some of the important switching characteristics of tungsten lamps, neon lamps, and LED's are examined in this part of the experiment.

Wire the tungsten lamp driver–photoconductive cell circuit of Fig. E2–41. Set the DTM square wave at a frequency of 1 Hz. Connect the square wave output to the tungsten lamp driver and also to light *Q* on the BIM. Connect the OA output to the connector on the OA card marked DIG IN. This connects at the OA output the binary level interface shown in dotted lines in Fig. E2–41. Connect a wire from the connector on the OA card marked DIG OUT to light *W* on the BIM.

First, disconnect point *A* from the 100 kΩ OA input resistor, and *use the oscilloscope to observe the signal at point A.*

Is the source/detector rise time longer than its fall time? Briefly explain why.

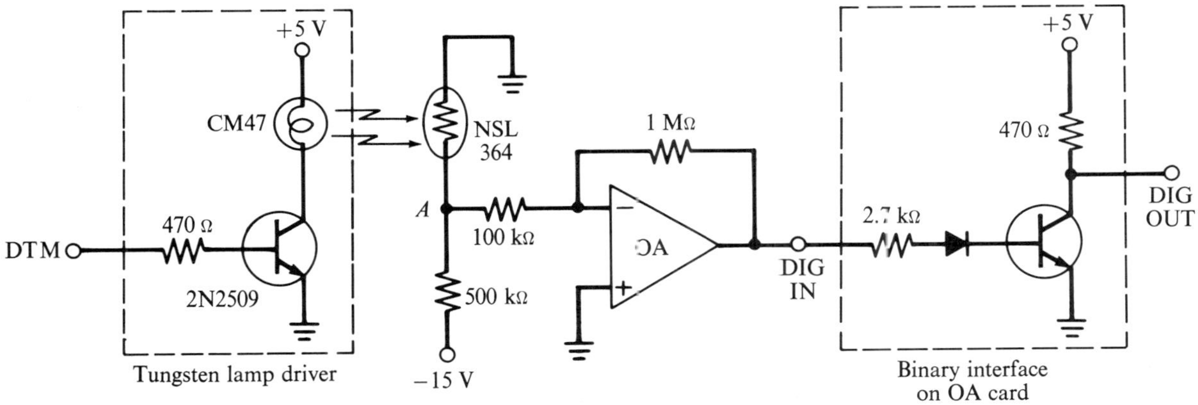

Fig. E2–41 Tungsten lamp driver and photoconductive cell detector.

Reconnect point A to the 100 kΩ OA input resistor. Observe lights Q and W on the BIM while increasing the DTM square wave frequency. As the frequency is increased, the system will no longer transfer information. This can be observed when light W (detector output) no longer follows light Q (DTM input). *Determine the maximum operating frequency, f_{max}, of this optoelectronic switch from the DTM frequency dial setting at the point where light W no longer follows light Q.*

f_{max} = ________________________

A similar experiment can be performed with a neon lamp–phototransistor combination. The neon lamp and driver contained in the BIM will be used. Connect the circuit of Fig. E2–42. Bring the phototransistor as close as possible to light X and try to minimize the amount of room light seen by the phototransistor.

Set the DTM frequency to 10 Hz. *Record the amplitude of the signal displayed on the oscilloscope.*

dc level ________________________

ac signal ________________________

Increase the DTM frequency until the signal is one-half its former amplitude. *Record this frequency as f_{max} below.*

f_{max} = ________________________

Explain why ac coupling on the oscilloscope is necessary.

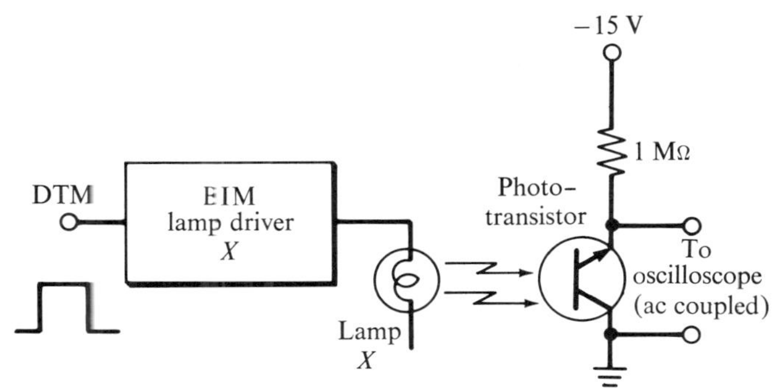

Fig. E2–42 Phototransistor detector.

Disconnect light X and connect the square wave output to the anode of a light-emitting diode. Wire the LED cathode to ground. *Measure the maximum operating frequency on the oscilloscope.*

f_{max} = ________________________

Note that if anything were to happen to the light driving circuit, no damage would be done to the measurement system and subsequent circuits. This type of isolation is highly desirable in many applications (interfacing an environmental probe to a computer, for example), and commercial optical isolators are available. They use LED's as sources; LED's are known to be usable at higher frequencies than other sources. Explain your data in light of these facts. *Do LED's have other advantages besides speed?*

Compare the response characteristics of the source-detector combinations studied above.

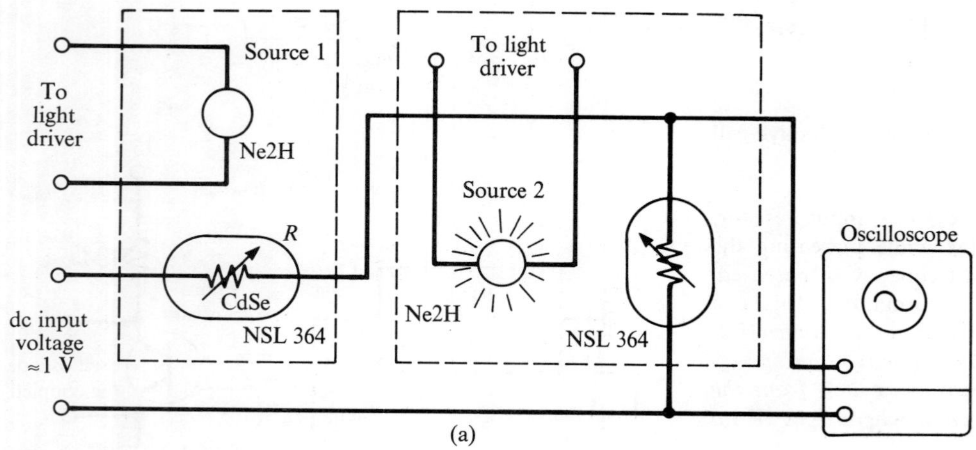

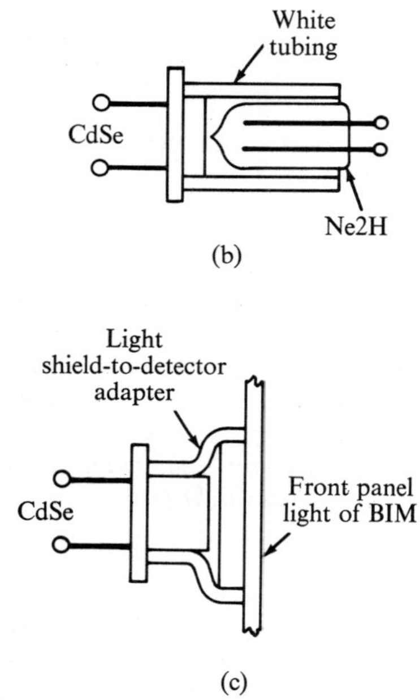

Fig. E2–43 Optoelectronic chopper: (a) schematic; (b) pictorial showing a neon lamp and CdSe photoconductive cell mounted in white rubber tubing; (c) pictorial showing use of front panel from BIM and special adapter to accommodate light shield and CdSe detector.

E2–25.2 Optoelectronic Chopper

Optoelectronic circuits can also be used to switch analog signals. The circuit shown in Fig. E2–43a is a series-shunt optoelectronic chopper. The two light sources are driven out of phase; i.e., when source 1 is ON, source 2 is OFF. Assemble the circuit. Use the neon lamps on the BIM for light sources, and use the DTM as the input to the light drivers. The DTM has complementary square wave outputs; the outputs from connectors 23–25 on the DTM are ON when the outputs from 26–28 are OFF, and vice versa.

Set the DTM frequency to 1 Hz and the VRS output to 1 V. Try to minimize the amount of room light which reaches the photoconductive cells by using the shielding arrangement shown in Fig. E2–43b and c.

Draw the waveforms observed below.

Change the VRS output to 5 V and observe the waveforms.

Explain the new waveforms below.

Raise the DTM frequency. *At what frequency is the amplitude of the chopped waveform cut in half?*

Convert the series-shunt chopper of Fig. E2–43a to a series chopper by replacing source 2 and its photoconductive cell with a 10 kΩ resistor.

Draw the waveforms for VRS voltages of 1 V *and* 5 V.

Next make a shunt chopper by rewiring source 2 and its photocell, and replacing source 1 and its photocell with a 10 kΩ resistor.

Repeat the same measurements.

Explain the differences in the performance of the series-parallel, series, and shunt optoelectronic choppers.

Experiment 2–26 *Diode Clipping and Clamping Circuits*

Equipment: sine-square generator or function generator
6.8 V Zener diode
silicon signal diodes
oscilloscope
breadboard connection system

Several diode clipper circuits are connected and tested to illustrate how they can slice off sections of a waveform. A diode clamp is also connected, and its ability to restore the dc component in an input square wave after it has been coupled through an *RC* circuit is investigated.

E2–26.1 Zener Clipping Circuit

Clip a 1 kHz sine wave output signal of the sine-square generator between −0.6 and +6.8 V using the Zener diode in the circuit shown in Fig. E2–44. Use an input signal of 10 V or greater peak amplitude.

Observe and record below the input and output signals, including voltage levels.

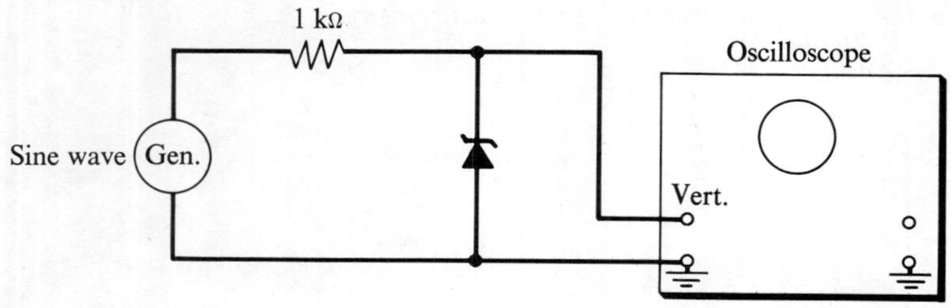

Fig. E2–44 Zener clipping circuit.

E2–26.2 Silicon Diode Clipper

Design and test a clipping circuit which will clip between +0.6 and −1.2 V using silicon signal diodes in place of the Zener diode. *Draw the circuit diagram below and record the waveforms.*

E2–26.3 Diode Clamping Circuits

Connect the square wave output of the sine-square generator or function generator through an *RC* network as shown in Fig. E2–45. Leave out the diode shown in dotted lines and set the square wave generator at 1 kHz.

Observe and record the input and output waveforms and measure the amplitudes. Use dc coupling on the scope and note that the dc level of the input signal is lost at the output of the *RC* coupling network.

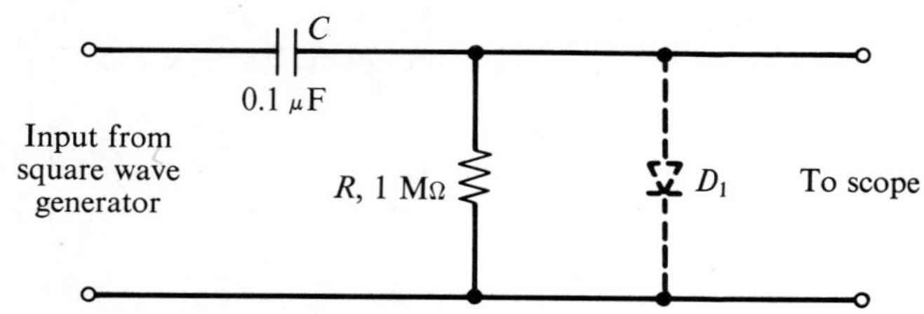

Fig. E2–45 Diode clamping circuit.

Now connect the diode across *R* and *observe, measure, and record the output signal and the dc level.*

Reverse the diode and *again record the output signal. Explain how the diode clamp works.*

Experiment 2–27 *Waveform Generators*

Equipment: ADD
- dual OA card and single OA card or three single OA cards
- 2N3393 npn transistor
- capacitors
- resistors
- silicon signal diode
- multiple connector card
- oscilloscope

This experiment illustrates that a general purpose function generator can be constructed from a basic operational amplifier linear sweep generator. The function generator produces variable frequency sine, square, and triangular waveforms. A linear sweep generator with automatic reset is first constructed and tested. Then the circuit is modified to produce the various other waveforms. Some characteristics and limitations of the function generator are then studied.

E2–27.1 Repetitive Linear Sweep Generator

Construct the linear sweep generator with automatic reset shown in Fig. E2–46. Use one OA for the sweep generator and the second for the comparator. Use one of the 100 kΩ potentiometers on the DTM for the threshold-adjust potentiometer.

Connect the sweep generator output, v_o, to CH. 1 of the dual-trace oscilloscope and also to the external trigger input. Connect the comparator output v_{comp} to CH. 2 of the oscilloscope. Set the oscilloscope triggering controls to EXT. TRIGGER, and the vertical mode control to ALT. Position the two traces so that both outputs can be observed. *Record the waveforms observed on the scope and note the time relationship between the two waveforms.*

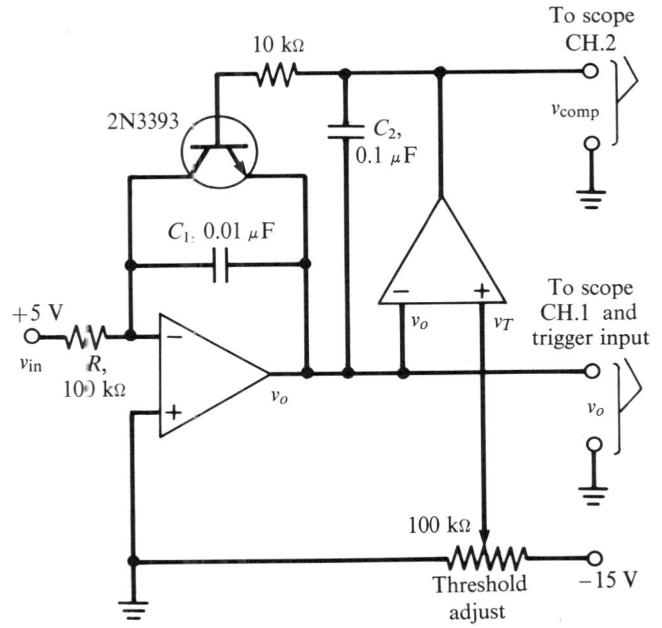

Fig. E2–46 Repetitive linear sweep generator.

Vary the threshold adjustment and note any changes in output. At what point does capacitor C_1 fail to discharge to 0 V?

Remove capacitor C_2 and note any changes in the waveforms. Explain below.

Reinsert capacitor C_2 and verify the relationship

$$\text{sweep rate} = -\frac{v_{in}}{RC_1}$$

for various values of C_1 from 0.1 to 0.01 μF. *Note*: the threshold voltage should be set at a low value to allow C_1 to discharge at the higher sweep rates. *Record the results below.*

C_1	Sweep rate observed	Sweep rate calculated	Difference, %

E2–27.2 Triangular and Square Wave Generator

A triangular wave can also be produced by reversing the polarity of the integrator input voltage. Construct the triangular and square wave generator of Fig. E2–47. Leave out the diode wave-shaping circuit shown in dotted lines. Use the DTM potentiometers for the two variable resistors. Connect the square wave output v_{sqr} to the scope's CH. 1 input and to the external trigger input. Connect the triangular wave output v_{tr} to CH. 2. Observe both traces simultaneously. *Record the waveforms observed on the scope and note the time relationship between the two waveforms.*

Change the threshold adjustment and record any changes in the output waveforms.

Vary potentiometer R and record what happens. Explain below.

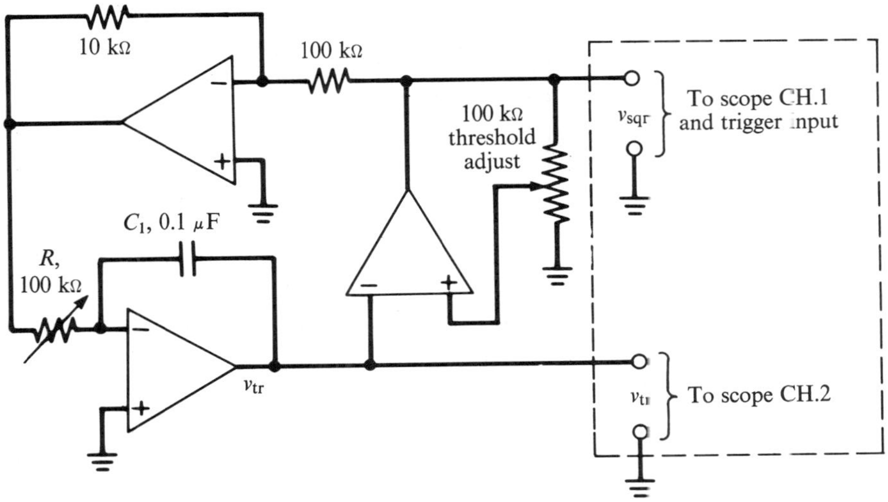

Fig. E2–47 Triangular and square wave generator.

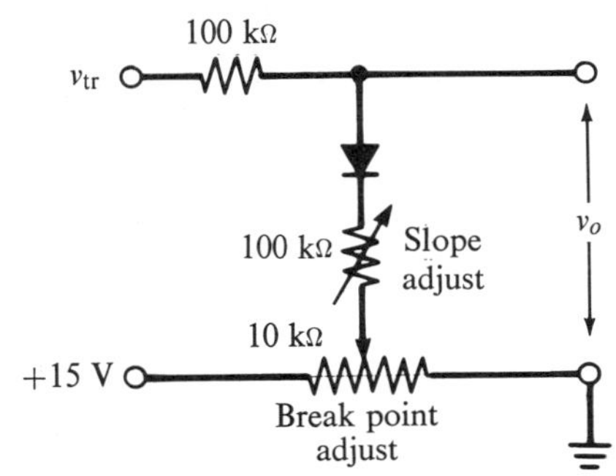

Fig. E2–48 Diode wave-shaper.

E2–27.3 Sine Wave Output

A sine wave can be produced from the triangular wave output of the function generator of Fig. E2–47 by adding the diode wave-shaping circuit shown in Fig. E2–48. Connect this circuit and observe the output on the oscilloscope. Note the effect of changes in the slope and breakpoint adjustments. *Record the waveform.*

Explain below how a more nearly sinusoidal waveform could be produced from the triangular wave output.

Experiment 2–28 *Multivibrators*

Equipment: ADD
one OA card or dual OA card
two signal diodes
capacitors
resistors
multiple connector card
oscilloscope

In this experiment various multivibrator circuits are constructed and tested. The basic astable, monostable, and bistable multivibrators are constructed with operational amplifier circuits. The limitations and advantages of these circuits are noted.

E2–28.1 The OA Astable Multivibrator

Construct the self-starting astable multivibrator shown in Fig. E2–49. Note that without the capacitor in the circuit, a square wave will not be produced at the output. The astable circuit of Fig. E2–49 has two metastable states, one of which corresponds to positive saturation of the amplifier, the other to negative saturation, usually ±10 V.

The frequency of oscillation is given by

$$f = \frac{1}{2RCln\left(1 + 2\frac{R_1}{R_2}\right)}.$$

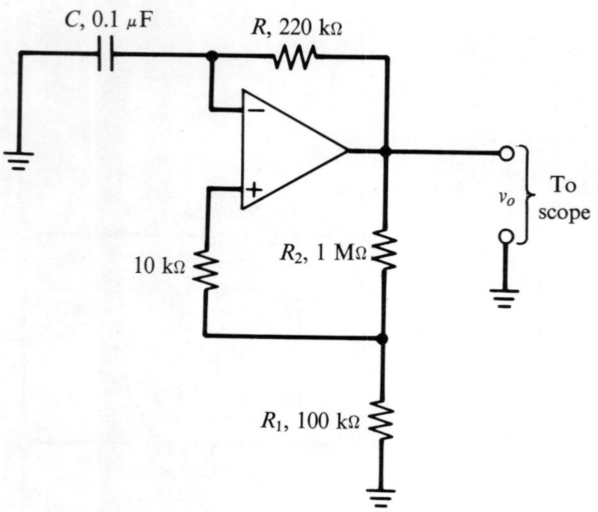

Fig. E2–49 Single OA astable multivibrator.

Verify this relationship for values of C from 0.1 *to* 0.01 μF.

C	Observed f	Calculated f	Difference, %

Record the waveform at the output, and note the value of peak amplitude and the position of 0 V.

At what values of C and the corresponding frequency does oscillation stop?

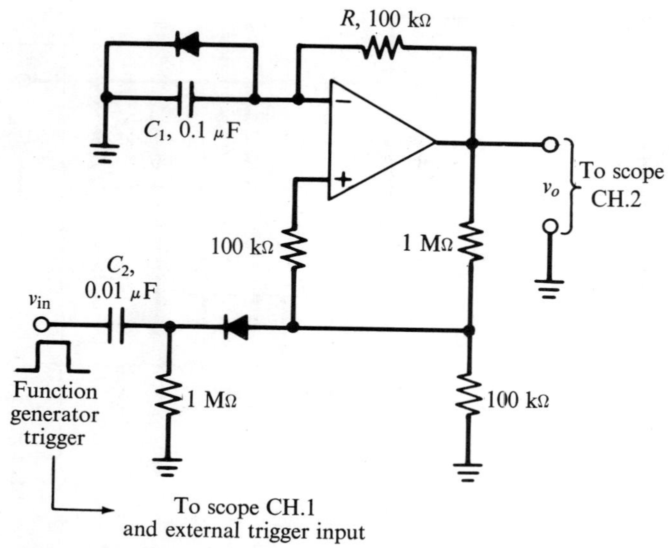

Fig. E2–50 Monostable multivibrator.

E2–28.2 The OA Monostable Multivibrator

Construct the monostable multivibrator of Fig. E2–50. Apply a square wave from the function generator of almost 100 Hz to the input to trigger the "single shot" of the monostable. Switch the scope triggering to EXT. TRIGGER and connect the square wave to the external trigger input and to CH. 1. Connect the output of the monostable multivibrator to CH. 2 and set the vertical mode control to ALT. Position the two traces so that both the input and output waves can be observed simultaneously.

Record the waveforms displayed on the scope, and note the time relationships between them.

Increase the frequency of the triggering pulse. Does the length of the monostable pulse change?

At what frequency will the monostable multivibrator cease to trigger?

E2–28.3 The Bistable Multivibrator

Construct the bistable multivibrator shown in Fig. E2–51. Connect a pushbutton to the input and observe the output states on the scope.

The bistable multivibrator has two stable states; in this circuit they correspond to positive and negative saturation of the operational amplifier. The bistable will remain in a given state until a trigger pulse is introduced at the input. A pulse of opposite polarity applied to the input will reverse the polarity of the voltage at the positive input. The output will quickly switch to its other limit. The positive feedback through R will keep the output in that state until another input pulse is applied.

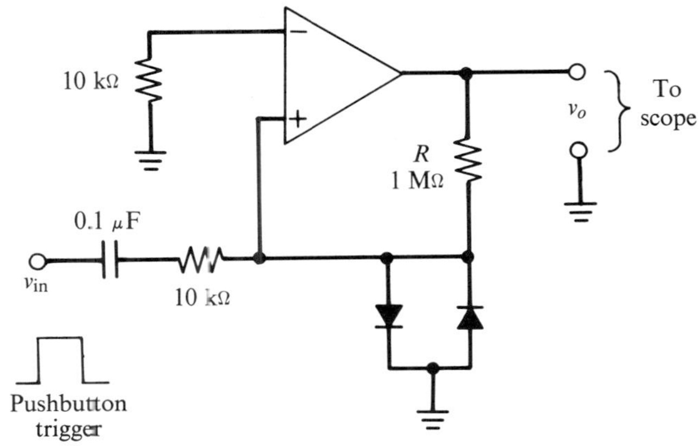

Fig. E2–51 Bistable multivibrator.

When the output of a bistable multivibrator is made compatible to logic level signals, it becomes the basic digital storage unit. *Push the button and observe the bistable multivibrator change states on the scope.*

Experiment 2–29 *Control of Power in a Load with Solid State Switches*

Equipment: LVPS or TPS
HVPS or UPS
SCR
triac
Zener diode
ADD
OA card
transistor
diode
oscilloscope
variable resistor

Several circuits employing solid state power switches are constructed. This experiment illustrates how SCR's and triacs can be used to accomplish power supply protection, and several modes of light source control. A circuit is studied as an overvoltage protector. Then a lamp output is controlled by a triac circuit. A circuit is constructed to turn the lamp off after a precise amount of radiation has been received by the detector, and to control the power applied to the lamp in order to keep the intensity constant.

E2–29.1 Control of Direct Current

Quite often it is critical that a power supply voltage not rise above a certain level. Modern power supplies seldom fail, but overvoltage protection is still commonly incorporated into many power supplies.

The circuit of Fig. E2–52a is called a "crowbar." When the voltage rises above the Zener diode's breakdown voltage, the Zener conducts and triggers the SCR. The SCR switches from the forward blocking state to the forward conducting state and provides a low resistance path to ground. The current will rise and blow the fuse; the current will then drop below the holding current, and this will turn off the SCR.

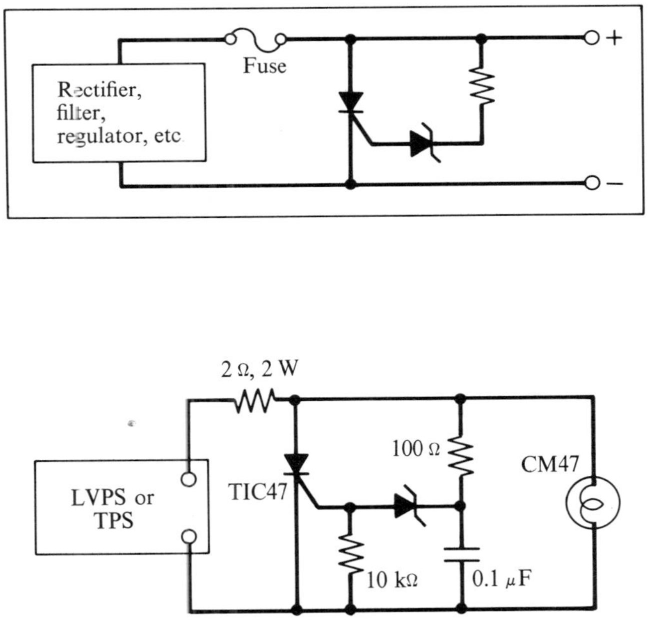

Fig. E2–52 Power supply protection circuits: (a) general overvoltage protector; (b) specific experimental circuit.

Wire the circuit shown in Fig. E2–52b. The 2 Ω resistor is inserted to limit the current drawn by the SCR when it conducts in order to protect the power supply and the SCR.

Use a 6.8 V Zener diode to limit the voltage. Turn the LVPS or TPS voltage control to 0 and the current limit to maximum and then turn the power supply ON. Slowly in-

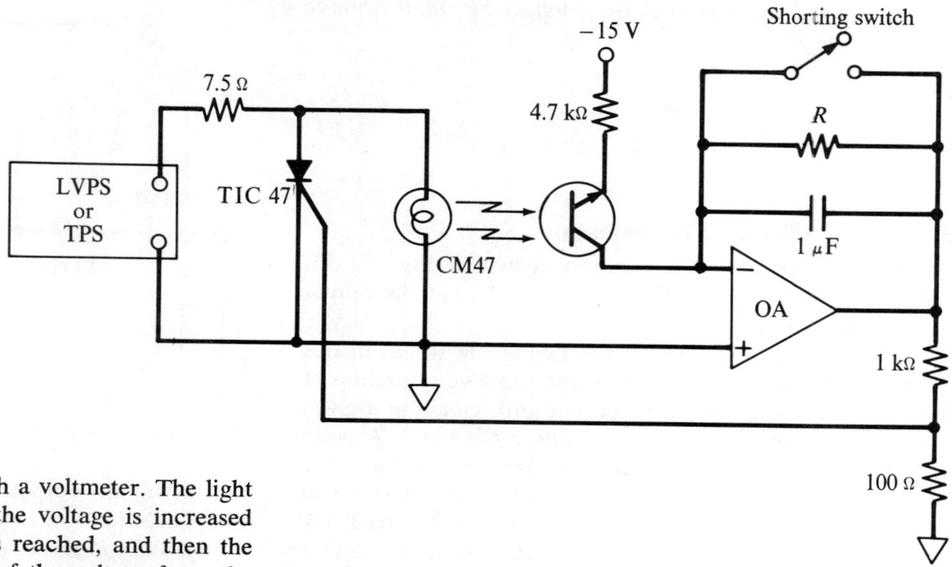

Fig. E2–53 Integrated light-activated circuit.

crease the voltage and monitor it with a voltmeter. The light will start to glow more brightly as the voltage is increased until the Zener breakdown voltage is reached, and then the SCR will conduct and remove most of the voltage from the external load. If an LVPS is used, the current limit light will go ON and the voltage must be reduced in order to turn the SCR OFF.

Record the trigger voltage.

What is the function of the 100 Ω *resistor?*

The capacitor was added in Fig. E2–52 to decrease the circuit's sensitivity to transients. The LVPS or TPS may provide a spike when it is turned ON and the spike can turn ON the SCR.

Measure the maximum voltage the LVPS can be set to that will not cause the supply to "crowbar" when turned ON.

With capacitor ______________________

Without capacitor ______________________

A similar circuit can also be used to turn the power off when signaled by an external device. Figure E2–53 shows a circuit which allows a precise amount of light to strike a detector. Increased radiation then shuts off the light. The OA and capacitor integrate the photocurrent. The feedback resistor will discharge the capacitor, and should be chosen so that the capacitor is discharged at the same rate as the background photocurrent is integrated. The resistor should be experimentally determined, and should be in the 100 kΩ to 1 MΩ range.

Construct the circuit in Fig. E2–53.

What are the functions of the 1 kΩ *and* 100 Ω *resistors?*

Vary the distance from the light to the detector to simulate varying light levels. *What is the longest* ON *time that can be achieved?*

Optional: A circuit similar to that in Fig. E2–53 has been designed so that an electronic flash camera can measure the reflected light and turn the flash off when the appropriate level is reached. The circuit, however, shorts out the power supply and wastes charge and battery life. *Propose a circuit which can shut off a lamp but not short the power supply to ground.* It is permissible to use operational amplifiers.

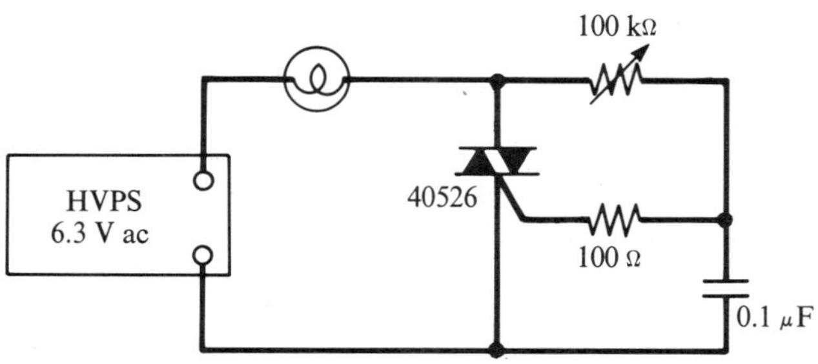

Fig. E2–54 Tungsten lamp ac controller.

E2–29.2 Control of Alternating Current

Alternating current can be controlled by a triac. A tungsten light dimmer is shown in Fig. E2–54. The capacitor is not necessary, but allows smoother control over the current. Build the circuit and *answer the following question.*

What quadrants and gate currents are used?

Can the light be varied smoothly? Why?

The circuit just wired can be used as part of a stabilized light control system. The complete system is shown in Fig. E2–55.

When the light output increases, the photocurrent and OA output both increase. This increases the base current to the 2N3393, and the transistor turns **ON**, so that the collector voltage drops, and the triac gate current decreases. The triac now conducts less, and the lamp output decreases.

Wire the circuit of Fig. E2–55. Use the 100 kΩ variable resistor on the DTM to set a partially conducting (half **ON**) initial state.

What happens if the light is blocked?

Bring the lamp close to the detector. Explain the oscillation.

Is an oscillation expected if the lamp is moved away from the detector? Why?

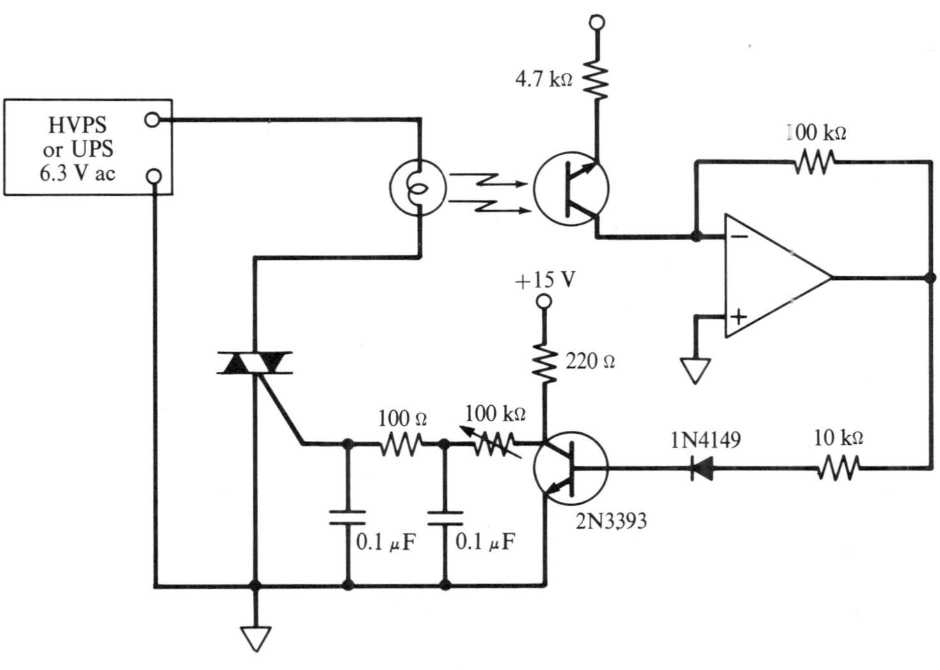

Fig. E2–55 Optically regulated light source.

Instrumentation Power Supplies

Section 2–4

It was shown in previous sections that electronic circuits, devices, and systems all require specific voltages V and the transfer of charge Q between the various circuit elements and networks so as to perform the functions for which they were designed. In other words, a supply of electrical energy W is required for suitable operation, and W (joules) $= Q$ (coulombs) $\times V$ (volts). However, it is not sufficient for an energy source to supply the required charge over an arbitrary period of time. It is necessary that the power P (the time rate of supplying the energy) be sufficient for continuous operation of the many circuits and devices in a system; the electrical power P (watts) $= W$ (joules)$/t$ (seconds). Since the voltage V (volts) is equal to W (joules)$/Q$ (coulombs), and the current I (amperes) is equal to Q (coulombs)$/t$ (seconds), the power $P = VI$. One watt of electrical power is thus one volt-ampere or one joule per second.

The sources of electrical power that are used to meet the specific voltage and current requirements for electronic measurement and control systems are called **instrument power supplies.** The major type of instrument power supply uses the ac electrical power that is fed into buildings from power companies and converts it into the required voltages, ensuring that these voltage levels are maintained in accordance with the current demands. By combining diodes, resistors, capacitors, and other devices discussed in the previous sections, it is possible to convert the ac line power into dc power at the required voltages. Therefore, the line-operated dc power supply circuit is a form of control circuit for manipulation of the ac line power so as to maintain the desired dc output voltages and currents for which it was designed.

The power company that supplies the ac electrical line power might have developed its electrical power in any of several ways. In the classical methods a water turbine is used to drive an electrical turbine (hydroelectric power plant), or a fossil fuel (coal, oil, gas) is burned to convert water into steam and then steam turbines move electrical conductors through a magnetic field so as to generate the ac voltage and current. More recently atomic energy (at nuclear power plants) is used to convert water into steam to drive the steam turbines which operate the electrical generators. However, when we plug instruments into a wall socket it does not make any difference what was the original type of energy used to provide the electrical power; the only important thing is that the correct ac electrical power is available at that socket. Therefore, the first part of the discussion in this section will assume the availability of ac electrical line power and will concentrate on its conversion to the dc voltages and currents required by modern electronic measurement and control systems. For example, transistor and integrated circuits usually require dc operating voltages of about 3 to 30 V and currents of a few milliamperes to several amperes. Transducers such as photomultiplier tubes and ionization detectors require dc power supplies that provide hundreds to thousands of volts but very low currents, usually less than 1 mA. The stability of the instrument power supply is frequently very important because it can determine the stability of the entire instrument. Therefore, the general types of power supply regulation devices are discussed in this section.

In recent years portable electronic instruments that do not require ac electrical line power have become increasingly important (remote pollution monitors, space flights) and popular (portable television). Most of these instruments utilize some form of electrochemical cell or batteries to provide the necessary voltages and currents. For these instruments, it is important to know something about the basic energy source in order to recognize the range of usefulness and limitations of many types of portable instruments with self-contained electrical energy sources. Both the disposable dry cells and rechargeable batteries are discussed. Also, the principles of solar cells and fuel cells are presented.

2–4.1 CONVERSION OF AC LINE VOLTAGE TO DC VOLTAGE

The block diagram in Fig. 2–179 illustrates the basic components and functions usually required to convert an ac to a useful dc voltage. First, a transformer is used to convert the ac line voltage to about the desired voltage. Next, a rectifier circuit converts the ac to a pulsating dc voltage. To smooth out the pulsating dc voltage, a filter network is employed. In modern instruments some type of regulator circuit is usually added to improve the stability and regulation and to reduce further the ripple of the output voltage. The critical characteristics of a dc power supply are its rated output voltage, its internal resistance, its output voltage variation (noise), and its current limits. Any voltage source, being made of imperfect conductors, has some internal resistance. Since the current in the voltage source is the same as

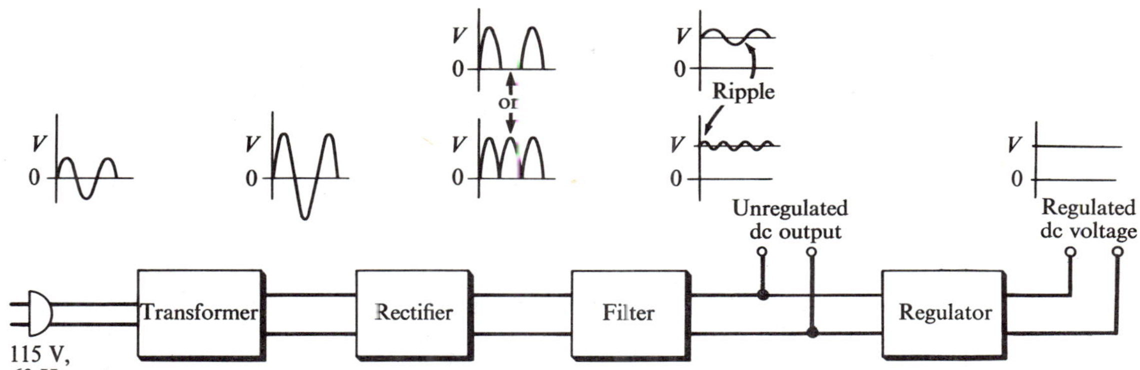

Fig. 2–179 Block diagram showing conversion of ac input to dc output voltage.

the current in the external circuit (for power supplies, the external circuit is called the load), an *IR* potential drop will be developed in the voltage source. Since this potential drop is in opposition to the no-load voltage source output potential difference, the output voltage will decrease as the load current increases.

An ideal dc voltage power supply would have (1) a constant output voltage regardless of variations in the current required by the load (good regulation or low output resistance); (2) a constant output voltage regardless of variations in temperature, ac line voltage, age of power supply, etc. (good stability); and (3) no noise voltage of line or other frequency superimposed on the dc output (low ripple). In addition to these characteristics, the dc output voltage and the current capability of the power supply must meet the operational requirements of the electronic devices.

Transformers

A transformer is used to provide an efficient multiplication or division of the ac power line voltage. The line voltages in the United States are usually 110 V or 220 V, 60 Hz, and in Europe 220 V, 50 Hz. A schematic diagram and pictorial of a transformer are given in Fig. 2–180. The operation of the

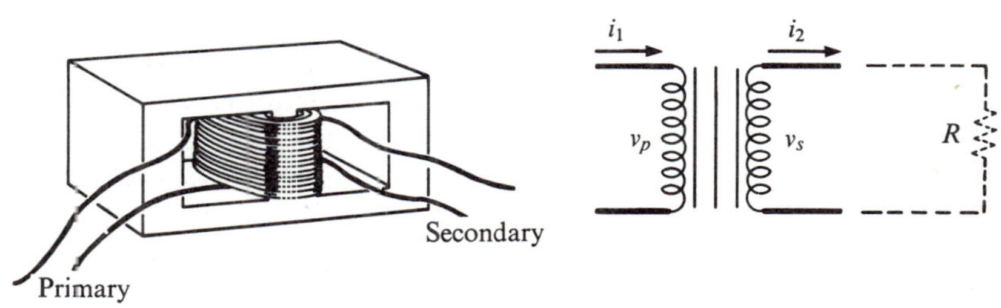

Fig. 2–180 Pictorial and schematic representation of transformer.

transformer is based on the principles of inductance introduced in Section 2–2.4. A changing current in the primary winding produces a changing magnetic flux in the secondary coil, which produces an induced emf across that coil. The voltage produced in the secondary winding is equal to the number of turns, N_s, times the rate of change of magnetic flux ϕ. The relationship, earlier given as Eq. (2–64), is repeated here:

$$v_s = -N_s \frac{d\phi}{dt}. \tag{2–131}$$

Since the primary and secondary are intimately wound on a core of high magnetic permeability, the magnetic flux in the primary and the magnetic flux in the secondary windings are equal. According to Eq. (2–66) the rate of change of flux is

$$\frac{d\phi}{dt} = \frac{-v_p}{N_p}, \tag{2–132}$$

where v_p and N_p are the voltage on the primary and the number of turns in the primary, respectively.

Combining Eqs. (2–131) and (2–132) to eliminate $d\phi/dt$, we get

$$\frac{v_s}{v_p} = \frac{N_s}{N_p}. \tag{2–133}$$

Therefore, the voltage induced in the secondary winding is proportional to the voltage applied to the primary. The voltages referred to are, of course, ac voltages because the transformer depends on a constantly changing flux. By choosing a transformer with an appropriate turns ratio, one can convert the ac line voltage to a higher or lower value. A single transformer can have several windings or several taps in one winding to satisfy several different voltage requirements. The transformer also provides a degree of isolation of the secondary load from the primary supply, since there is no direct connection between the two. Isolation transformers are used where there is a problem of conflicting common voltage levels between the primary and secondary circuits.

Rectifier Circuits

Since a diode conducts current effectively in only one direction, the current in a circuit which is in series with a diode must be dc. The simplest rectifier circuit is shown in Fig. 2–181. On the positive half-cycle of the ac voltage, the diode can conduct, allowing current to pass through R_L. The resistor R_L is the "load" or the circuit which is to be supplied with direct current. On the negative half-cycle, the diode is reverse biased and, therefore, nonconducting. This rectifier circuit is called a "half-wave" rectifier because only half of the ac current wave is present in the load circuit.

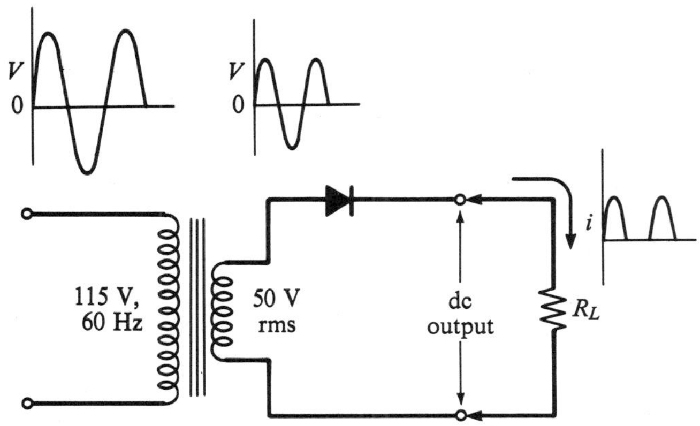

Fig. 2–181 Half-wave rectifier circuit.

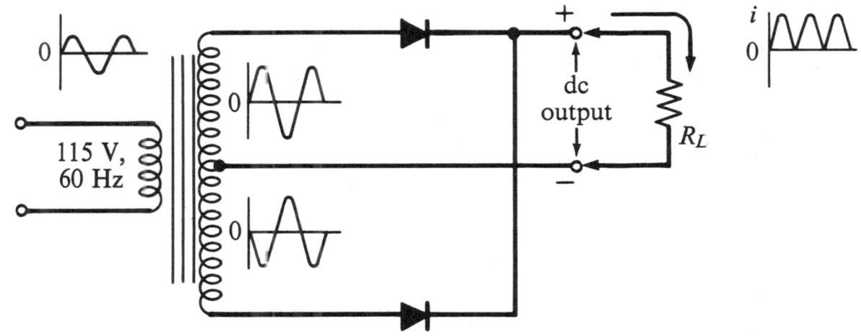

Fig. 2–182 Full-wave rectifier circuit.

In considering rectifier circuits, it is important to consider the ratings of the rectifiers, which include: (1) the maximum average forward current rating, which is approximately $\frac{1}{2} \times V_{av}/R_L$, because the diode conducts only half the time; and (2) the **peak inverse voltage** (PIV), which is the maximum voltage of the reverse or nonconduction polarity that should be applied to the rectifier. The diode of Fig. 2–181 has to withstand a peak inverse voltage of $50 \times 1.4 = 70$ V. If a capacitor is connected across the output of the rectifier circuit in Fig. 2–181 the diode PIV must be 2×70 V $= 140$ V. The effective resistance of a conducting diode is not constant but depends on the current, but an approximate knowledge of the forward resistance allows one to calculate the power loss in the rectifier (I^2R_{fs}).[6]

Full-Wave Rectifiers. For many applications it is desirable to have a rectifier circuit which supplies current during both half-cycles of the ac power, and thus provides a more continuous current to the load. A "full-wave" rectifier circuit is shown in Fig. 2–182. This circuit is essentially two half-wave rectifiers in parallel with inputs that have a phase difference of 180°. The voltage output of the full-wave rectifier is equal to the voltage developed by each half of the transformer secondary. For a 30 V peak output from a full-wave rectifier, one would use a $(60/1.4) + 1 = 44$ V center-tapped transformer. The extra 1 V is to compensate for the drop in the rectifiers. Note that the rectifiers must withstand an inverse voltage of twice the peak value of the source. For the case above, the peak inverse voltage would be $1.4 \times 44 = 62$ V.

A way to obtain full-wave rectification which does not require a center-tapped transformer is shown in Fig. 2–183. This circuit is called the "bridge rectifier." Trace the current path through the rectifiers. On the positive half-

Note 6. Peak, Average, and RMS Current and Voltage.

In calculating power the rms values are used. The relationships between peak current, I_p, peak-to-peak current, $I_{p\text{-}p}$, average current, I_{av}, and root mean square current, I_{rms}, are

$$I_{p\text{-}p} = 2I_p,$$

$$I_{rms} = 0.707I_p,$$

$$I_{av} = 0.637I_p.$$

See Appendix C for the derivation of these basic relationships.

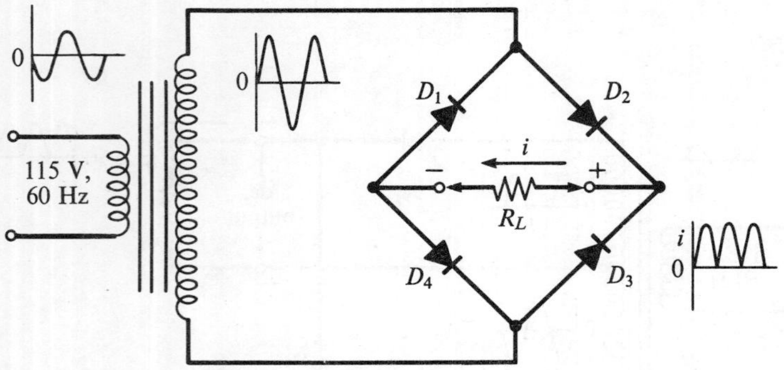

Fig. 2–183 Bridge rectifier circuit.

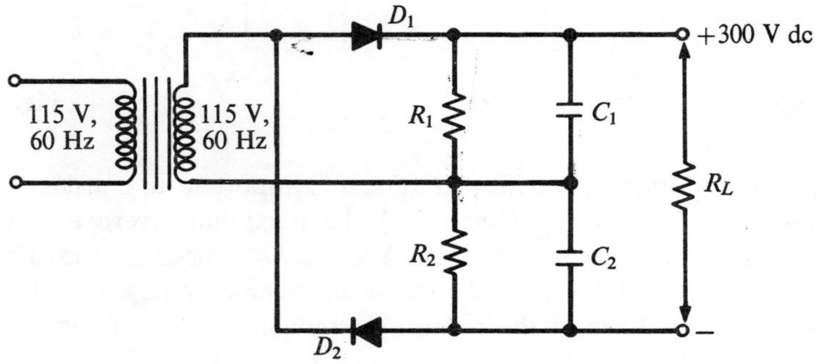

Fig. 2–184 Voltage doubler.

cycle, D_2 and D_4 conduct. On the negative half-cycle, D_1 and D_3 conduct. In each case, the direction of electron flow through the load R_L is the same.

Having two rectifiers in series with the load drops the peak inverse voltage that each rectifier must withstand to the peak value of the supply voltage.

Voltage-Doubler Rectifier. Two rectifiers can be connected to a single ac source and wired so that their outputs are in series as in Fig. 2–184. The output voltage available from such a circuit is twice that available from the ac source with a half-wave or bridge rectifier. For this reason, this kind of circuit is called a "voltage-doubler" rectifier. On the positive half-cycle, capacitor C_1 is charged to the peak value of the supply voltage (in this case $115 \times 1.4 = 160$ V). On the negative half-cycle, C_2 is charged to the same potential. Since C_1 and C_2 are in series across the load, the output voltage is twice the peak voltage of the ac source. The capacitors are essential to the operation of the circuit, because they maintain the voltage developed during one half-cycle so that the voltage developed during the next half-cycle can be added to it. The capacitors C_1 and C_2 have a filtering action which is described in the next section. Since current is drawn from the transformer during both half-cycles, this voltage-doubler circuit is considered to be full-wave. The peak inverse voltage applied to each rectifier is twice the peak value of the supply voltage, in this case 320 V. Where the peak inverse voltage rating of the diode is insufficient, two diodes may be put in series so that their peak inverse voltage ratings are additive.

Filtering the Rectified Voltage

The output voltages of the rectified circuits discussed in the previous section vary with time as shown in Fig. 2–185. The output can be considered as a voltage that varies about the average dc voltage. The average dc voltage is called the dc component of the output. For the half-wave rectifier output, the lowest and predominant frequency of the ac component is the frequency of the ac line as shown by the dotted line in Fig. 2–185a. From Fig. 2–185b it can be seen that the lowest and predominant frequency of the output from a full-wave rectifier is twice the ac supply frequency. For most applications, it is necessary to reduce the magnitude of the ac component of the rectifier output to a value which is very small compared with the average dc potential. The electrical device that accomplishes this task is called a **filter.** The effectiveness of the filter is called the **ripple factor,** r, which is defined as the rms value of the ac component or ripple divided by the average dc voltage; that is, $r = I_{ac}/I_{dc} = V_{ac}/V_{dc}$.

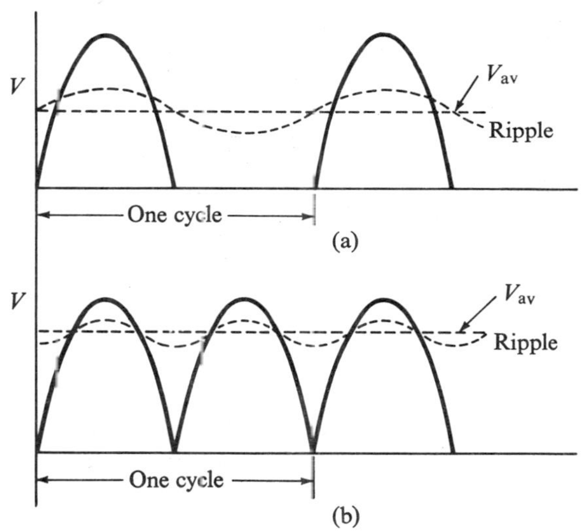

Fig. 2–185 Rectified voltage and fundamental ripple frequency: (a) half-wave and (b) full-wave rectifier outputs.

A rather effective filter is simply a large capacitance in parallel with the load R_L. The capacitor may be thought of as storing charge when the ac component is positive and as discharging through the load when the ac component is negative. Another way to think of the capacitor filter is to consider

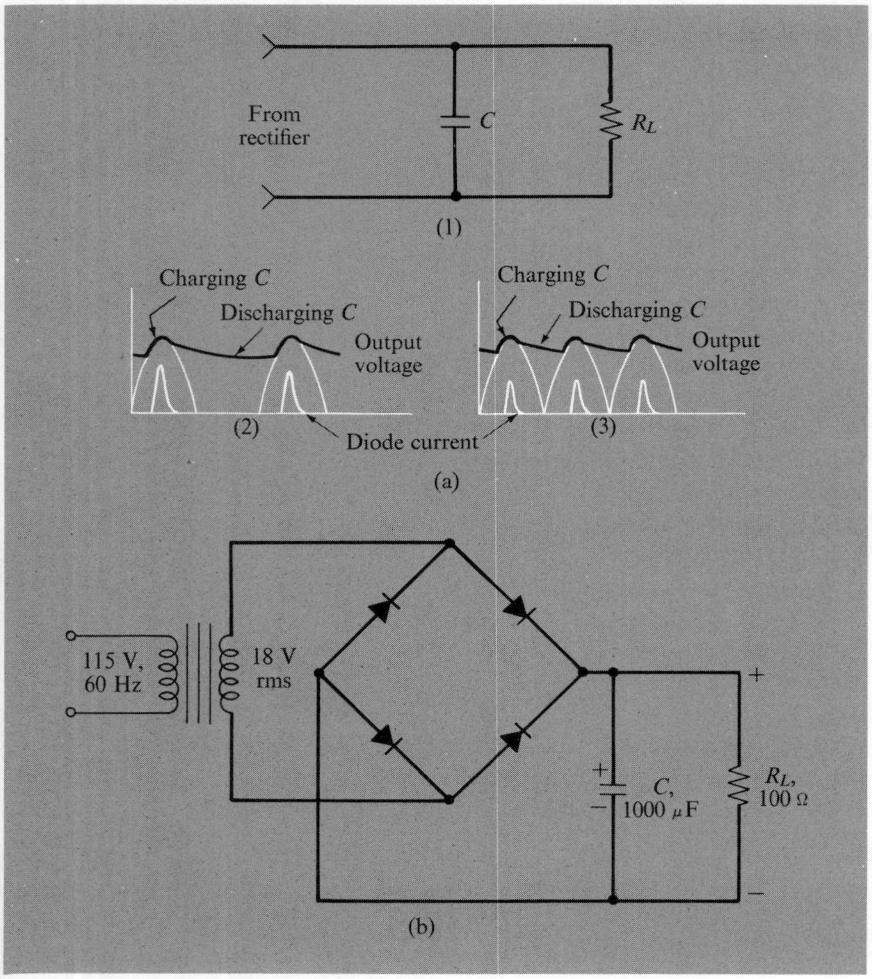

Fig. 2–186 (a) Capacitor filter: (1) circuit; (2) half-wave input; (3) full-wave input. (b) Low-voltage, capacitor-filtered power supply.

that it', resistance for the ripple frequency is very low compared with R_L, and thus it diverts most of the alternating current away from the load circuit.

A more detailed picture of the action of a capacitor filter is presented in Fig. 2–186a. The capacitor charges to the peak value of the input voltage. If R_L were infinite, the voltage across the capacitor would quickly reach a constant value equal to the peak value of the alternating current supplying the rectifier. In the practical case where R_L is not infinite, the capacitor begins to discharge through R_L as soon as the input voltage decreases to that voltage to which the capacitor has charged. This results in the output waveshapes shown in Fig. 2–186a.

It can be seen from Fig. 2–186a that the magnitude of the ripple voltage will be decreased if R_L, C, or the frequency is increased. The expression for the ripple factor, $r = 1/(2\sqrt{3}fCR_L)$, bears this out. Here f is the frequency of the main ac component, equal to the line frequency for half-wave rectifiers and twice the line frequency for full-wave rectifiers. It can also be seen that, as the ripple increases in magnitude, the average dc output will decrease and be approximately

$$V_{dc} = 1.4V_{rms} - \frac{I_{dc}}{2fC}, \tag{2–134}$$

where V_{rms} is the rms value of the rectifier supply voltage and I_{dc} is the average dc current through R_L. The regulation of the capacitor filter improves also with larger values of C and f.

Since $I_{dc} \approx V_{dc}/R_L$, by substituting for I_{dc} in Eq. (2–134), we obtain

$$V_{dc} = 1.4\ \mathrm{V}_{rms}\,[1 - 1/(2fCR_L)]. \tag{2–135}$$

Equations (2–134) and (2–135) are based on assumptions with respect to waveshape and the load. They are not very suitable if the load is sufficient to cause the output voltage to decrease by more than about 20% from the peak voltage (1.4 V_{rms}).

The output voltage and ripple voltage are determined here for the low voltage power source in Fig. 2–186b. Note that, since this is a full-wave rectifier, $f = 2 \times$ line frequency $= 120$ Hz. Inserting the specific values for f, C, and R_L, we obtain

$$V_{dc} = 1.4 \times 18\,[1 - 1/(2 \times 120 \times 1000 \times 10^{-6} \times 100)] = 24.2\ \mathrm{V}. \tag{2–136}$$

The ripple factor r is

$$r = \frac{1}{2\sqrt{3}fCR_L}$$

$$= \frac{1}{3.5 \times 120 \times 1000 \times 10^{-6} \times 100} = 0.024. \tag{2–137}$$

The rms ripple voltage is

$$V_{ripple} = rV_{dc} \approx 0.6\ \mathrm{V}_{rms} \quad (1.7\ \mathrm{V\ peak\ to\ peak}). \tag{2–138}$$

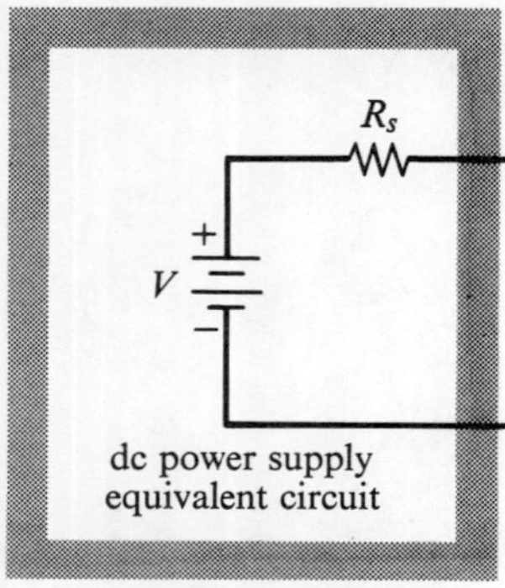

Fig. 2–187 Power supply equivalent circuit.

Another filter device is simply an inductance in series with the load. The inductance, by opposing the changes in current through it, tends to maintain a more constant current through the load. This results in a more constant output voltage. The current tends toward the average value under the influence of the inductance. The voltage across the load then becomes the average voltage output of the rectifier. For a full-wave rectifier and a choke filter, $V_{\text{dc}} = 2/\pi V_{\text{peak}} = 2\sqrt{2}/\pi V_{\text{rms}} \approx 0.9 V_{\text{rms}}$. This expression assumes that the dc resistances of the transformer, rectifier, and choke are negligible compared with R_L. Note that the output voltage is quite independent of the load resistance, making the choke-filtered supply well regulated. The output voltage, however, is considerably lower than with the capacitor filter.

To be effective with a half-wave rectifier, the inductance value must be rather large. The sharp increase in current which occurs at the beginning of the conduction cycle in a half-wave rectifier causes such a large voltage to be induced in the coil that a severe strain is put on the insulation of the windings. For this reason, it is not practical to use chokes to filter the output of a half-wave rectifier.

Combinations of chokes and capacitors make the best passive filters, but the current trend is toward active filtering and regulating circuits for power supplies in precision instruments. Passive and active filters are considered in detail in Modules 3 and 4.

Voltage Regulation

Any power supply can be represented by the Thévenin equivalent circuit (see Section 2–1) shown in Fig. 2–187 with the load R_L connected at its output. The voltage v_o across the load is

$$v_o = V - iR_s. \tag{2–139}$$

Therefore, anything that changes the current i changes the voltage v_o. Obviously, if the load changes, the current changes, and the effect on voltage v_o depends on the relative values of R_L and R_s. Also, anything that changes V will change v_o. Therefore, if the line voltage changes, the voltage across the load changes. In other words, to regulate fully (control) the voltage v_o, it is necessary to compensate for changes of V and/or changes of load current i.

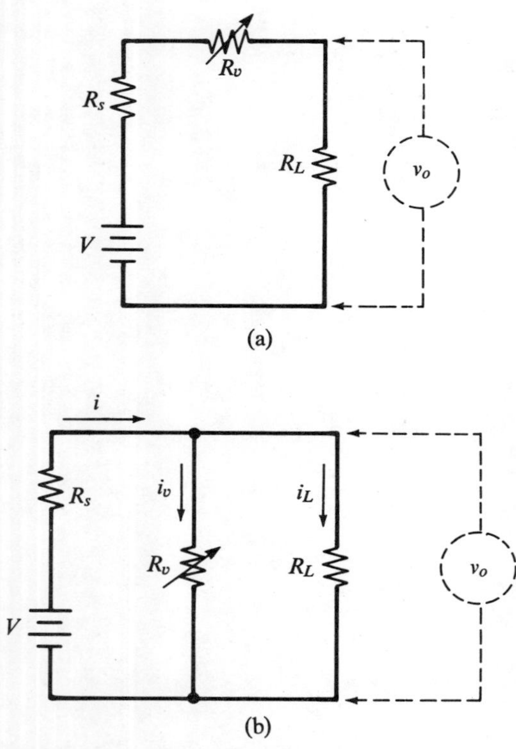

Fig. 2–188 Types of voltage regulators: (a) series type; (b) shunt type.

One method of providing voltage regulation is to introduce a variable resistor R_v in series with the load, as shown in Fig. 2–188a. The output voltage is now

$$v_o = V - i_L(R_s + R_v), \tag{2–140}$$

and in order to provide regulation of the output voltage v_o, the resistance R_v is varied automatically to compensate for changes of V or i_L. There are various control circuits that can automatically change the effective resistance of a series element by monitoring the output voltage and comparing it to a fixed reference voltage. This type of **series regulator** can provide excellent voltage regulation. With the recent developments in integrated circuits the cost of providing series regulation is low. Because of the low cost and excellent voltage regulation, the series type regulator is now very widely used. Specific circuits for series regulators are discussed in Section 2–6.

Another type of voltage regulator is the **shunt regulator,** illustrated in Fig. 2–188b. One of its most important applications is for low-power voltage reference sources where Zener diodes are used as the automatically variable shunts. In the shunt regulator, part of the total supply current i goes through the load, $R_L(i_L)$, and part through a variable resistance, $R_v(i_v)$. Since $v_o = V - iR_s$ and since $i = i_L(R_v + R_L)/R_v$, it follows that

$$v_o = V - \left(i_L \frac{R_v + R_L}{R_v}\right) R_s. \qquad (2\text{–}141)$$

It can be seen that a device that would effectively vary R_v so as to maintain v_o constant at all times would be a valuable regulator. This is the role of a Zener diode in certain regulated supplies. A Zener-regulated supply is shown in Fig. 2–189.

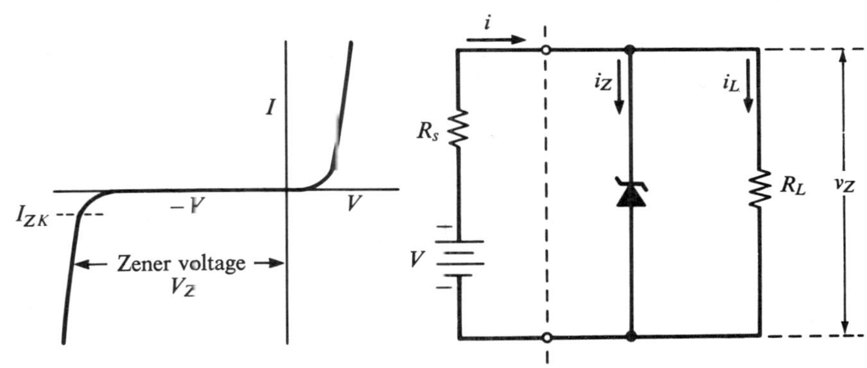

Fig. 2–189 Zener-regulated supply and characteristic curve.

Since the Zener voltage remains essentially constant for a wide range of currents, its effective resistance $R_v = V_Z/i$ changes to compensate for changes of supply voltage V or load current. The diode continues to regulate until the circuit conditions require the Zener current to fall to I_{ZK} in the region of the knee on the diode I-V curve. The power dissipation rating determines the upper limit for the Zener diode current ($P_{max} = I_{Z(max)}V_Z$).

2–4.2 BATTERIES

Despite the availability of high-quality units for the conversion of line-voltage alternating current to direct current in all voltage and current ranges, batteries find a widespread application in modern scientific instruments. The use of batteries as power sources for instruments is essential in remote places, such as outer space, underseas, and in mines, where a central line voltage is not available. Batteries are called for when mobility, portability, and extremely high reliability are required. For some applications, a battery is less expensive and simpler than a power supply that would have the equivalent stability. Also the battery is free from line frequency noise.

A battery consists of several electrochemical "cells" connected to provide the necessary voltage and current capacity. However, it is not unusual to hear the term "battery" used in reference to a single cell. For many years there were only two kinds of batteries readily available that were suitable for light- or moderate-duty applications. These were the familiar carbon-zinc dry cell and the lead-acid storage battery. Today the designer or user of batteries has a choice of half a dozen different kinds of batteries, all commercially available in a wide variety of sizes and voltages. Each of these battery types has characteristics which make it particularly valuable for certain uses. Some new kinds of batteries come in the same case sizes and voltage ratings as the familiar carbon-zinc dry cell, thus allowing the user to choose the best type for his particular application. In this section, the construction, characteristics, and general areas of usefulness in instruments will be discussed for each of the common battery types.

The Carbon-Zinc Dry Cell

This is still the most generally used of all the so-called dry cells. Its structure is shown in Fig. 2–190. During discharge, the zinc (Zn) metal of the can goes into the electroylte as zinc chloride, hydroxide, amine, or oxychloride, leaving two electrons in the zinc can for every atom of zinc dissolved. The MnO_2 in contact with the carbon (C) electrode changes to $Mn_2O_3 \cdot H_2O$, requiring one electron from the carbon for every molecule of MnO_2 converted.

This chemical action establishes a voltage between the two electrodes of about 1.5 to 1.6 V. The carbon electrode is positive, and electrons flow in an external circuit from the zinc can to the carbon rod. Thirty individual cylindrical cells could be connected in series to provide a standard 45 V battery, but this would waste space. Therefore, dry batteries are generally made by stacking a carbon plate, a layer of electrolyte paste, and a zinc plate, alternately, as many times as necessary to give the desired voltage. The most common dry batteries have voltages of 1.5, 3, 6, 7.5, 22.5, 45, 67.5, and 90 V.

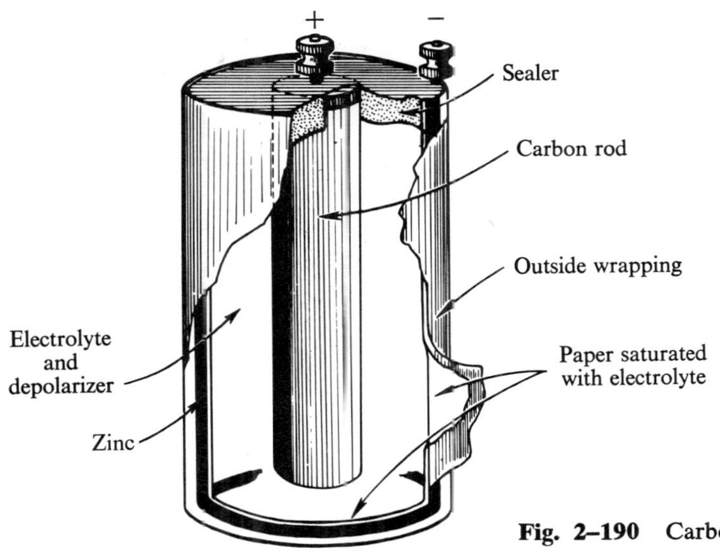

Fig. 2–190 Carbon-zinc dry cell.

The carbon-zinc dry cell is one of several kinds of primary batteries, that is, nonrechargeable, "one-shot" batteries.

The service life of a battery is the number of operating hours during which a fresh battery will satisfactorily operate the actual circuit under normal operating conditions. The service life of a battery can vary widely depending on the following factors: the quality of the battery, the length of time it has been stored before use, its temperature during the storage period, the rate at which it is discharged, the number and duration of the off periods, its temperature during discharge, and the lowest voltage for satisfactory operation of the circuit. After actuation all batteries discharge at some rate internally. This limits the time a dry cell can be stored before use. At room temperature a dry cell could be stored for about a year, but increasing the temperature shortens the shelf life considerably.

It is very important to use a battery which is physically large enough to handle the current drain comfortably. The discharge rate is commonly measured by the number of operating hours necessary to wear out the battery, that is, the service life. For a carbon-zinc cell, a service life of less than 10 hours is considered a heavy drain. In addition to the nuisance of frequent battery changing, the number of milliampere-hours (the capacity) of the battery falls off severely as the rate of discharge increases. Furthermore, the voltage of a battery under heavy drain is low because of the *IR* drop in the cell and the polarization at the electrodes. If a carbon-zinc dry cell must be used under conditions of heavy drain, periodic rest periods will definitely extend the capacity and thus the service life of the battery.

Carbon-zinc dry cells do not work very effectively at temperatures below freezing. Even at temperatures where the cell can still supply a current, the service life may be very low. The potential of the carbon-zinc cell decreases continuously during use. This decrease is due partly to the formation of compounds which interfere with the operation of the battery and partly to the fact that substances in the electrolyte may be involved, to some extent, in the electrode reaction. When the latter is true, the change in electrolyte composition during discharge has a direct effect on the electrode potentials.

The dissolution of the zinc tends to weaken the structure of the cell. Furthermore, during discharge or storage a pressure of evolved hydrogen gas builds up. This can lead to a rupture of the zinc and leakage of the corrosive electrolyte into the instrument. *Instruments using these dry cells should therefore not be stored with the batteries installed.*

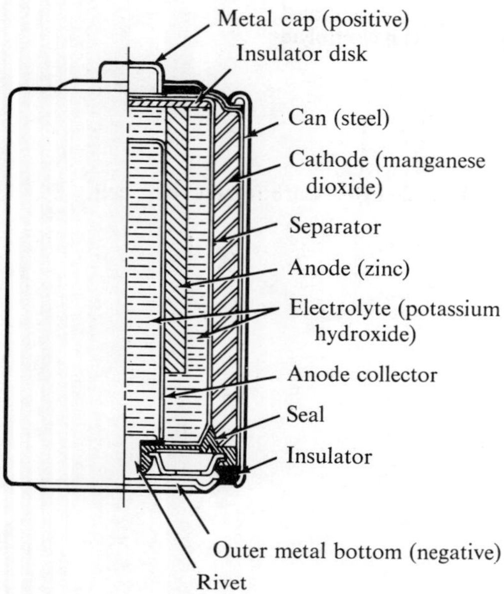

Fig. 2–191 Alkaline-manganese cell.

The Alkaline-Manganese Battery

Commonly called the alkaline battery, the alkaline-manganese battery is quite similar in mechanism to the carbon-zinc dry cell. The electrode reactions are basically the same. The arrangement of the electrodes is different, and the electrolyte is strongly basic. The structure is shown in Fig. 2–191. The changes have resulted in quite a number of improvements over the ordinary carbon-zinc cell. The capacity has roughly doubled. The shelf life has increased. The capacity does not decrease under heavy drain. The internal resistance is lower, and the available current is higher. The low-temperature operating limit has been reduced to 40°F. At roughly four times the price, the alkaline battery will not replace the carbon-zinc cell completely. It does have obvious advantages for heavy drain applications, where the capacity of the conventional cell is greatly decreased. Its longer shelf life and lower operating temperature make it a much better choice for an emergency power source. Its low internal impedance gives it an advantage for circuits where large fluctuations in the current demand occur.

The Mercury Battery

This is a third common kind of primary cell. It has a zinc amalgam for one electrode and mercuric oxide–carbon for the other. During discharge the zinc oxidizes to zinc oxide, yielding two electrons per atom, while the mercuric oxide is reduced to mercury by accepting two electrons per molecule from the carbon electrode. The structure is shown in Fig. 2–192a. The potential developed by this cell is 1.35 V and is remarkably reproducible from one cell to the next. The capacity of this cell is somewhat greater than even that of the alkaline cell, but this capacity is not maintained under heavy drain or at low temperatures (only 58% capacity at 40°F). The

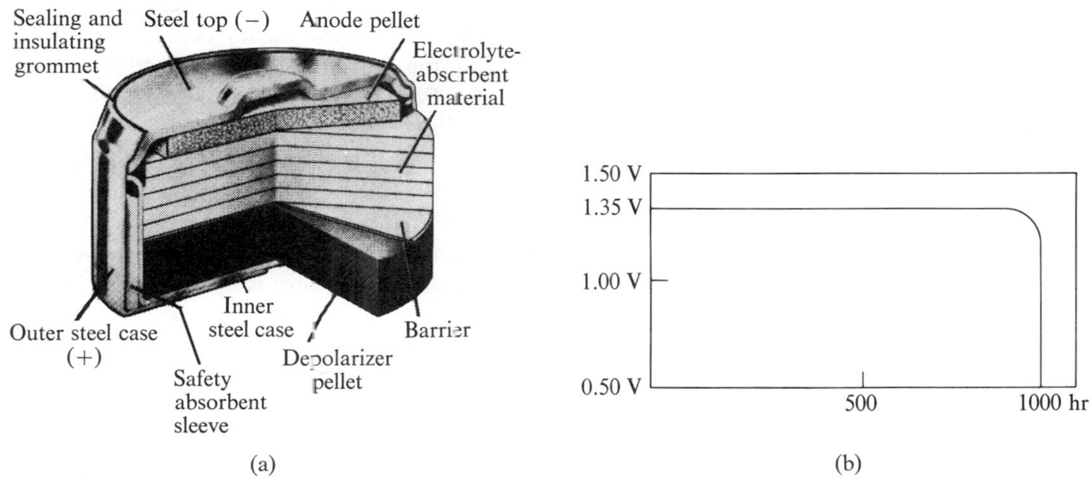

Fig. 2–192 Mercury battery: (a) pictorial; (b) discharge curve.

internal resistance is very low, as with the alkaline battery. In addition to the reproducibility of cell voltages, the great virtue of the mercury battery is its virtually constant voltage during discharge. This is achieved because of the electrode structure and composition and because the electrolyte does not change composition during discharge. The voltage discharge curve (Fig. 2–192b) is so flat that the mercury battery has been used in some older instruments as a secondary voltage standard. No gas is evolved in the mercury battery, and its structure does not readily deteriorate. Because of its high price (about five times that of the alkaline battery for a comparable size), the mercury battery is likely to be used only where a constant potential during discharge is important. In many applications the circuit will not operate satisfactorily at reduced voltages. This requirement shortens the service life of the less expensive batteries so much that the mercury battery is practical economically. Mercury batteries are available in many voltages from 1.35 to 42 V.

The Lead-Acid Storage Battery

The familiar car battery, this is a series combination of several $Pb/H_2SO_4/PbO_2$ cells. Under discharge the lead (Pb) in the lead electrode is converted to insoluble lead sulfate, releasing two electrons to the lead electrode per molecule of lead sulfate created. The lead dioxide in the PbO_2-Pb electrode is reduced to lead sulfate by accepting two electrons per molecule of lead sulfate formed. The chemical reaction which provides the electrical energy is

$$Pb + PbO_2 + 2H_2SO_4 \rightarrow 2PbSO_4 + 2H_2O.$$

When fully charged, the potential of each cell is 2.06 to 2.14 V. The lead-acid battery is a **secondary** cell; i.e., it can be recharged. When a reverse or charging current is forced through the cell, the $PbSO_4$ is converted to Pb and PbO_2 respectively, so that the cell is returned to very nearly its original state.

The reaction shows that sulfuric acid is converted to water on discharge. This dilution of the sulfuric acid causes the cell voltage to decrease during discharge. The decrease in voltage is slow at first but becomes quite rapid during the last one-third of the service life. The internal resistance of the lead-acid battery is so low that it can be ignored at normal discharge rates. The high current free from ac ripple makes it the most economical power source for certain applications. However, the lead-acid battery is bulky and heavy and requires a considerable amount of care if it is to give proper service.

The Nickel-Cadmium Battery

The nickel-cadmium battery has come into widespread use in recent years. One reason for its growth in popularity has been the development of the sealed nickel-cadmium cell. This is a completely sealed unit which requires no attention other than charging. Two forms of this cell are shown in Fig. 2–193a. Under discharge the cadmium (Cd) is oxidized, supplying electrons, while at the positive electrode, nickel oxide is reduced to a lower oxidation state by accepting electrons. The open-circuit potential of this cell is 1.3 V. The electrolyte is not involved in the electrode reaction, and thus the potential is fairly constant over the service life. The capacity of the nickel-cadmium battery is reduced very little even at very high discharge rates. The batteries can be stored charged or discharged without harm. The sealed units will not leak. They have the lowest self-discharge rate of any secondary cell. Sealed nickel-cadmium batteries are made in many sizes and voltages which make them interchangeable with the primary batteries. To all the areas of the application of conventional batteries, nickel-cadmium batteries offer the advantages of high current capability, long service life, reasonably constant potential, and the possibility of recharge. These characteristics are particularly suited to instruments where heavy drain and frequent use would require frequent battery replacement. The price is about 25 times that of a similar-sized carbon-zinc dry cell, and there is some concern about cadmium pollution in the environment.

During the charging process, the nickel oxide is reoxidized to its higher oxidation state and the cadmium oxide is reduced. The cell is constructed with some oxidized cadmium so that the nickel electrode is the first electrode to reach full charge. If the cell is overcharged, the oxygen gas produced at the nickel electrode diffuses to the cadmium electrode, where it is reduced.

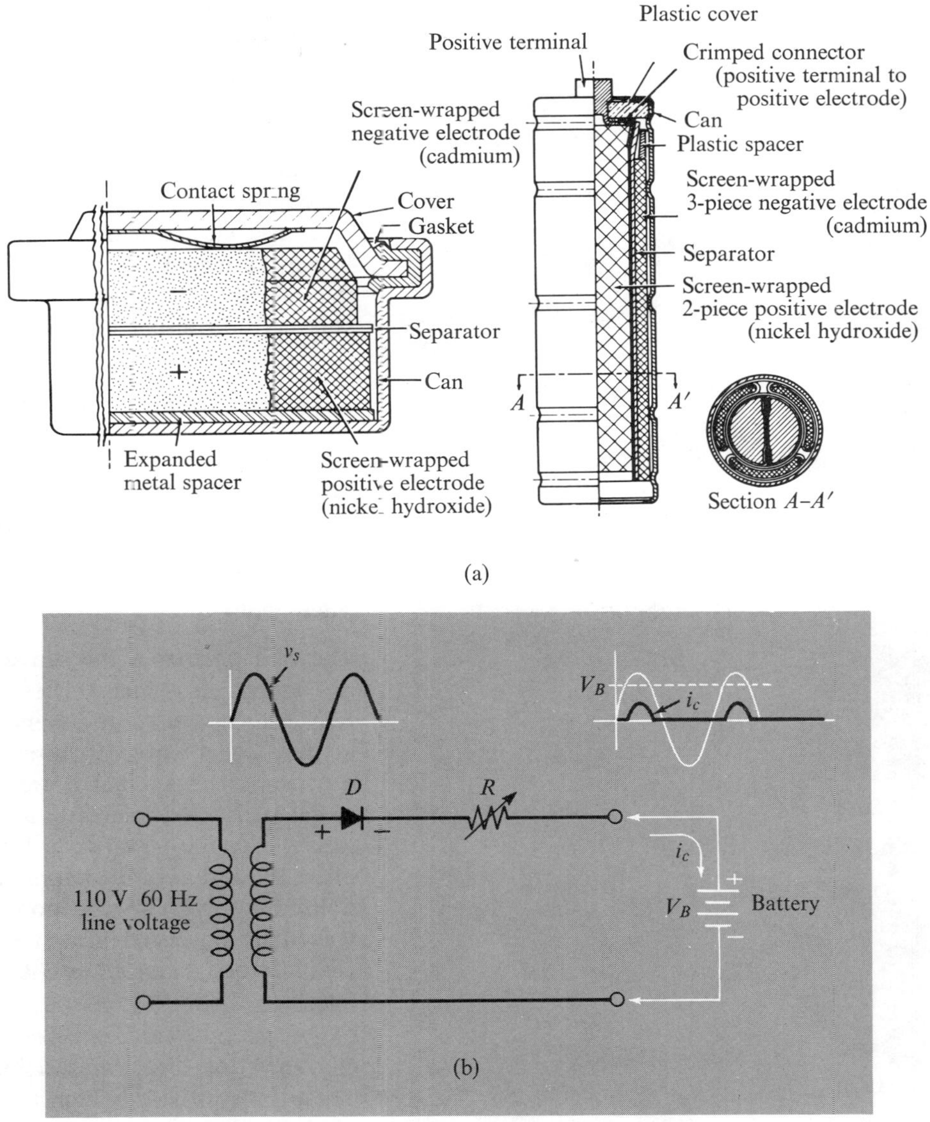

Fig. 2–193 Rechargeable nickel-cadmium cell: (a) sealed button cell at left and cylindrical cell at right; (b) battery charger for rechargeable cells.

This provides a "chemical short circuit" for excess charge and prevents the buildup of a large gas pressure.

Battery Charger

Rechargeable batteries such as the nickel-cadmium battery can be readily charged from the ac power line by use of a half-wave rectifier circuit as shown in Fig. 2–193b. The battery is charged by reversing the electrode reactions which occur when the battery is a voltage source. A dc charging current, i_c, is obtained by introducing a diode in series with the battery and applying a sufficiently high voltage, v_s, at the secondary of the transformer so that the diode conducts during part of the ac cycle. When the magnitude v_s of the sine wave exceeds the battery voltage, V_B, the diode conducts and the charging current, i_c, flows in the correct direction so as to force the electrochemical reactions to occur which recharge the battery. When the magnitude of v_s is less than V_B the diode is reverse biased, and this effectively opens the circuit so that no discharge current occurs. By varying the resistance R, the rate of charging the battery can be adjusted.

2–4.3 SOLAR AND FUEL CELLS

Although not generally applicable now, both solar cells and fuel cells hold considerable promise for the direct conversion of sunlight and chemical energy, respectively, into electrical power. These cells would be especially useful for instrumentation in remote locations.

Solar Cells

The radiation from the sun to the earth each day represents an amount of energy that is many times our yearly consumption of electrical energy. It is estimated that on a summer day the solar radiant power is 1 kW/m^2 at the earth's surface. Unfortunately there is no efficient way of directly converting this radiant power into electrical power. Of course, the solar radiation is used indirectly to obtain electrical power. The sun's radiation causes evaporation of water from the earth's oceans and other major bodies of water, and the rain returns the water to the mountains, where it can be utilized by hydroelectric power plants. Also, the sun's radiation was involved in the chemical and biological processes that resulted over many years in the formation of coal and oil, which are used in fossil fuel power plants.

In recent years, **semiconductor solar cells** have been used with fair efficiency for direct conversion of solar radiation into electrical power. These cells provide a high power capacity per unit weight and have been used successfully on space vehicles and communication satellites. A representation of a silicon photovoltaic cell is shown in Fig. 2–194a. A 1 × 2 cm silicon wafer has been used to build up single-panel arrays of many thousands of wafers so as to provide a capacity of several hundred watts.

A very thin layer of p-type semiconductor is formed on an n-type silicon strip (or vice versa) to form a pn junction and to provide the light-sensitive face. A narrow conducting strip serves as the collector term-

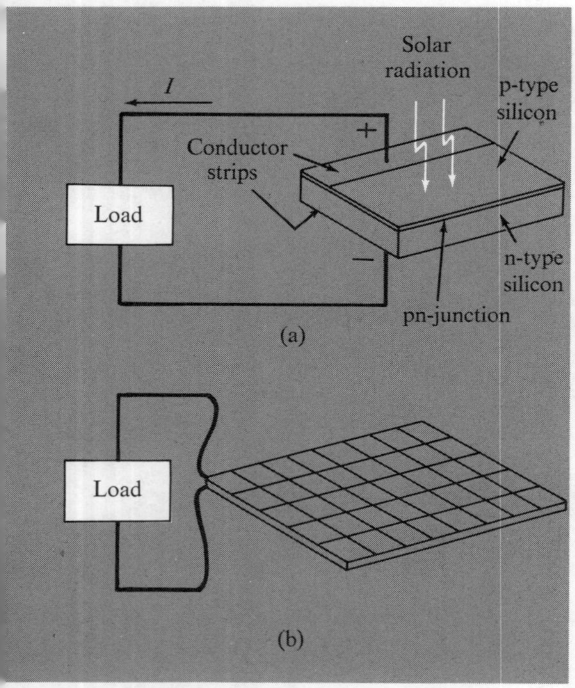

Fig. 2–194 Solar cells: (a) silicon cell; (b) an array of forty 1 × 2 cm silicon cells.

inal. The bottom of the cell is nickel plated and tinned. The mechanism of conversion from solar radiation to electrical power can be understood by recalling the discussion of the pn diode in Section 2–3. The diffusion of majority carriers from both the n- and p-type materials causes a combination of holes and electrons that sets up a barrier potential across the junction. The p- and n-type materials contain many valence electrons that are not affected by the impurity doping. However, a photon of light can interact directly with a valence electron and provide sufficient energy to promote it to a conductance electron. The free electron and the hole are now attracted by the barrier potential at the pn junction, and will travel in the opposite direction to the majority semiconductor carrier. That is, the photoelectrons will travel toward the positively charged n-type silicon and the holes toward the negatively charged p-type silicon so as to provide a current I through a load as shown in Fig. 2–194a.

A series-parallel array of forty 1×2 cm cells as shown in Fig. 2–194b could provide about 1 W electrical power for 10 W of incident sunlight. The satellite arrays generally provide about 10 W output per pound. In view of the huge supply of silicon and sunlight, there should be increased application of solar cells as the technology improves.

Fuel Cells

The direct conversion of chemical energy in a fuel into electrical power is now practical on a small scale by operation of hydrogen-oxygen fuel cells. The Gemini spacecraft utilized H_2-O_2 fuel cells of the type shown in Fig. 2–195. There are two separate chambers and two porous electrodes separated by an electrolyte. The hydrogen gas in the top chamber diffuses through electrode A and reacts in the electrolyte (in the presence of a catalyst) to produce hydrogen ions and free electrons ($2H_2 \rightarrow 4H^+ + 4e^-$). The hydrogen ions move through the electrolyte to electrode B and there they combine with oxygen and electrons to form water ($4H^+ + 4e^- + O_2 \rightarrow 2H_2O$). The reactions at electrodes A and B cause a flow of current I through the load, as shown in Fig. 2–195.

The theoretical efficiency of greater than 90% and already attained efficiencies of over 80% with laboratory cells, low cost, and quiet operation without exhaust are major reasons for the great interest in this type of direct chemical-to-electrical converter.

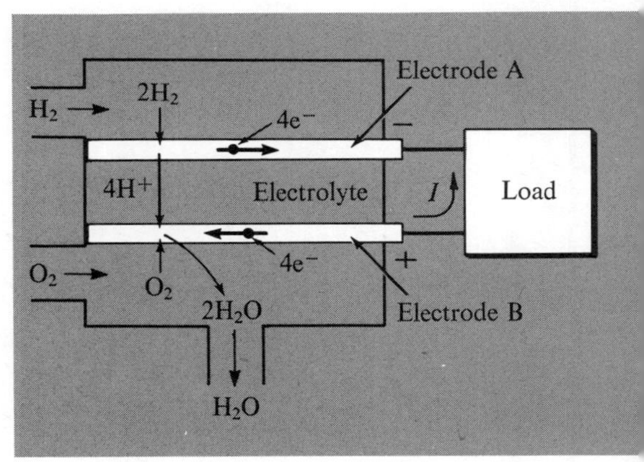

Fig. 2–195 Hydrogen-oxygen fuel cell.

2–4.4 ELECTRICAL POWER DISSIPATION

When charge moves through a resistive element, the power loss in the element appears as heat. The product of the voltage drop across the resistive element and the current through it is equal to the power dissipated in the device, as expressed in Eq. (2–142):

$$P \text{ (watts)} = V \text{ (volts)} \times I \text{ (amperes)}. \qquad (2\text{–}142)$$

Since $V = IR$, Eq. (2–142) can be rewritten as

$$P = I^2R \qquad (2\text{–}143)$$

or

$$P = \frac{V^2}{R}. \qquad (2\text{–}144)$$

In building power supplies or in connecting loads at the output of power supplies, it is important not to exceed the wattage rating of resistors or other devices. For example, suppose that a 47 kΩ resistor is rated at 1 W. The maximum current through this resistor is thus $\sqrt{1/47{,}000} = 5 \times 10^{-3}$ A, or 5 mA. The maximum voltage that can be applied to the resistor is $\sqrt{1 \times 47{,}000} = 230$ V. A resistor that is dissipating its **maximum rated power** can be hot enough to cause burns and must be well ventilated. For cooler and therefore more stable resistance values, a maximum rated power two to ten times greater than the power dissipation is recommended.

Some devices, such as light bulbs, are rated according to the **power normally consumed** rather than the maximum rated power. For these, the power rating reveals other electrical characteristics.

Experiment 2–30 *Instrumentation Power Supplies*

This experiment illustrates how a Zener-regulated power supply can be built from a rectifier circuit, a filter circuit, and a Zener regulator circuit. Half-wave and full-wave rectifier circuits are first studied and then a capacitor filter is added to a full-wave bridge rectifier. The effects of the *RC* time constant of the filter on the output waveform are noted. A Zener regulator is then added and its characteristics and limitations are studied.

PROBLEMS

1. By what percentage will the secondary voltage of a transformer be reduced if the line voltage decreases from 115 V ac to 95 V ac?
2. A transformer on a power pole transforms the voltage from an efficient transmission value of 4400 V to a safe working value of 220 V. What turns ratio must the transformer have?

3. What is the minimum peak inverse voltage (PIV) rating for a diode to be used in a half-wave rectifier circuit if a transformer is used which has 10 V rms on the secondary, and a large filter capacitor is placed across the load?

4. (a) What are the minimum PIV ratings for diodes to be used in a full-wave rectifier circuit if a transformer is used which has 5 V rms on either side of the center tap and if a large filter capacitor is placed across the load? (b) What are the minimum PIV ratings for the diodes and the voltage rating for the transformer to be used in a bridge rectifier circuit which supplies the same voltage to the same filtered load as in part (a) above?

5. Various rectifier circuits are being considered for use with a transformer which has 25 V rms on the secondary. If it is assumed that the transformer and the diodes are ideal and the load is negligible, what are the maximum output voltages available for the following rectifier circuits with a capacitor filter: (a) half-wave; (b) full-wave bridge; (c) voltage doubler?

6. Consider the power supply of Fig. P–6 and assume that the diodes and transformer are ideal. Calculate (a) the output voltage; (b) the current through the 5 kΩ load; (c) the rms ripple voltage.

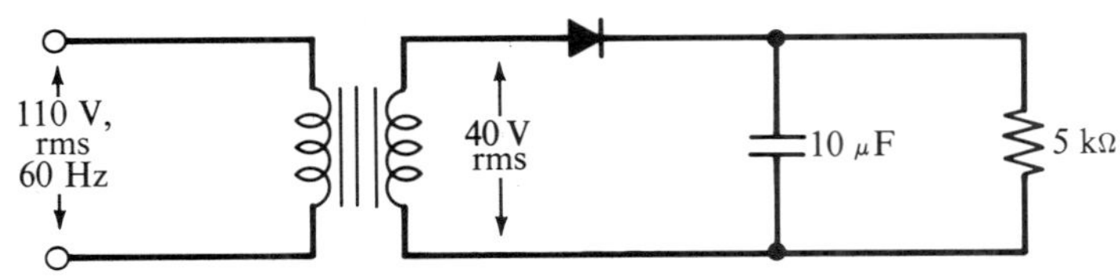

Fig. P–6

7. The power supply of Fig. 2–186b is considered for various values of the filter capacitor C. Calculate the output voltage and ripple voltage for $C = 10$ μF, 100 μF, 500 μF, and 1000 μF.

8. A Zener diode is used in the low-voltage power supply shown in Fig. 2–189. The Zener voltage is 5.1 V and the Zener diode can dissipate a maximum power of $P_{max} = 400$ mW. Calculate the required value of R_s if $V = 10$ V, $R_L = 100$ Ω, and the power dissipation is to be 100 mW.

9. A 50 Ω resistor has a maximum power rating of $\frac{1}{2}$ W. What is the maximum current through this resistor and the maximum voltage which can be applied without exceeding the power rating?

Experiment 2–30 *Instrumentation Power Supplies*

Equipment: HVPS or LVPS
breadboard system
four rectifier diodes
resistors
10 kΩ potentiometer
Zener diode (≈6 V)
capacitors
VOM
oscilloscope

This experiment illustrates how a Zener-regulated power supply can be built from a rectifier circuit, a filter circuit, and a Zener regulator circuit. Half-wave and full-wave rectifier circuits are first studied and then a capacitor filter is added to a full-wave bridge rectifier. The effects of the *RC* time constant of the filter on the output waveform are noted. A Zener regulator is then added and its characteristics and limitations are studied.

E2–30.1 Half-Wave Rectifier

Because the power supply patch card will be used in subsequent experiments, it is recommended that this card be inserted into the breadboard chassis and the power cable connected to one of the supplies that provides 6.3 V output, such as the EUW-15 and EU-40 supplies, as shown in Fig. E2–56.

The 6.3 V ac secondary terminals of the transformer of the power supply go to the power patch card (PPC) terminals marked FILAMENT (25–28 for one and 17–20 for the other), and the center tap goes to the terminals marked COMMON, numbers 9–12, except for the EUW-15 power supply, which has the center tap connected to connectors 3, 4, and 5, marked HV on the power patch card. These connections are shown in Fig. E2–56.

Caution: When using the EUW-15 supply, disconnect the high voltage (HV) jumper (if connected) so that the high voltage will not be connected to the power patch card for this set of experiments, and keep the HV switch OFF. Always turn the complete power supply OFF when connecting circuits or replacing the power card. Note that for the EUW-15 supply the HV goes to terminal 3 of the octal socket, which goes to PPC terminals 9–12 marked COMMON, and that the power supply common goes to the HV terminals of the PPC.

Insert the power card, multiple connector card, and the breadboard card in the three-card chassis (EU-51), as illustrated in Fig. E2–57, and connect the half-wave rectifier circuit shown in the schematic diagram in Fig. E2–58.

Record below the output waveform for the half-wave rectifier circuit and measure the amplitude on the oscilloscope. Reverse the connections of the diode and again record the waveform and compare the two cases.

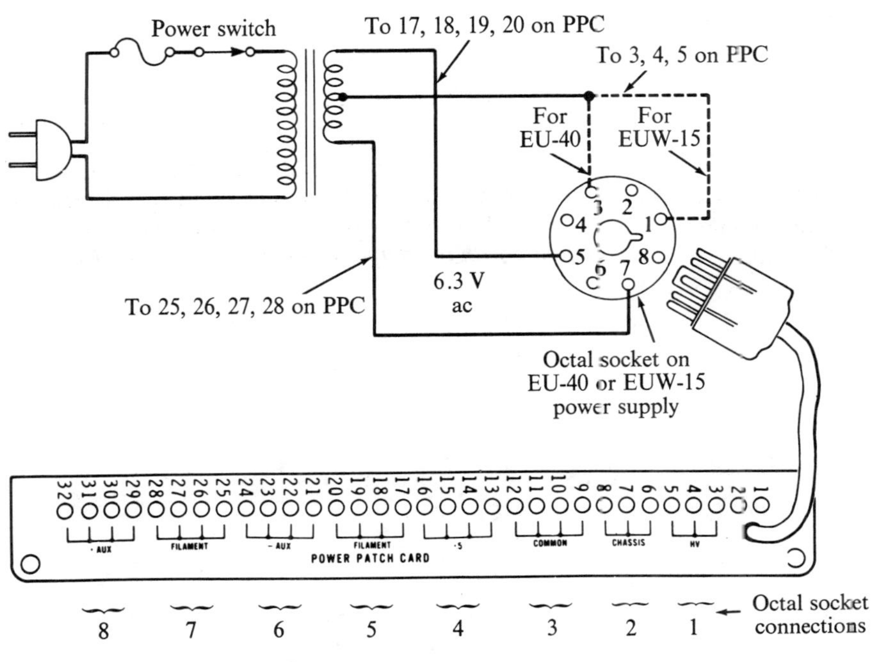

Fig. E2–56 Connections from octal socket of EUW-15 supply to power patch card (PPC).

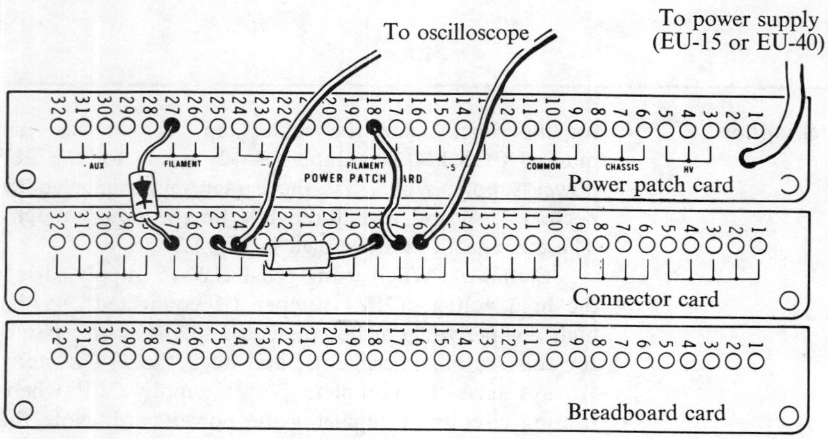

Fig. E2–57 Arrangement of cards in breadboard chasis.

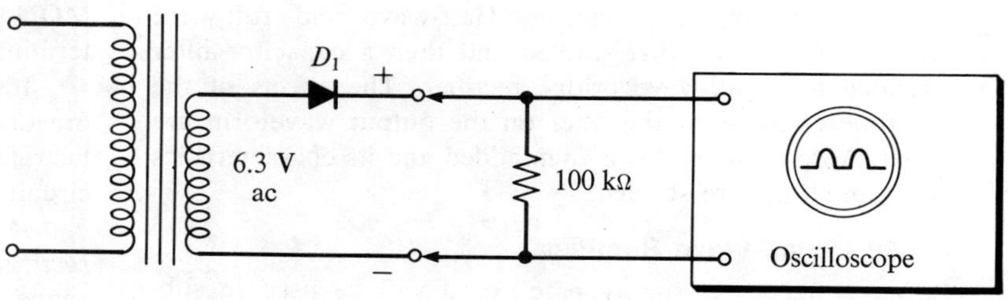

Fig. E2–58 Half-wave rectifier.

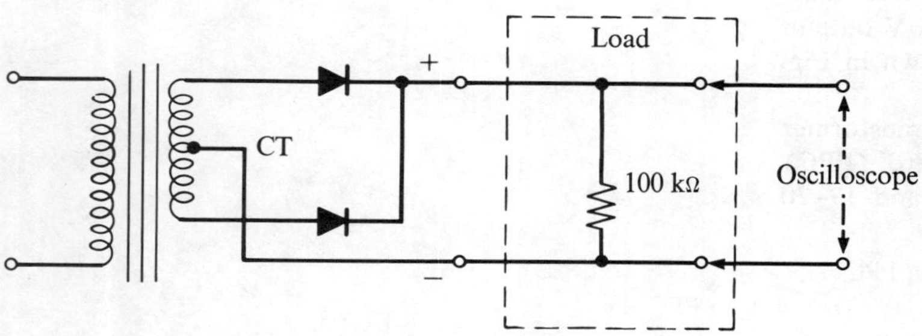

Fig. E2–59 Full-wave rectifier.

E2–30.2 Full-Wave Rectifiers

Wire the full-wave rectifier shown in Fig. E2–59. *Record the output waveform below and the measured amplitude. Compare the results with the half-wave rectifier results.*

Wire the full-wave bridge rectifier as shown in the schematic diagram of Fig. E2–60a and the pictorial of Fig. E2–60b. *Record the output waveform and magnitude and compare with the output of the rectifier in Fig. E2–59.*

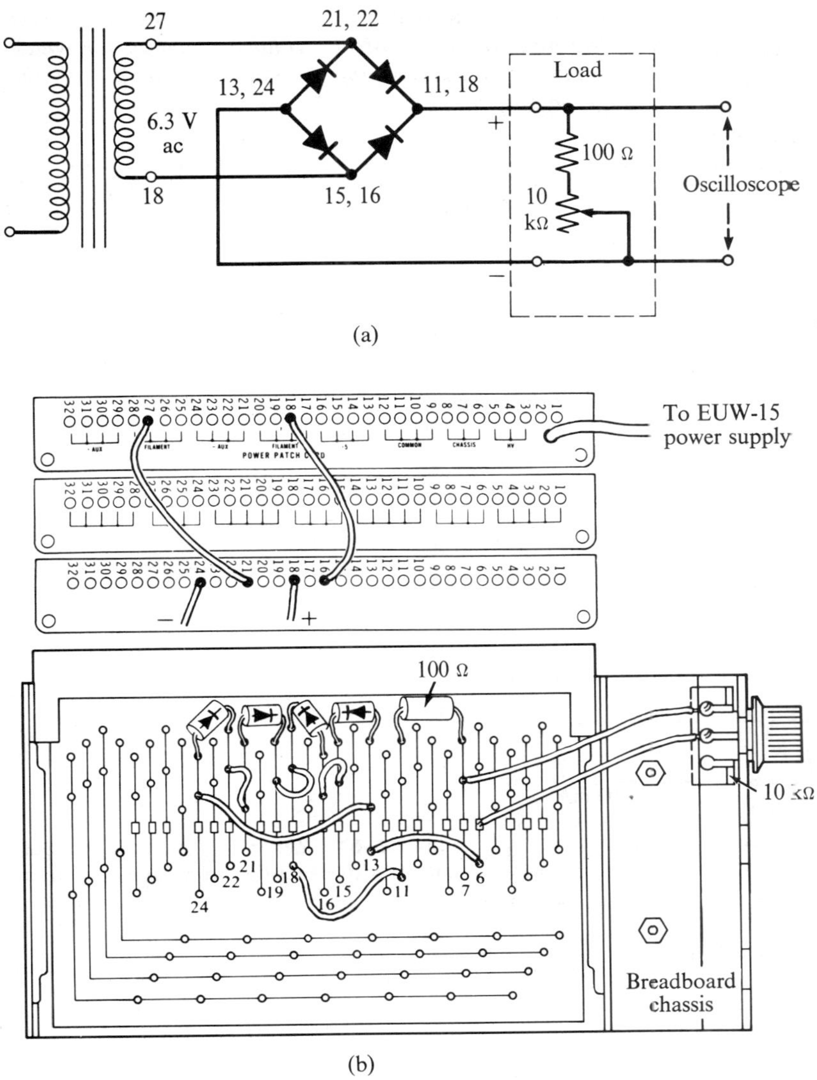

Fig. E2–60 Bridge full-wave rectifier: (a) schematic diagram; (b) pictorial of connections to breadboard card.

Note: Keep the bridge rectifier circuit connected for subsequent experiments.

E2–30.3 Capacitor Filter for Full-Wave Rectifier

Connect a 20 μF capacitor across the output of the bridge rectifier and observe on the oscilloscope the effect of changing the load current from 1 to 50 mA. Be certain to connect the electrolytic capacitor with the correct polarity (see Fig. E2–61 for polarity). *Measure the ripple frequency and ripple and dc voltages and determine the ripple factor.*

Change the filter capacitor to 100 μF and *compare its effectiveness to that of the* 20 μF *capacitor.*

E2–30.4 Zener Diode Regulator

Connect the Zener diode voltage regulator at the output of the bridge rectifier as shown in Fig. E2–61. *Record below the effect on the Zener regulator of changes of load and input voltage, and the reduction of ripple. Determine the range of load for which the regulation is within 2% of the average output voltage.*

If a Variac is available, plug the supply into the Variac and *determine the range of line voltage variation for which the output of the Zener supply remains within 2% of the average dc output.*

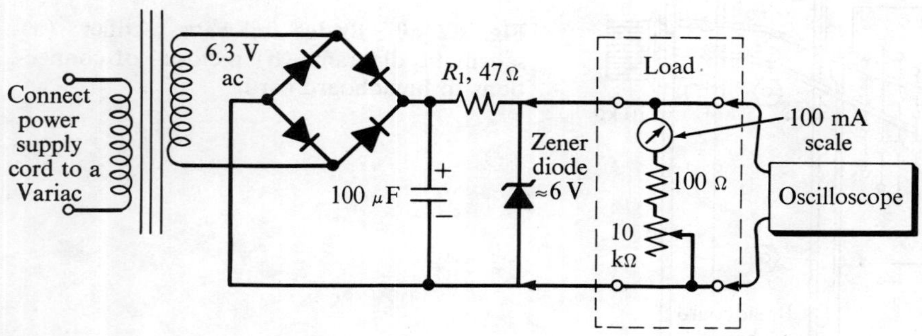

Fig. E2–61 Zener regulator at the output of bridge rectifier with capacitor filter.

Gain Control of Voltage and Current

Section 2–5

The sequential or simultaneous control of events by electronic switching, often at fantastically high speeds, and the control of electrical signals with resistive, capacitive, and inductive networks, were described in the previous sections of this module. Combinations of these basic control operations were shown to provide important functions such as electronic waveform generation and shaping, integration, differentiation, and log conversion. For many of these previously described applications, amplifiers were assumed to be available with the required characteristics, and were used as **gain blocks.** That is, in those cases where voltage and/or current gain of an electrical signal were necessary, an amplifier functional block was inserted that could meet the gain requirements and it was represented in diagrams by a triangular symbol. Fortunately, there are now on the market many inexpensive general purpose amplifiers that come in small prepackaged units and fulfill specific amplification (current and/or voltage gain) needs. Unfortunately, however, all amplifiers have limitations and these limitations can cause problems unless they are recognized and allowed for.

From our own experience we have found that the misuse of amplifier functional blocks can be most easily avoided by the application of some basic and practical understanding of how the control of voltage and/or current gains is achieved with amplifiers. Therefore, many of the basic considerations of ideal and practical amplifiers are presented. In Modules 3 and 4, certain of these considerations, such as frequency response of amplifiers, are discussed in more detail. Some of the most frequently encountered basic amplifiers are presented here so as to acquaint the reader with the terminology and methods of circuit modeling. Several of these basic amplifiers are then combined to form a dc instrumentation amplifier or gain block of the general type that has been used (and represented with a triangular symbol) to perform many of the measurement and control functions previously reported.

2–5.1 IDEAL AMPLIFIERS

Both voltage and current amplification are important in measurement and control applications in instrumentation systems. Often the voltage signal from a transducer or other source is so small that it needs to be increased in magnitude by a **voltage amplifier** before subsequent information processing. In other cases the magnitude of the voltage is sufficient but the current signal needs to be increased by a **current amplifier,** which also provides the necessary **power gain,** so that the signal can control a motor, relay, meter, or other device. Although gain blocks with ideal characteristics are not available, it is often helpful to consider what constitutes an ideal device for a specific purpose. Therefore, in this section both ideal voltage and current amplifiers will be considered and illustrated with circuit models.

An Ideal Voltage Amplifier

First let us examine the characteristics of an ideal voltage amplifier (voltage-controlled source), in which the output voltage v_o is directly proportional to the input voltage v_{in} at every instant, and μ is the proportionality constant (called the voltage amplification factor). The input and output of the amplifier are related as shown in Eq. (2–145), and represented in the circuit model of Fig. 2–196.

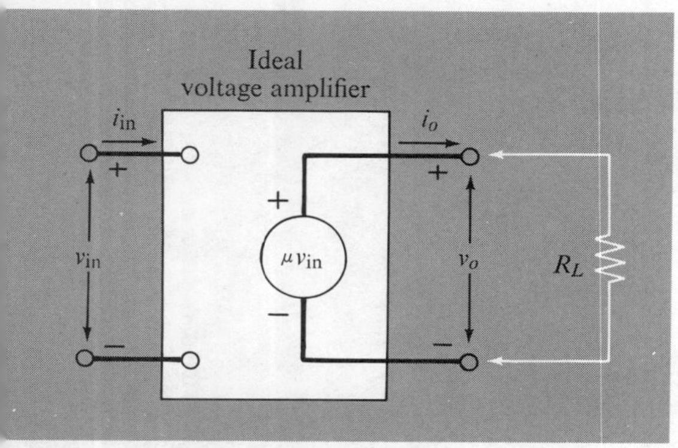

Fig. 2–196 Circuit model of an ideal voltage amplifier.

$$v_o = \mu v_{\text{in}}. \tag{2–145}$$

There are several significant conclusions that can be reached from considering the ideal voltage amplifier. The input terminals are isolated from the output terminals. That is, the electrical connections at the output are completely independent of the electrical connections at the input, as illustrated in Fig. 2–196. The input resistance is infinite (open circuit), so that the input current $i_{\text{in}} = 0$. Therefore the input power is zero and there is no loading of the input voltage source. The output power, P_o, is

$$P_o = \frac{v_o^2}{R_L} = \frac{(\mu v_{\text{in}})^2}{R_L}. \tag{2–146}$$

Therefore in the ideal voltage amplifier the ratio of output power to input power (i.e., power gain) is infinite.

Ideally, as is also implied in Eq. (2–145), there are no restrictions on the output current. In other words, as the load changes from no load (open circuit) to a short circuit the output voltage remains constant, and this implies that the output resistance is zero.

When the output load is fixed at R_L, it can be seen from Eq. (2–146) that the output power depends on the input voltage. Therefore, if μ is fixed and v_{in} is small, the output power and the voltage might be insufficient for operation of a specific load. In this case several voltage amplifiers can be used in cascade, the output of the first amplifier being connected to the input of the second amplifier, and the load connected to the output of the second amplifier.

An Ideal Current Amplifier

A current-controlled current source or **current amplifier** would be ideal if the output current i_o remained proportional to the input current i_{in} at every instant. The proportionality constant β is called the current amplification factor, and the relationship is expressed in Eq. (2–147):

$$i_o = \beta i_{in}. \tag{2–147}$$

The circuit model of an ideal current amplifier is shown in Fig. 2–197. The input current is applied through a zero input resistance so that the input voltage is zero and the input power ($P_{in} = i_{in}^2 R_{in}$) is zero. Also, the input terminals are perfectly isolated from the output terminals. It can be seen from Eq. (2–147) that the output current is independent of the output voltage and load, and that the output power ($P_o = i_o^2 R_L$) varies with the load. Therefore the output power gain of an ideal current amplifier is infinite.

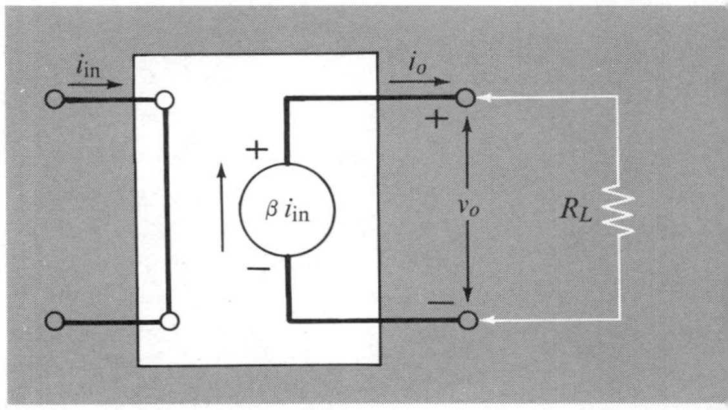

Fig. 2–197 Circuit model of an ideal current amplifier.

Having considered the basic characteristics of ideal voltage and current amplifiers, we will now turn our attention to various amplifier types and become familiar with basic amplifier terminology. Although practical amplifiers fall short of being ideal, field-effect transistor amplifiers approximate ideal voltage amplifiers, while bipolar junction transistor amplifiers can often be treated as ideal current amplifiers.

2–5.2 BASIC AMPLIFIERS

There are many types of basic amplifiers, only a few of which are presented in this section. However, by describing a basic junction field-effect transistor (JFET) amplifier and a basic bipolar junction transistor (BJT) amplifier it is possible to introduce much of the electronic amplifier terminology and circuit modeling concepts that are important in reading the electronics literature. These basic presentations are followed by discussions of the emitter-follower amplifier and the difference amplifier, both of which are frequently found in all types of instrumentation systems. With this brief

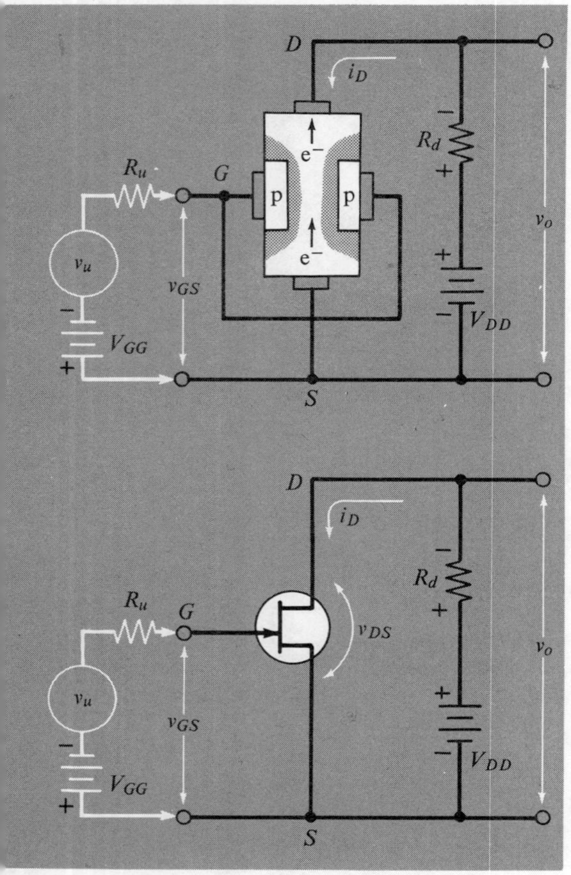

Fig. 2–198 Basic n-channel JFET amplifier: (a) pictorial diagram; (b) schematic diagram.

background in amplifiers it is possible to understand how the basic devices are combined to form a general purpose operational amplifier or gain block that has almost unlimited application for measurement and control functions. This combination of devices to form a general purpose gain block is the subject of the next section.

A Basic JFET Amplifier

The basic structure of a JFET was described in Module 1, and its application as a switch was presented in Section 2–3 of this module. In Fig. 2–198 a silicon n-channel JFET is shown connected in a basic common-source amplifier configuration. The term **common-source** implies that the source terminal of the JFET is common to both the input and output terminals of the amplifier.

The majority carriers (electrons for the n-channel silicon FET shown in Fig. 2–198) flow from the source S to the drain D because of the applied voltage V_{DD}. The gate (G) junctions are usually reverse biased (in this example by a voltage source V_{GG}) and only very small currents (10^{-8} to 10^{-13} A) exist at the gate terminals. Therefore, the voltage drop across the resistance R_u of the input source is usually negligible, and the instantaneous voltage between gate and source, v_{GS}, is essentially equal to the sum of the signal input voltage and the bias voltage, $v_{GS} = v_u + V_{GG}$. The gate voltage v_{GS} controls the current, i_D, between the drain and source of the JFET. Thus v_{GS} also controls the voltage drop, $i_D R_d$, across the drain load resistor R_d, which is connected between the drain terminal of the JFET and the positive lead of the drain voltage supply V_{DD}.

The output voltage, v_o, is equal to the difference between the supply voltage V_{DD} and the voltage drop across R_d caused by the drain current, i_D, so that

$$v_{DS} = v_o = V_{DD} - i_D R_d. \tag{2–148}$$

The linear equation (2–148) relates the output voltage and device current, and this can be drawn as a current-voltage line on a set of JFET characteristic curves. This method of drawing a load line on a set of characteristic curves was first introduced in Section 2–3 for switching circuits. In this case it provides a means of graphically observing how an input signal is amplified by a specific JFET, as illustrated in Fig. 2–199. It should be noted that the input signal varies about an average value of V_{GS}, so that when input signal $v_u = 0$, there is an average dc drain current I_D which causes an average dc drain-source voltage V_{DS}. These average values are known as the **quiescent voltages and current**, and they describe an average or **quiescent operating point** about which the signal varies. The quiescent values are designated by capital letters.

A JFET Small-Signal Circuit Model

In analyzing the operation of practical amplifier circuits it is convenient to use circuit models. In developing these models for amplifiers one generally assumes that a linear equivalent circuit is a valid approximation over small portions of the operating range. That is, the circuit model will describe the relationship of the small changes in the amplifier current and voltage caused by a small input signal.

For the JFET amplifier the application of an input signal causes the instantaneous circuit values of v_{GS}, i_D, and v_{DS} (Fig. 2–199) to differ from the quiescent values V_{GS}, I_D, and V_{DS}. The differences between the actual values and the quiescent values are called the signal values and are given the symbols v_{gs}, i_d, and v_{ds}. In other words $v_{gs} = v_{DS} - V_{DS}$, where v_{gs} is the change in gate-to-source voltage which causes a change in drain current $i_d = i_D - I_D$, which causes an output signal $v_{ds} = v_{DS} - V_{DS}$.

By noting that the actual circuit quantities v_{GS}, i_D, and v_{DS} are interdependent variables, and choosing i_D as the dependent variable, the relationship among i_D and v_{GS} and v_{DS} can be written as a general function as shown in Eq. (2–149):

$$i_D = f(v_{GS}, v_{DS}). \tag{2–149}$$

If the differential[7] of Eq. (2–149) is taken, Eq. (2–150) results:

$$di_D = \left(\frac{\partial i_D}{\partial v_{GS}}\right)_{v_{DS}} dv_{GS} + \left(\frac{\partial i_D}{\partial v_{DS}}\right)_{v_{GS}} dv_{DS}. \tag{2–150}$$

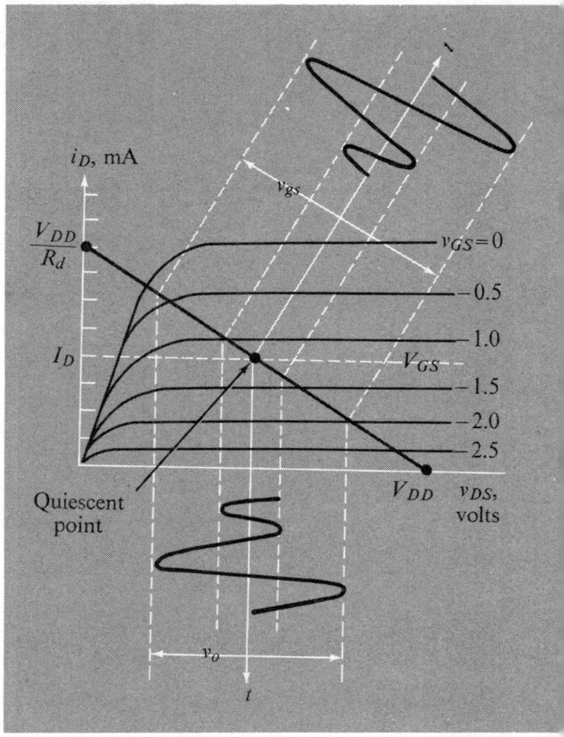

Fig. 2–199 Graphical illustration of amplification by JFET amplifier.

The quantity $(\partial i_D/\partial v_{DS})_{v_{GS}}$ is the reciprocal of the drain-to-source dynamic resistance, or $1/r_{ds}$. The quantity $(\partial i_D/\partial v_{GS})_{v_{DS}}$ is called the **mutual conductance** or **transconductance** and given the symbol g_m. It has the units 1/ohms, or mhos. Since the quantities di_D, dv_{GS}, and dv_{DS} are the small-parameter changes from quiescent values of drain current, gate-to-source voltage, and drain-to-source voltage, respectively, they can be represented by i_d, v_{gs}, and v_{ds}. Therefore Eq. (2–150) can be written as

$$i_d = g_m v_{gs} + \frac{v_{ds}}{r_{ds}}. \tag{2–151}$$

Note 7. Interpretation of Eq. (2–150). A knowledge of differential equations is NOT necessary to understand this formula or the derivation. Equation (2–150) reads simply: The change in i_D is equal to the change in i_D resulting from a change in v_{GS} at constant v_{DS} times the change in v_{GS}, plus the change in i_D resulting from a change in v_{DS} at constant v_{GS} times the change in v_{DS}.

Equation (2–151) suggests that the signal drain current i_D is the sum of the currents in two current paths between the source and the drain. The main current is $g_m v_{gs}$, and the other is a small current caused by a slight decrease or increase in channel length with increasing or decreasing drain voltage that results in a small change in drain current equal to v_{ds}/r_{ds}. The slope m of a drain characteristic curve in the region of pinch-off is equal to i_d/v_{ds}, which is $1/r_{ds}$ and is given the symbol g_d. The circuit model can thus be drawn as shown in Fig. 2–200a or b. It is called a **current-**

source circuit model because the input signal v_{gs} results in an output current $g_m v_{gs}$.

For the amplifier circuit of Fig. 2–201a, a circuit model can be drawn as shown in Fig. 2–201b. This circuit model includes the input source v_u and its internal resistance R_u and the output load resistance R_d. Since $i_d \approx g_m v_{gs}$, the output voltage $v_o \approx -g_m v_{gs} R_d$. The negative sign is used, as previously, to indicate a 180° phase reversal. That is, when the input voltage increases, the output voltage decreases proportionately, and vice versa, as readily seen in Fig. 2–199.

The voltage gain, A_v, is thus

$$A_v = \frac{v_o}{v_{gs}} = -\,g_m R_d. \qquad (2\text{–}152)$$

The concept of the small-signal model is very useful in analyzing electronic circuits and it is used extensively. The model represents the device and/or circuit in terms that can be readily manipulated and visualized.

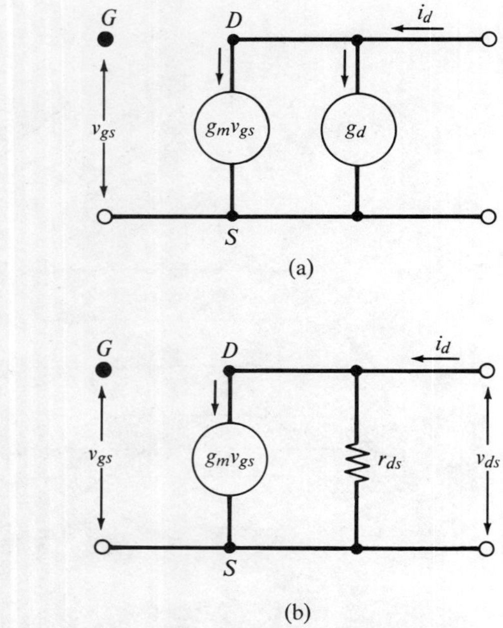

Fig. 2–200 Circuit models for JFET amplifier.

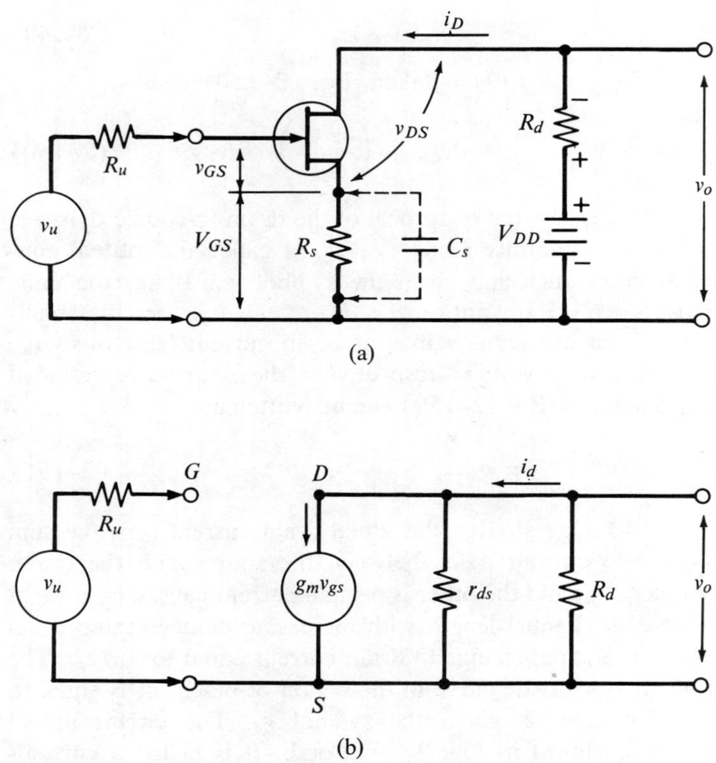

Fig. 2–201 Basic JFET amplifier with input voltage source: (a) circuit schematic; (b) circuit model.

However, it must be remembered that these models are restricted to a limited range of operating conditions. It should be noted that the terms **circuit model** and **equivalent circuit** are often used interchangeably, but that in recent literature the term circuit model is gaining preference because it shifts emphasis from equivalence to approximation of the circuit.

Self-Bias of JFET Amplifier

In the amplifier circuit shown in Fig. 2–198 the gate-to-drain operating point was established with a bias voltage source V_{GG}. Although it is possible to use a bias source, it is not convenient or economically practical. Instead a resistance R_s is generally put in the source circuit as shown in Fig. 2–201a. The current through R_s with no input signal applied is then I_D, because the gate-to-drain current i_G is negligible. Therefore there is a voltage drop $I_D R_s$ across the JFET source resistor which provides the quiescent bias voltage V_{GS} between gate and source. When an input signal voltage v_u is applied to a self-biased amplifier, the drain current, $i_D = I_D + i_d$, changes with the input signal. This would cause a change in the bias voltage and decrease the effective input signal. To maintain a relatively constant quiescent bias voltage a capacitor C_s is connected across the resistor R_s. The capacitor C_s is normally of very high capacitance so that it acts as a very low impedance[8] for even relatively low frequency signals. Thus the signal current, i_d, is effectively bypassed or short-circuited around R_s, so that the current in R_s is I_D and the current in C_s is i_1. The bias voltage across the parallel combination is then essentially constant, $V_{GS} = I_D R_s$.

When the circuit model is drawn for ac signals, the resistor R_s is omitted and the circuit analysis is the same as for Fig. 2–201a except for the added restriction that the signal must be ac. The circuit model for the JFET basic amplifier is shown in Fig. 2–201b.

Note 8. Capacitive Reactance.
Capacitive reactance, *X_C, is the opposition or* **impedance** *of a capacitor to alternating current and is measured in ohms. It is dependent on frequency f and the capacitance, so that $X_C = 1/2\pi fC$. A thorough discussion of impedance concepts and measurements is included in Module 3.*

A Basic BJT Amplifier

A basic BJT transistor common-emitter amplifier is shown in Fig. 2–202. One great difference between the BJT and the FET amplifiers is the

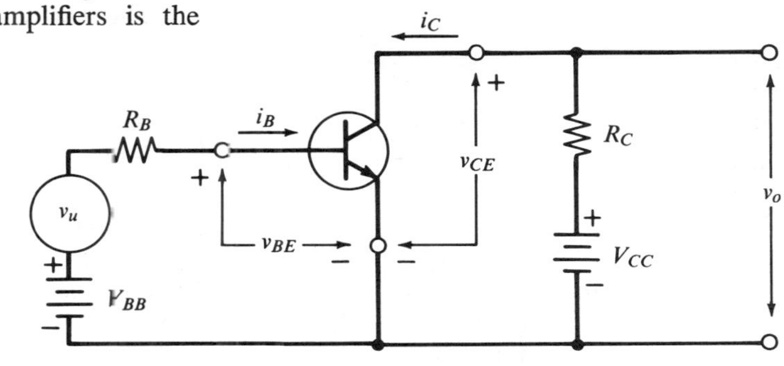

Fig. 2–202 Basic BJT amplifier.

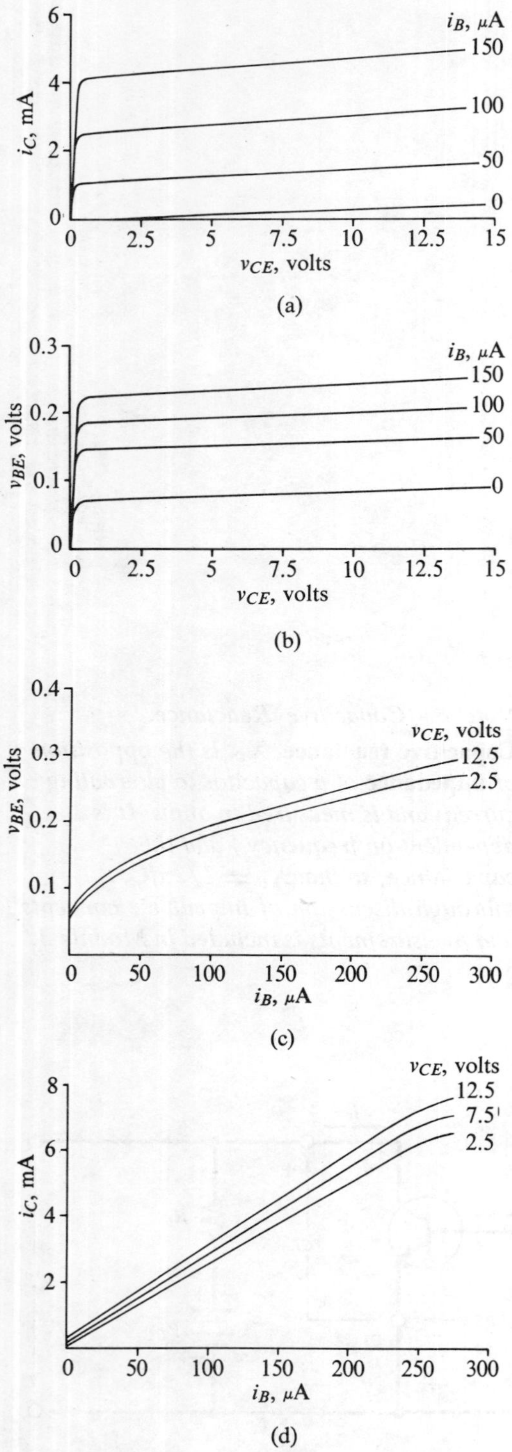

appreciable base current for the BJT circuit, which is required for control of the emitter-collector current; in contrast, the gate current for the FET is extremely small. The base current, I_B, and base-emitter voltage, V_{BE}, influence the charge distribution and thus the operating characteristics of the BJT. Hence the characteristics of the output circuit (emitter to collector) are dependent on certain parameters of the input circuit. Similarly, conditions in the output circuit will affect the characteristics of the input circuit. The interdependence of the input and output circuits complicates the analysis of the BJT in comparison to the FET described in the previous section.

The characteristic curves of the BJT are readily obtained, as discussed previously. Theoretically, eight different three-parameter characteristic curves can be determined for the four quantities i_B, v_{BE}, i_C, and v_{CE} shown in Fig. 2–203. At least two types of curves are necessary for a complete characterization. Four types of curves are shown in Fig. 2–203.

Circuit Model for BJT Amplifier

To derive a circuit model for the BJT, we must consider all four variables i_B, v_{BE}, i_C, and v_{CE}. Two variables are chosen as independent. It is more logical to choose the two most easy to control, which are i_B and v_{CE}. The other two variables are dependent. Thus we have the equations

$$v_{BE} = f_1(i_B, v_{CE}) \tag{2–153}$$

and

$$i_C = f_2(i_B, v_{CE}). \tag{2–154}$$

The differentiation of the first will yield the input equivalent circuit, and the differentiation of the second the output equivalent circuit. From the first equation,

$$dv_{BE} = \left(\frac{\partial v_{BE}}{\partial i_B}\right)_{v_{CE}} di_B + \left(\frac{\partial v_{BE}}{\partial v_{CE}}\right)_{i_B} dv_{CE}. \tag{2–155}$$

The partial differential $(\partial v_{BE}/\partial i_B)_{v_{CE}}$ is the slope of the v_{BE} vs. i_B characteristic curve (Fig. 2–203c) and is defined as h_{ie}. The symbol h is for hybrid, since the parameters developed by these equations are called hybrid parameters. The i indicates input, and e is for the common-emitter configuration. Note that h_{ie} has the units of resistance and is sometimes referred

Fig. 2–203 Common-emitter characteristic curves: (a) collector characteristic curves; (b) reverse transfer characteristic curves; (c) input characteristic curves; (d) forward transfer characteristic curves.

to as the input resistance. The partial differential $(\partial v_{BE}/\partial v_{CE})_{i_B}$ is the slope of the v_{BE} vs. v_{CE} characteristic curve (Fig. 2–203b) and is defined as h_{re}. This parameter indicates the effect of the output voltage on the input voltage and hence is a sort of amplification factor in reverse. Here h stands for hybrid parameter, r is for reverse, and e represents the common-emitter configuration.

With the substituted parameters, the input circuit equation is

$$v_{be} = h_{ie}i_b + h_{re}v_{ce}, \tag{2–156}$$

where v_{be}, i_b, v_{ce}, and other symbols with lower-case subscripts are defined as the signal values. This equation suggests a resistance h_{ie} in the base lead which develops an iR drop of $i_b h_{ie}$ in series with a voltage source $h_{re}v_{ce}$. The schematic representation of the transistor input circuit model is shown as part of Fig. 2–204. The term $h_{re}v_{ce}$ in Eq. (2–156) shows the dependence of the input circuit on the conditions existing in the output circuit (v_{ce}). The reverse amplification factor h_{re} is a measure of the degree of interaction between the input and output voltages.

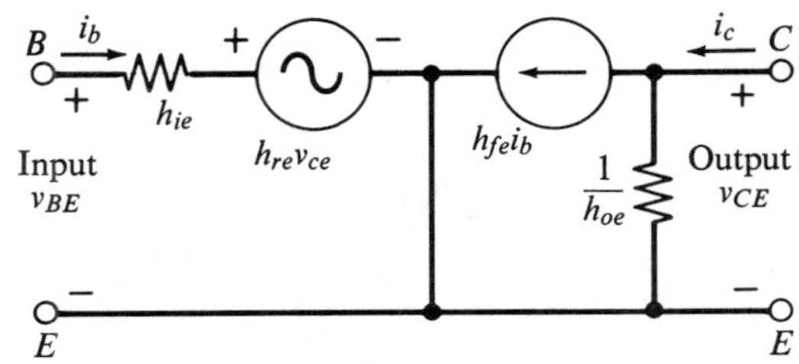

Fig. 2–204 BJT amplifier circuit model with hybrid parameters.

Differentiating the output equation, Eq. (2–154), we obtain

$$di_c = \left(\frac{\partial i_C}{\partial i_B}\right)_{v_{CE}} di_B + \left(\frac{\partial i_C}{\partial v_{CE}}\right)_{i_B} dv_{CE}. \tag{2–157}$$

The partial differential $(\partial i_C/\partial i_B)_{v_{CE}}$ is the slope of the i_C vs. i_B characteristic curve (Fig. 2–203d). It is the current amplification factor symbolized by h_{fe}, the f standing for forward and e for the common-emitter configuration. The partial differential $(\partial i_C/\partial v_{CE})_{i_B}$ is the slope of the i_C vs. v_{CE} characteristic curve (Fig. 2–203a). It is a conductance, has the units mhos, and is symbolized by h_{oe}. Again h is for hybrid, and the o indicates output and e the common-emitter configuration. Therefore

$$i_c = h_{fe}i_b + h_{oe}v_{ce}. \tag{2–158}$$

The collector current is the sum of two current paths, a conductance h_{oe} between the emitter and collector, and the amplified input current $h_{fe}i_b$. In the equivalent circuit the conductance h_{oe} is shown as a resistance of $1/h_{oe}$ ohms. The resulting equivalent circuit is shown in Fig. 2–204. Other equivalent circuits and other parameters result from a different choice of the independent variables.

A BJT Common-Emitter Amplifier

A BJT amplifier which has the emitter lead in common with the input and output circuits is shown in Fig. 2–205. The dc analysis of this circuit is very similar to that of the JFET. First the load line is determined by

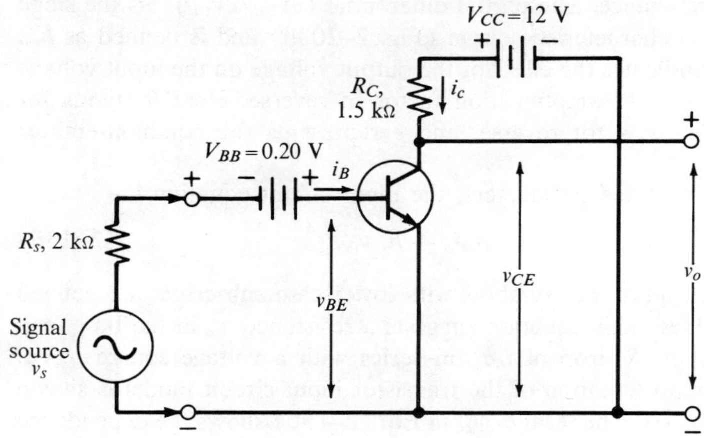

Fig. 2–205 Transistor amplifier circuit.

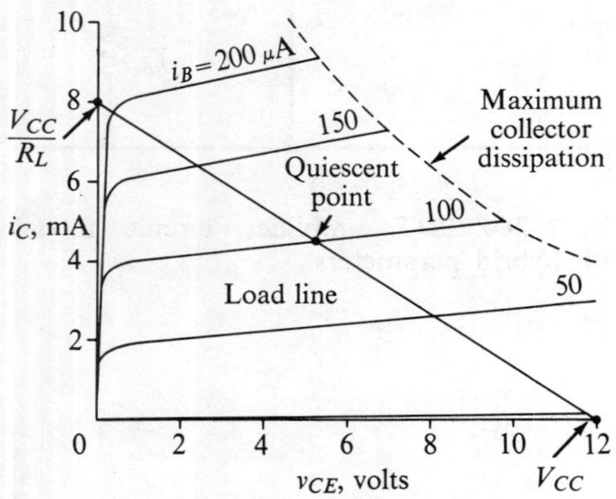

Fig. 2–206 DC analysis for the circuit of Fig. 2–205.

connecting V_{CC} on the voltage axis with V_{CC}/R_L on the current axis of the collector-characteristic curve for the transistor used (Fig. 2–206). Note that the load line should fall well within the maximum collector dissipation line. The maximum collector dissipation $P_{\max}$ is a characteristic of the transistor and is usually given in the transistor manuals or catalog descriptions. The maximum collector dissipation hyberbola is drawn by connecting all the points satisfying the equation $i_C v_{CE} = P_{\max}$. The quiescent base current can be estimated by considering the input circuit of Fig. 2–205. Assume that the base-to-emitter junction is simply the transistor input resistance h_{ie}. If $v_s = 0$ (quiescent state), $i_B = V_{BB}/(R_s + h_{ie})$. Often R_s will be very much greater than h_{ie}, so that $i_B \approx V_{BB}/R_s$. In this case, then, $i_B \approx 0.20\ \text{V}/2\ \text{k}\Omega = 100\ \mu\text{A}$. Thus the choice of bias voltage will depend on the nature of the signal source. Other methods of biasing are used, but none of them is independent of the input signal circuit to the extent that the JFET is.

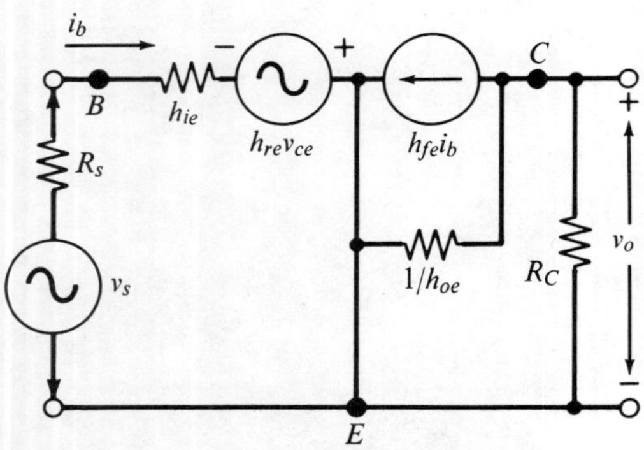

Fig. 2–207 Common-emitter amplifier circuit model with hybrid parameters.

Transistor Amplifier Characteristics. The common-emitter amplifier circuit model is shown in Fig. 2–207. Equations for the various amplifier characteristics can be derived from the equivalent circuit and the mathematical relations used to obtain that circuit. The derivations are more involved than those for the JFET amplifier because of the greater complexity of the transistor circuit model. Some of the characteristics of the common-emitter amplifier are given below. The actual derivations can be found in some of the books listed in the bibliography.

The voltage gain is

$$A_v = \frac{-h_{fe}R_C}{(h_{ie}h_{oe} - h_{fe}h_{re})R_C + h_{ie}}. \qquad (2\text{–}159)$$

For the transistor amplifier circuit shown in Fig. 2–205, the gain can be calculated if the parameters for the transistor are known. Assume that

$$h_{fe} = 55,$$
$$h_{ie} = 2720\ \Omega,$$
$$h_{oe} = 14 \times 10^{-6}\ \Omega^{-1},$$
$$h_{re} = 3.23 \times 10^{-4}.$$

Then

$$A_v = \frac{-55 \times 1500}{(2720 \times 14 \times 10^{-6} - 55 \times 3.33 \times 10^{-4})1500 + 2720}$$

$$= \frac{-8.3 \times 10^4}{(3.8 \times 10^{-2} - 1.8 \times 10^{-2})1500 + 2720}$$

$$= \frac{-8.3 \times 10^4}{30 + 2720} = -30.$$

The term $h_{ie}h_{oe} - h_{fe}h_{re}$ is found so often in transistor equations that it is given a special symbol, Δh_e. For this transistor, $\Delta h_e = 0.02$. It can be seen from the above equation that the term $\Delta h_e R_C$ in the denominator does not influence the result significantly. Thus the voltage gain could be approximated by the equation

$$A_v \approx \frac{-h_{fe}R_C}{h_{ie}} \qquad (2\text{–}160)$$

unless Δh_e or R_C is unusually large. The minus sign indicates that the output voltage is 180° out of phase with the input voltage. Recall that phase inversion is also a characteristic of the analogous JFET circuit (the common-source amplifier). Assume that R_C of Fig. 2–205 is a much better conductor than $1/h_{oe}$, that is, $R_C \ll 1/h_{oe}$. Assume also that the generator voltage $h_{re}v_{ce}$ is small compared with the voltage $i_b h_{ie}$. Now the model could be drawn as in Fig. 2–208. From this circuit it can be seen that $v_{\text{in}} = i_b h_{ie}$ and that $v_o = -h_{fe}i_b R_C$, so that the gain $A_v = v_o/v_{\text{in}} = -h_{fe}i_b R_C/i_b h_{ie} = -h_{fe}R_C/h_{ie}$, which is Eq. (2–160). Thus the assumptions leading to Eq. (2–160), which were not obvious previously, are actually $R_C \ll 1/h_{oe}$ and $h_{re}v_{ce} \ll i_b h_{ie}$.

The current gain A_i is given by the formula

$$A_i = \frac{-h_{fe}}{h_{oe}R_C + 1}. \qquad (2\text{–}161)$$

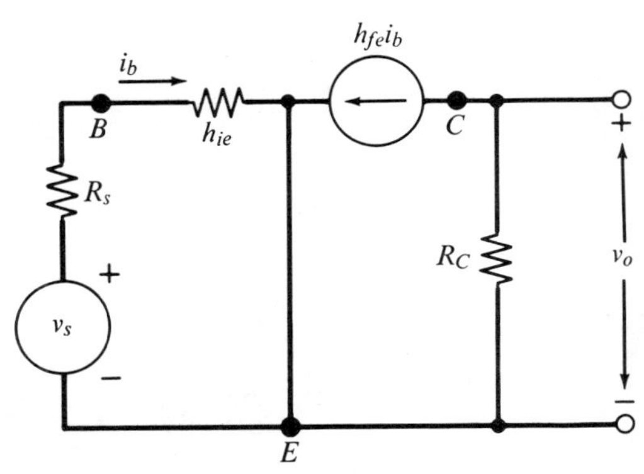

Fig. 2–208 Transistor amplifier simplified equivalent circuit.

The current gain for the amplifier of Fig. 2–205 is then

$$A_i = \frac{-55}{14 \times 10^{-6} \times 1500 + 1} = \frac{-55}{2.1 \times 10^{-2} + 1} = -55.$$

The term $h_{oe}R_C$ is usually so small compared with 1 that it can be neglected. Thus

$$A_i \approx -h_{fe}. \tag{2–162}$$

Again the minus sign is an indication of the phase reversal between the input and output signals. The assumption made in Eq. (2–162), that is, $h_{oe}R_C \ll 1$, is the same as the previous assumption, $R_C \ll 1/h_{oe}$. In calculating this current gain, the output current is defined as the current through R_C, not the current which can be drawn from the output terminals. If an additional load R_L is connected to the output terminals of the amplifier, the output current will be split between R_C and R_L; thus the effective value of the parallel combination of R_C and R_L should be used in the above equations.

The power gain A_p is simply the product of the current and voltage gains. When the assumptions made in Eqs. (2–160) and (2–162) are valid,

$$A_p \approx \frac{(h_{fe})^2 R_C}{h_{ie}}. \tag{2–163}$$

For the amplifier under analysis,

$$A_p \approx (-30)(-55) \quad \text{or} \quad \frac{(-55)^2(1500)}{2720} = 1660.$$

The power gain does not include power losses in the signal source or in transferring the output signal to a load other than R_C. It is simply the ratio of the signal power dissipated in R_C to the signal power dissipated in h_{ie}. Equation (2–163) verifies this definition.

The input resistance R_{in} is found from the formula

$$R_{\text{in}} = \frac{h_{ie} + (h_{oe}h_{ie} - h_{fe}h_{re})R_C}{1 + h_{oe}R_C} = \frac{h_{ie} + \Delta h_e R_C}{1 + h_{oe}R_C}. \tag{2–164}$$

Substitution of the parameters shows R_{in}, in this case, to be

$$R_{\text{in}} = \frac{2720 + 0.02 \times 1500}{1 + (14 \times 10^{-6}) \times 1500} = \frac{2720 + 30}{1 + 2.1 \times 10^{-2}} = 2700\ \Omega.$$

Since $\Delta h_e R_C \ll h_{ie}$ and $h_{oe}R_C \ll 1$,

$$R_{\text{in}} \approx h_{ie}. \tag{2–165}$$

It is obvious that the assumption $h_{re}v_{ce} \ll i_b h_{ie}$ made in Fig. 2–203 leads to this same conclusion. Note also that $R_{\text{in}} = R_C(A_i/A_v)$.

The output resistance R_o can be calculated from

$$R_o = \frac{h_{ie} + R_s}{h_{oe}h_{ie} - h_{re}h_{fe} + h_{oe}R_s} = \frac{h_{ie} + R_s}{\Delta h_e + h_{oe}R_s}. \qquad (2\text{–}166)$$

For the circuit of Fig. 2–205,

$$R_o = \frac{2720 + 2000}{0.02 + 14 \times 10^{-6} \times 2000} = \frac{4720}{0.048} = 10^5\ \Omega.$$

In this case the terms in the numerator and the denominator are both of the same order of magnitude. It can be seen that, if R_s is small compared with h_{ie},

$$R_o \approx \frac{h_{ie}}{\Delta h_e} \qquad (2\text{–}167)$$

or that, if R_s is much larger than h_{ie},

$$R_o \approx \frac{1}{h_{oe}}. \qquad (2\text{–}168)$$

As Eq. (2–168) indicates, the output resistance, R_o, is in this case the resistance that the transistor as a power source presents to the load R_C. The output resistance of the entire amplifier circuit as seen by a device attached to the output terminals would actually be R_o in parallel with R_C.

The common-emitter BJT amplifier is thus shown to possess the properties of voltage gain, current gain, phase reversal of the input signal, low input resistance, and high output resistance. Similar analyses can be made for the common-collector and common-base amplifiers.

Emitter-Follower Amplifier

A very practical and important circuit that has the basic common-collector configuration is shown in Fig. 2–209. The input and output terminals are connected to the transistor collector through the collector supply voltage V_{CC}. However, the three terminals are in common for ac signals because, from the viewpoint of the varying signal levels, the collector voltage is a fixed constant value just as surely as if it were grounded. It is considered to be in common with the common input and output terminals because the potential difference is a constant.

This circuit has many desirable characteristics such as high input impedance, low output impedance, and excellent linearity over a wide dynamic range, and is therefore frequently used as a buffer stage or an

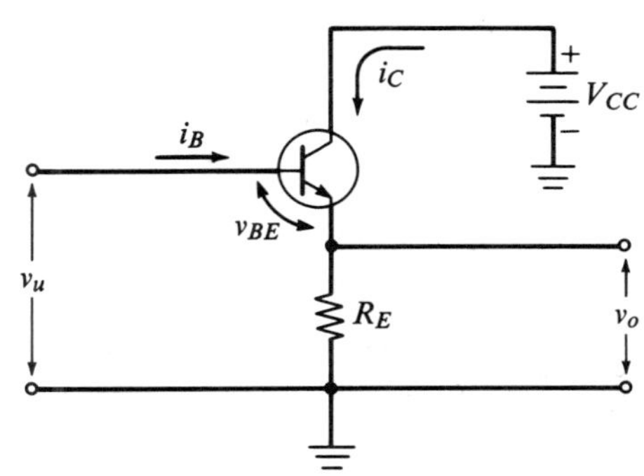

Fig. 2–209 Emitter-follower circuit.

output circuit. The wide applicability of this simple linear amplifier circuit makes it worthwhile to gain an understanding of its characteristics.

When a positive voltage v_u is applied to the base, the transistor conducts, causing a voltage drop v_o across R_E which increases the transistor emitter voltage. If the input voltage v_u then decreases, i_C decreases, and so does the emitter potential. Thus the emitter voltage, v_o, "follows" the input voltage. Therefore this type of common-collector configuration is usually called an "emitter follower." The characteristics of this circuit are analyzed in greater detail as follows.

When the input signal voltage, v_u, is greater than the turn-on voltage for the base-emitter junction, v_{BE}, the output voltage, v_o, is

$$v_o = v_u - v_{BE} \tag{2–169}$$

and the ac voltage gain A_v of the circuit is

$$A_v = \frac{dv_o}{dv_u} = \frac{d(v_u - v_{BE})}{dv_u} = 1 - \frac{dv_{BE}}{dv_u}. \tag{2–170}$$

Since dv_{BE}/dv_u is always positive, it can be seen from Eq. (2–170) that the voltage gain is always less than 1, but it approaches unity in the linear amplification region, where dv_{BE}/dv_u is small. However, the input v_u cannot exceed the supply voltage V_{CC} by much before the collector-base junction becomes forward biased and the voltage between collector and base $v_{CB(\mathrm{ON})}$ is only a few tenths of a volt. The input would thus be clamped at $V_{CC} + v_{CB(\mathrm{ON})}$.

If the transistor of Fig. 2–209 is not saturated,

$$i_C = h_{fe} i_B. \tag{2–171}$$

Since $i_E = i_C + i_B$, it follows that

$$i_E = h_{fe} i_B + i_B = i_B(h_{fe} + 1). \tag{2–172}$$

Therefore, from Eq. (2–172) the current gain A_i is

$$A_i = \frac{di_E}{di_B} = h_{fe} + 1. \tag{2–173}$$

Input Resistance. One of the important characteristics of the emitter follower is its relatively high input resistance, which means that it will not "load" the input signal source significantly. In this section the effective resistance to an ac signal is considered. It is defined as

$$R_{\mathrm{in}} = \frac{dv_u}{di_B}. \tag{2–174}$$

If the input voltage signal, v_u, is applied directly to the base terminal, it can be seen from Eq. (2–169) that $v_u = v_o + v_{BE}$, and since $v_o = i_E R_E$ it follows that

$$v_u = i_E R_E + v_{BE}. \qquad (2\text{–}175)$$

By substituting the equivalent of i_B from Eq. (2–172), $i_B = i_E/(h_{fe} + 1)$, and of v_u from Eq. (2–175) into Eq. (2–174), it follows that

$$R_{\text{in}} = \frac{d(i_E R_E + v_{BE})}{d(i_E/(h_{fe}+1))} = \left(R_E + \frac{dv_{BE}}{di_E}\right)(h_{fe}+1). \qquad (2\text{–}176)$$

If the emitter current i_E is larger than a few milliamperes, dv_{BE}/di_E will typically be negligible compared to R_E, and R_{in} will be

$$R_{\text{in}} \approx R_E(h_{fe}+1) \qquad (2\text{–}177)$$

Equation (2–177) shows that the dc input resistance, R_{in}, is greater than the emitter load, R_E, by the factor $h_{fe} + 1$.

Output Resistance. The emitter-follower stage is often connected between an inverter circuit and the external load, because of its ability to act as an impedance transformer. This is illustrated by the circuit of Fig. 2–210a. A low output resistance for a voltage source means that relatively heavy loads can be connected to the output of the source without changing the output voltage. The low output resistance of the emitter follower is one of its most important characteristics. The output resistance of the emitter follower is seen to be only a fraction $[1/(h_{fe}+1)]$ of the input source resistance $(R_u = dv_u/di_B)$ by considering the following equations obtained by substitution from Eqs. (2–169) and (2–172):

$$R_o = \frac{dv_o}{di_E} = \frac{d(v_u - v_{BE})}{di_B(h_{fe}+1)} = \frac{1}{h_{fe}+1} \times \frac{d(v_u - v_{BE})}{di_B}$$

and, if v_{BE} is negligible compared to v_u,

$$R_o = \frac{R_u}{h_{fe}+1}. \qquad (2\text{–}178)$$

Since the maximum output voltage $v_{o(\max)}$ can be given as $v_{o(\max)} = V_{CC} - i_B R_C - v_{BE}$, it can also be readily shown, by substitution for $i_B = i_E/(h_{fe}+1)$ and for $I_E = v_{o(\max)}/R_L$, that

$$v_{o(\max)} = \frac{V_{CC} - v_{BE}}{1 + R_C/[R_L(h_{fe}+1)]}. \qquad (2\text{–}179)$$

If v_{BE} is assumed to be 0.6 V and the supply voltage $V_{CC} = 5$ V, $R_C = 1$ kΩ, and $h_{fe} = 50$ for the selected transistor Q_2 in Fig. 2–210a,

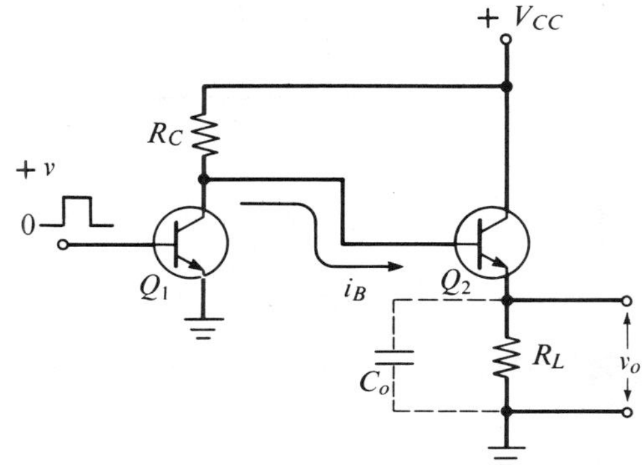

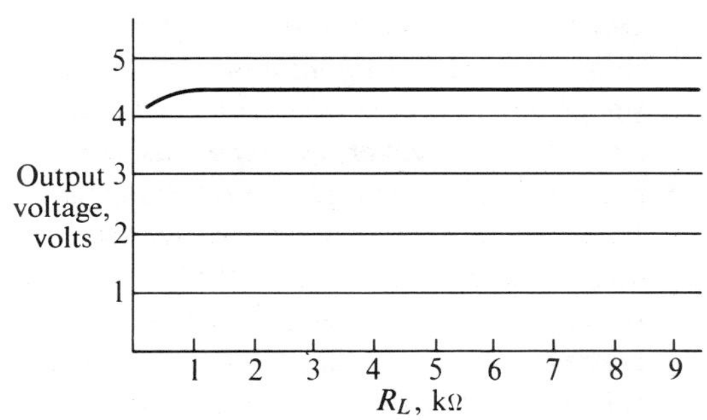

Fig. 2–210 Emitter follower as resistance transformer between inverter and output load: (a) circuit; (b) plot showing small change of output voltage vs. load resistance.

then it can be seen that the second term in the denominator in Eq. (2–179) is negligible for a wide range of load resistance R_L. A plot of output voltage vs. load resistance for this example is shown in Fig. 2–210b.

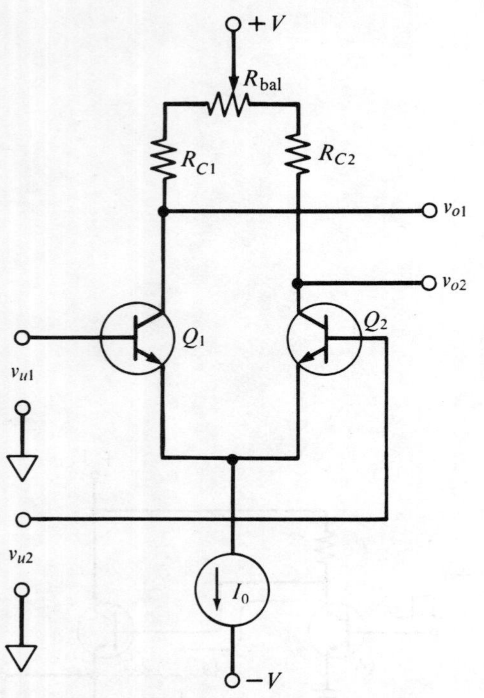

Fig. 2–211 Basic BJT differential amplifier.

A Basic Differential Amplifier

A difference or differential amplifier is one whose output is a function of the potential difference between its two inputs, and is relatively nonresponsive to voltages of equal magnitude (common-mode voltages v_{cm}) present at both inputs. It is a very useful amplifier because it is capable of rejecting spurious common-mode signals that are coupled to both inputs either from pickup of radiated noise or conduction from the input source. Also, a well-designed differential amplifier is relatively insensitive to environmental changes of temperature or other conditions.

It was shown in Section 1–3.4 of Module 1, *Electronic Analog Measurements and Transducers,* that the very important voltage comparator is a differential amplifier designed specifically for null comparison measurements. It has well-balanced differential inputs and controlled output limits. Although the ideal differential amplifier would provide an output proportional only to the difference in voltage at the two inputs regardless of the magnitude of the input voltages, it is necessary to settle for something less than ideal. The commonly used measure of the perfection of input balance was also shown in Section 1–3.4 to be the **common mode rejection ratio** (CMRR):

$$\text{CMRR} = \frac{\Delta v_o/\Delta v_d}{\Delta v_o/\Delta v_{cm}}. \tag{2–180}$$

The numerator of Eq. (2–180) is the change in output voltage Δv_o for a difference signal Δv_d at the inputs, and the denominator is the change in output voltage for a change Δv_{cm} in the common mode signal voltage that is applied to both inputs.

Often the common mode voltage rejection (CMRR) is expressed as a certain number of decibels (dB),[9] where

$$\text{CMRR} = 20 \log (\text{CMRR}) \text{ dB}. \tag{2–181}$$

For example, if the CMRR is 100,000, then CMRR $= 20 \log 10^5 = 100$ dB.

A modern comparator amplifier is generally a combination of several types of amplifiers, and the combination is usually represented by the triangular symbol with two inputs. The following description is of the basic differential amplifier shown in Fig. 2–211, which is made with two matched BJT amplifier stages. The emitters of the two transistors are connected together and to a constant current source I_0.

Note 9. Decibel.

Decibel, *dB, is a term used primarily to relate the ratio of signal power output to signal power input, where dB* $= 10 \log (P_{out}/P_{in})$. *However, the term has also been extended through the relationship* $P = V^2/R$ *to express relative voltage levels where the resistance R for the related levels is the same. In this case the expression given in Eq. (2–181) is used. It is incorrect to use Eq. (2–181) if the resistance for the related levels is not the same but this has become common practice when referring to OA voltage gains.*

To understand the operation of the difference amplifier, consider first how the output at v_{o2}, which is influenced by both input signals, originates. The input signal at v_{u1} gives rise to an output at v_{o2} which is amplified, but not inverted, since transistor Q_1 acts as an emitter-follower stage for this signal and Q_2 acts as a common-base stage. Transistor Q_2 provides voltage gain for the signal at v_{u1}, while Q_1 provides high input resistance and low output resistance to drive the common-base stage. The input signal at v_{u2} gives rise to an output at v_{o2} which is both amplified and inverted, because transistor Q_2 acts as a normal common-emitter stage for v_{u2}. Thus the output at v_{o2} is a function of the difference between the two input voltages and can be expressed as

$$v_{o2} = K_1 v_{u1} - K_2 v_{u2},$$

where K_1 and K_2 are the voltage gains experienced by v_{u1} and v_{u2} respectively. Similar reasoning applied to the output at v_{o1} leads to the conclusion that the signal at v_{u2} gives rise to an output which is amplified but not inverted, while the signal at v_{u1} gives rise to an output at v_{o1} which is amplified and inverted. Thus the output at v_{o1} can be expressed as

$$v_{o1} = K'_2 v_{u2} - K'_1 v_{u1},$$

where again K'_2 and K'_1 are the gains experienced by v_{u2} and v_{u1} respectively at v_{o1}.

There are several modes in which the difference amplifier can be used. In the **balanced** or **differential** mode, the output voltage v_o is taken from collector to collector so that

$$v_o = v_{o2} - v_{o1} = A_1 v_{u1} - A_2 v_{u2}, \tag{2–182}$$

where $A_1 = K_1 + K'_1$ and $A_2 = K_2 + K'_2$. If the circuit is perfectly balanced, the voltage gains experienced by the two signals are equal ($A_1 = A_2 = A$) so that

$$v_o = A(v_{u1} - v_{u2}). \tag{2–183}$$

Perfect balance requires that the amplification factors of the two transistors and the collector load resistors, R_{C1} and R_{C2}, be perfectly matched and track with any changes in temperature or other operating conditions. Also, any differences between the base-emitter voltages of the two conducting transistors can lead to an offset voltage at the output, which changes with temperature. For these reasons the two transistors are often fabricated on the same chip, encapsulated in a single case, or placed together on a single heat sink. A balance potentiometer (R_{bal} in Fig. 2–211) is often used in the collector circuit to compensate for any voltage offsets and drifts. A well-balanced difference amplifier is highly independent of supply voltage fluctuations and ripple, since these affect both outputs similarly and do not influence the difference output very strongly.

To a large extent the CMRR depends on the quality of the constant current source I_0. A perfect current source would maintain the gains and dc bias points of Q_1 and Q_2 constant and independent of the input signals. If the current source is nonideal, however, the gains and bias points will shift with a common mode input, which leads to a poor CMRR. The simplest current source is a common mode resistor connected between the negative supply voltage $-V$ and the two emitters. Large values of the common mode resistor make the current relatively independent of the input signals. Improved values of CMRR can be obtained by replacing the common mode resistor with a transistor constant current source or by using feedback to control the current. Almost all high quality difference amplifiers use these latter methods for high CMRR.

The difference amplifier can be used in a **single-ended output** mode. In this mode the input is differential as in Fig. 2–211, but only one output, referenced to the $+V$ supply, is employed. With a single-ended output only one of the collector load resistors is necessary.

A third mode in which the differential amplifier can be operated is the **phase inverter** mode. In this mode, only one of the inputs is used and each output is taken separately. If input 2 in Fig. 2–211 is shorted to the common, $v_{o2} = K_1 v_{u1}$ and $v_{o1} = -K'_1 v_{u1}$. If $K_1 \approx K'_1$, the outputs at the two collectors are of equal magnitude, but always of opposite phase.

A differential amplifier made with matched JFET's is illustrated in the next section as the input stage for a general purpose operational amplifier.

Summary

The basic amplifiers presented in this section all illustrate that amplification is essentially a control operation. That is, various electronic devices such as BJT's and JFET's are used as control elements in various configurations to control a power supply, as illustrated in the block diagram, Fig. 2–212.

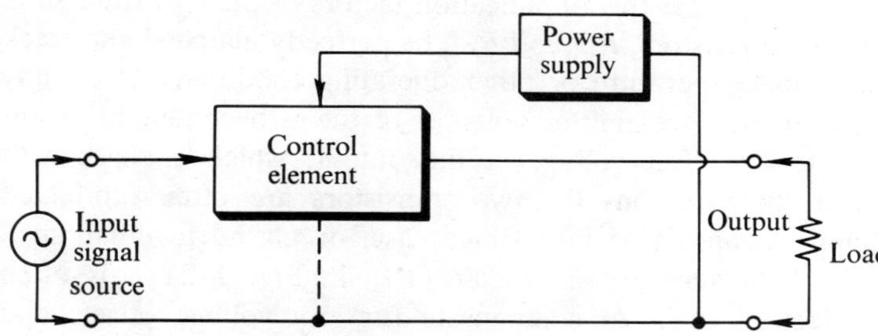

Fig. 2–212 Method of amplification.

The control element (BJT, FET, etc.) is actuated by the input signal, and it controls the magnitudes of the voltage and current that are taken from the power supply and made available at the amplifier output terminals. The controlled output voltage and current from the power supply will follow some known function related to the input signal. The amplification is a result of the small voltage and/or current required by the control element in order to control relatively large voltages and currents from the power supply.

The many amplifier configurations used in practice are aimed at meeting certain design requirements such as linearity (direct proportionality between input and output), high-frequency response (ability to provide the necessary control of power supply voltage and current even when the input signal varies at extremely high rates such as 100 MHz or more), or rejection of noise (as with the differential amplifier). The amplifier concepts, circuits, models, and terminology presented in this section should enable the reader to understand the basic operation of amplifiers and appreciate at least some of the limitations of amplifier circuits. Several basic amplifiers are usually combined in the production of amplifiers with special characteristics. An example of this is illustrated in the next section, where a differential JFET amplifier, a cascaded high-gain BJT amplifier, and an emitter-follower amplifier are combined to make a general purpose operational amplifier.

Experiment 2–31 *Difference Amplifiers*

A difference amplifier is constructed and studied as a differential balanced amplifier, as a phase inverter, and as a single-ended differential amplifier. The amplifier is constructed with both npn transistors and matched JFET's. Characteristics of the amplifiers are determined, and the common mode rejection ratios are determined.

2–5.3 INSTRUMENTATION DC AMPLIFIERS

Economical general purpose dc amplifiers, referred to as operational amplifiers (OA's), are now available which have outstanding characteristics for instrumentation applications. Operational amplifiers are utilized in hundreds of different ways to provide elegant solutions to measurement and control problems. Many applications have been presented in this module and others in the series.

The many functions that can be performed by using OA's with suitable feedback elements and in various combinations make it feasible for anyone with just a little experience to construct some very sophisticated instruments. It is, of course, not necessary to know what is inside the OA gain block (represented simply by a triangular symbol) in order to use an OA profitably. However, most users of OA's feel more comfortable in using them when they have some idea of how they are designed, and certainly the user can appreciate specifications and limitations better after investigation of the internal operation of a typical OA. Therefore the first part of this section is devoted to a look inside the triangular symbol. Then the combination of several OA's to form a very high quality dc instrumentation amplifier is presented, and the section concludes with a summary of specification considerations for OA's.

Integrated Circuit Operational Amplifier

The OA that we have chosen to investigate is constructed on a micro semiconductor chip. This integration of several basic amplifier circuits into a monolithic OA has resulted in units no larger than and costing no more than a single transistor. As previously indicated, the general purpose OA is a gain block with dc integrity, high input impedance, low output impedance, quite wideband frequency response with negative feedback, and other characteristics that will be described later. The ways by which these characteristics are obtained can be seen by referring to Fig. 2–213, which is a somewhat simplified circuit diagram of an integrated circuit OA.

The input circuit is a differential amplifier using matched p-channel JFET's as source followers. An external 10 kΩ potentiometer is used to balance the differential amplifier to compensate for any offset. The differential input provides for the inverting and noninverting inputs and excellent CMRR. The use of JFET's also provides input impedances of 10^9 Ω or higher and very low offset currents. This very high input impedance makes the OA generally applicable as a high quality integrator, for analog memory, or for other applications where high input impedance is necessary.

Two BJT transistors are coupled together to provide a second stage with high current gain. When two transistors are coupled together in this fashion, the pair is known as a **Darlington circuit.** The current gain of the pair is approximately equal to the product of the amplification factors ($\beta_1 \times \beta_2$) for the two transistors.

The output stage is a cascaded complementary emitter-follower circuit. It has the necessary very low output impedance (< 1 Ω with negative feedback) for the output stage, as described in the previous section.

Fig. 2–213 Integrated circuit operational amplifier.

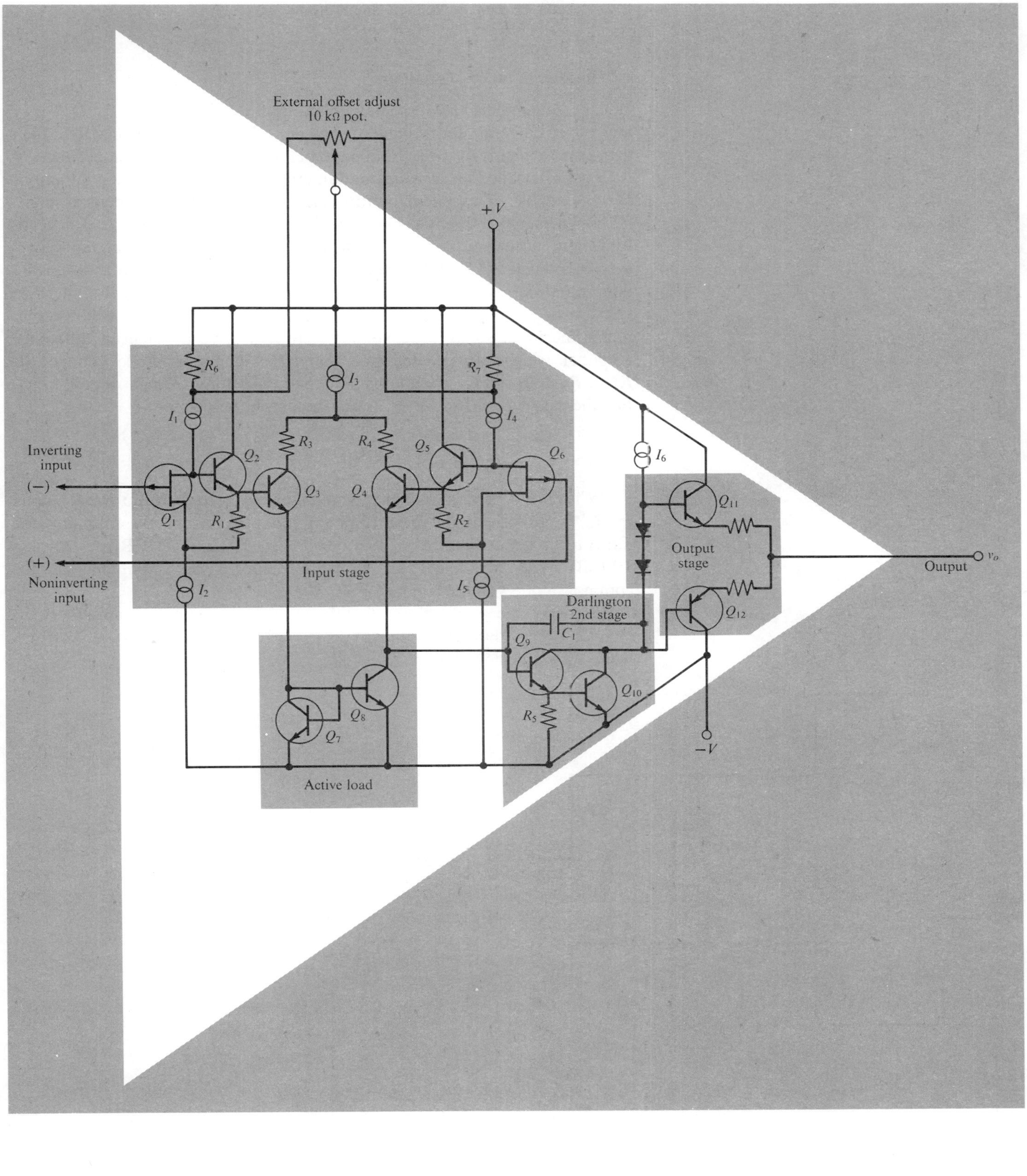

External offset adjust
10 kΩ pot.
+V
Inverting input
(−)
(+)
Noninverting input
Input stage
Active load
Darlington 2nd stage
Output stage
Output
−V

DC Instrumentation Amplifiers

When a dc amplifier has especially high input impedance ($> 10^9\ \Omega$), very low offset currents and voltages, negligible drift, very high CMRR, high gain over a relatively wide frequency band, and low output impedance, it is frequently called an **instrumentation amplifier.** Obviously, such an amplifier is capable of excellent performance in critical measurement systems. Although many differential OA's with FET inputs fulfill the general needs for many accurate instruments, it is possible to improve certain characteristics such as CMRR by combining three or more OA's so as to form a differential dc instrumentation amplifier, as illustrated in Fig. 2–214. Two OA's with very high input impedance are used as voltage followers to buffer the input signal. The voltage followers drive a balanced difference amplifier which can provide gain and reject common mode voltages. If $R_1 = R_2 = R_3 = R_4$, the gain is unity and the output voltage $v_o = v_2 - v_1$. For the case where $R_3/R_1 = R_4/R_2$ the output voltage is

$$v_o = (v_1 - v_2)\,\frac{R_3}{R_1}. \qquad (2\text{–}184)$$

With this type of amplifier it is necessary to keep the ratio R_3/R_1 equal to R_4/R_2 to obtain the best common mode rejection. By trimming the ratios to be better than 0.1% it is possible to obtain a CMRR greater than 100 dB. With the typical ± 15 V power supplies the amplitude limit for common mode voltage is about ± 10 V.

Some of the applications of differential instrumentation amplifiers include bridge, strain gauge, or any low level voltage measurement where a high CMRR is essential.

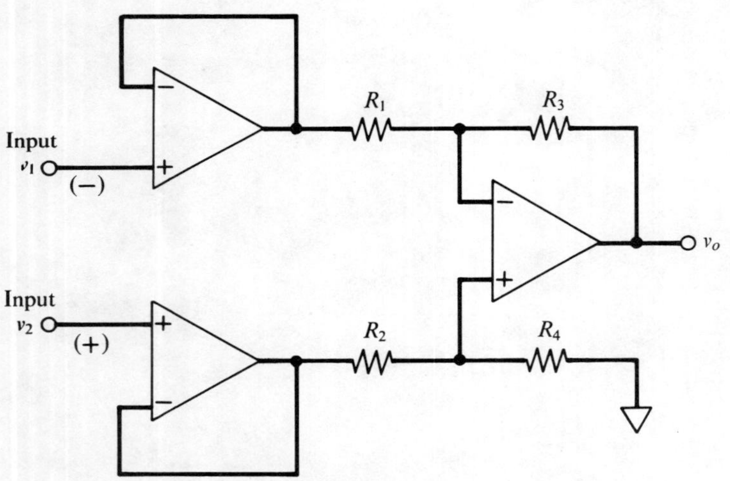

Fig. 2–214 Instrumentation differential amplifier.

Summary of OA Specifications

Every commercial OA has a set of specifications, and although many of these have been described in previous sections, they are summarized here as a useful reference. In addition, some of the other specifications and considerations that would not have meant much earlier can now be briefly described in light of the foregoing discussion of amplifiers.

Open-Loop Gain. The open-loop gain is the OA gain without any feedback. It is typically given in dB. For example, since gain (dB) = 20 log (v_o/v_u), an open-loop gain of 100 dB means that

$$\log \frac{v_o}{v_u} = \frac{100}{20} = 5,$$

so that

$$A = \frac{v_o}{v_u} = 100{,}000.$$

Frequency Response. It is important to realize that the open-loop gain, A, is a function of frequency and is relatively constant from dc to 10 Hz for most OA's and then falls off. As was previously shown in descriptions of various applications, it is important that A be very large for the basic equations to be reliable. Therefore, it is important to check the manufacturer's frequency response characteristics when using OA's for applications at frequencies above 10 Hz. Amplifier frequency response is discussed in detail in Module 3.

Rated Output. The power supply voltages used with many modern OA's are ±15 V with a range of tolerance from about ±12 to ±18 V. Therefore, the maximum available output voltage is often greater than ±10 V. Some IC OA's operate at voltages about ±5 V so that the voltage output swing is less than 5 V. The current output rating varies considerably with the specific OA, but it is typically 2–10 mA for general purpose OA's. Consequently, if the output voltage goes as high as 10 V and the output current rating is 10 mA, then the lowest value for the load resistor for rated output is 1 kΩ. If the current rating is 2 mA, the lowest possible value for the output load is 5 kΩ if the output voltage is to swing to its rated maximum.

Input Offset Voltage. As explained in the previous section, it is improbable that the input differential amplifier will be perfectly balanced, although perfect balance is a design goal of the manufacturer of dc amplifiers. The small, relatively constant, but temperature-dependent voltage that exists between the input terminals is called **input offset voltage.** It can be considered as a small voltage in series with one of the input leads. It will be amplified by the same factor as the input signal, and unless it is balanced out, the output v_o will not be zero with zero input. Usually an external pot is required to balance out the offset voltage. However, the offset voltage changes with temperature, supply voltage, and time. For example, manufacturer's specifications might be ± 50 μV/°C, ± 500 μV/V, and ± 100 μV/day respectively.

Input Bias Current. Even when the input voltage is zero, there is an input current in each input terminal. For JFET input stages the bias currents are the gate currents of the JFET's and any leakage currents within the amplifier. The manufacturer attempts to make this bias current as small as possible by selecting appropriate input control elements and by careful design and component layout. Typical bias current for a good instrumentation dc amplifier might be in the range 0.1–10 pA at 25°C. The bias current can be balanced by supplying a balance current to the summing point of an OA. However, it should be remembered that the bias current is quite sensitive to temperature changes. For example, for FET OA's the bias current doubles for about each 10°C change in temperature. The difference between the two input bias currents is called the **input offset current.**

Input Resistance. Modern FET amplifiers have extremely high input impedances. The value depends greatly on the design and will be generally in the range from 10^9 to 10^{14} Ω. An amplifier with an input resistance of 10^{14} Ω can keep the input current in the subpicoampere range for full common mode voltage swings of ± 10 V. The input capacitance in parallel with the input resistance is usually about 5 pF.

Input Noise. The input current noise of the amplifier is the limiting factor on signal resolution. It varies as a function of source impedance and frequency. Therefore, the specification of input noise is generally given as a function of frequency, and graphs of typical changes of input noise with source resistance are available from the manufacturer.

CMRR. The common mode rejection ratio of amplifiers classed as dc instrumentation amplifiers should be very good, typically 100 dB at ± 5 V and about 80 dB at ± 10 V.

Leakage Currents. When using FET amplifiers it is important to take special precautions in the layout of external wiring and feedback elements. That is, unless the circuit boards are free of fingerprints and other contamination, the leakage paths between the 15 V power supply conductors to the amplifier inputs might decrease the insulation resistance well below 10^{12} Ω. With a leakage path of 10^{12} Ω between either input and the 15 V supply there would be an input current of 15 pA. This can be much larger than the amplifier input bias current. Leakage paths are likely to undergo great variations with environmental changes and could cause erratic behavior.

Because of the difficulty in guarding input terminals from leakage paths when the OA's are soldered on printed-circuit boards, it is sometimes preferable to use **hardwiring.**

Shielding. The high resistance levels at which FET instrumentation amplifiers may be operated makes it important to shield the input terminals so as to reduce noise pickup. Shielded input cables are recommended and the feedback elements should be enclosed within a shield. When a shielded cable is flexed, it can generate input noise because of the cable conductor rubbing against the insulation. However, special low noise cables are available that have a graphite coating or conductive tape between the cable conductor and its insulation.

Any changes in capacitance at the input because of the input wiring shifting position can cause extraneous voltages at the input. The obvious remedy for this is to provide rigid wiring with leads as short as possible and spaced as far from other conductors as possible.

Slew Rate. Within the OA there are capacitances inherent in the semiconductor devices and leads used in its construction. The rate of change of voltage at each point in a circuit is limited by the available current to charge the capacitance at that point. The consideration is the same as for the response of the *RC* circuit to a step function, as discussed in Section 2–2. The maximum rate of change of voltage $(\Delta V/\Delta t)_{\max} = I_{\max}/C$. Therefore, for example, if a capacitance of 10 pF is to change voltage by 10 V in 1 nsec, then the available current at that point must be $I = (10/10^{-9})10^{-11} =$ 100 mA.

Settling Time. In today's world of high-speed data acquisition and processing, many measurements are made in a microsecond or less. Therefore the time required for an electronic feedback control circuit to perform its function with a specified accuracy is important and is referred to as **settling time.** For OA feedback circuits it is the time that elapses between the application of an ideal instantaneous step input and the instant at which the closed-loop amplifier output has entered and remained within a specified error band, usually symmetrical about the final value. The settling time includes any propagation delay in the circuit plus the time required to slew to the final value, recover from any overshooting of the final value associated with slewing, and finally settling within a certain specified accuracy for the final value.

Note that the settling time is associated with a closed-loop OA system, i.e., an OA circuit with feedback, and it cannot be predicted from open-loop specifications for the OA, such as slew rate or small-signal bandwidth.

Experiment 2–32 *Operational Amplifier Characteristics*

Two important characteristics of operational amplifiers which are easily measured are the common mode rejection ratio (CMRR) and the closed-loop output resistance (R_o). In this experiment these values are measured for two different types of operational amplifiers in order to compare and contrast these characteristics. Other OA characteristics can be measured if desired.

PROBLEMS

1. Circuit models of two semi-ideal amplifiers are shown in Fig. P–1. Fig. P–1a corresponds to a FET amplifier, while Fig. P–1b corresponds to a BJT amplifier. Note that the FET amplifier has a finite input resistance, as opposed to the ideal voltage amplifier, while the BJT amplifier has a non-zero input resistance. (a) Calculate the voltage gain of the FET amplifier of Fig. P–1a and the current gain of the BJT amplifier of Fig. P–1b. (b) Calculate the power gains of the two amplifiers. (c) Calculate the voltage gain of the BJT amplifier of Fig. P–1b.

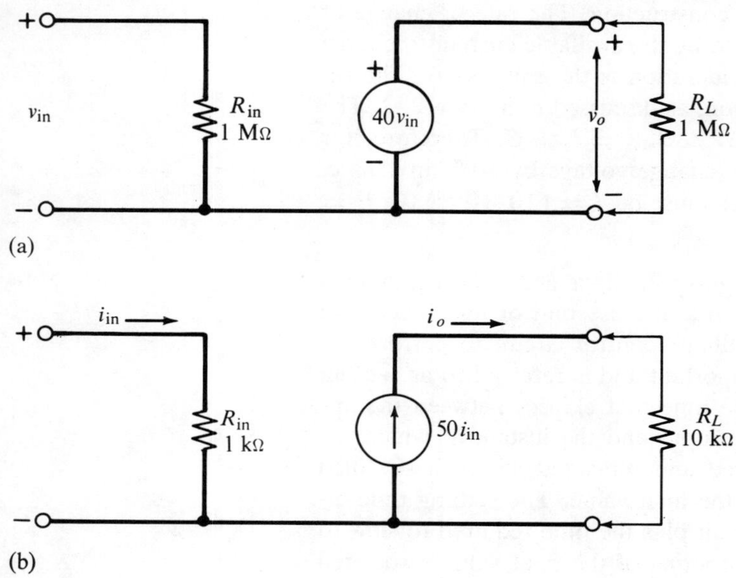

Fig. P–1

2. In the JFET amplifier of Fig. 2–201a, $R_s = 1$ kΩ, $R_d = 10$ kΩ, $C_s = 20$ μF, and $V_{DD} = 30$ V. For the particular JFET used, the transconductance $g_m = 2 \times 10^{-3}\ \Omega^{-1}$ and the drain-source dynamic resistance $r_{ds} = 100$ kΩ. (a) Assuming that the current v_{ds}/r_{ds} is negligible, calculate the amplifier

voltage gain. (b) For $v_{gs} = 1$ V, determine if v_{ds}/r_{ds} is significant compared to the current $g_m v_{gs}$.

3. Draw the small-signal model of the JFET amplifier of problem 2 including actual values of the small-signal components. Compare this model to the ideal voltage amplifier and describe how the JFET amplifier falls short of being ideal.
4. Determine the hybrid parameters h_{fe}, h_{ie}, h_{re}, and h_{oe} of the transistor whose characteristic curves are given in Fig. 2–203a through d.
5. The transistor whose hybrid parameters were found in problem 4 above is used in the amplifier circuit of Fig. 2–205. (a) Calculate the current gain, voltage gain, and power gain of the amplifier. (b) Calculate the amplifier input and output resistances.
6. Draw the small-signal model of the common-emitter amplifier of problem 5. Include specific values for all the small-signal circuit components. Compare the resulting model with that of the ideal current amplifier and describe how the BJT amplifier falls short of ideality.
7. The transistor used in the emitter-follower amplifier of Fig. 2–209 has an h_{fe} value of 60. A source is used with an equivalent resistance $R_u = 100\ \Omega$. The supply voltage $V_{CC} = 5$ V and $R_E = 5$ kΩ. (a) If $dv_{BE}/dv_u = 0.05$, calculate the emitter-follower voltage, current, and power gain. (b) Calculate the input and output resistance of the emitter-follower amplifier.
8. The output voltage of a difference amplifier changes by 5 V for an input difference signal of 250 μV, while the output voltage changes by 1 V for a common mode input signal of 500 mV. (a) What is the common mode rejection ratio (CMRR) of the difference amplifier? (b) Express the CMRR in decibels.
9. Using the differential input of an oscilloscope, it is desired to measure a 50 mV signal to within 1% accuracy. If the CMRR of the oscilloscope amplifier is 10^4, what is the maximum common mode signal which can be tolerated?
10. One circuit model of a difference amplifier is shown in Fig. P–10. The parameters are $\mu = 1000$, $R_{in} = 50$ kΩ, $R_o = 200\ \Omega$, and $R_L = 2$ kΩ. If the signal $v_{u2} = 0$, calculate the voltage gain, current gain, and power gain from the circuit model.

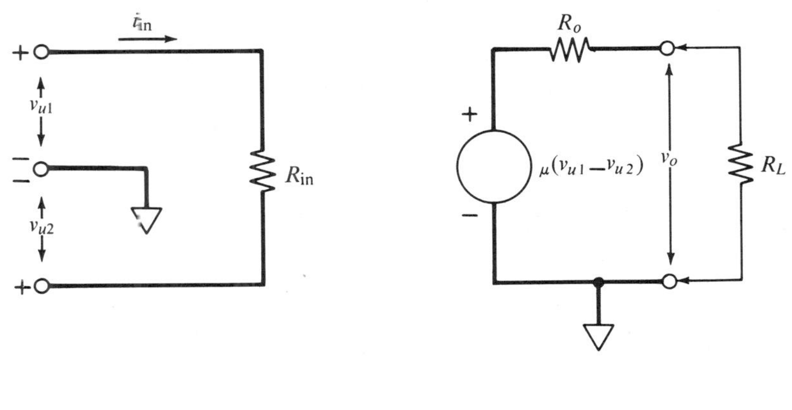

Fig. P–10

11. The same amplifier as in problem 10 is operated with a sine wave input at v_{u1} of 5 mV rms amplitude and $v_{u2} = 0$. Consider the load resistor R_L to be variable. (a) What is the maximum output power that can be drawn from the amplifier by adjustment of R_L? (b) What is the power gain at the value of R_L which gives maximum power?

Experiment 2–31 *Difference Amplifiers*

Equipment: ADD
breadboard card and chassis
two transistor sockets
one 6-pin socket
two 2N3393 npn transistors
one 2N3958 dual n-channel JFET
sine wave generator
capacitors
resistors
DVM
VRS
oscilloscope
can of freon

A difference amplifier is constructed and studied as a differential balanced amplifier, as a phase inverter, and as a single-ended differential amplifier. The amplifier is constructed with both npn transistors and matched JFET's. Characteristics of the amplifiers are determined, and the common mode rejection ratios are determined.

Construct the circuit of Fig. E2–62 without the capacitors C_1 and C_2 or either of the 100 kΩ base resistors. Connect input 2 (v_2) to a +5 V supply and input 1 (v_1) to the VRS. The amplifier is operating in the "balanced" mode. The output voltage is taken between v_{o1} and v_{o2}, so that

$$v_o = v_{o1} - v_{o2} = A_1 v_1 - A_2 v_2,$$

where A_1 and A_2 are the gains experienced by v_1 and v_2 respectively.

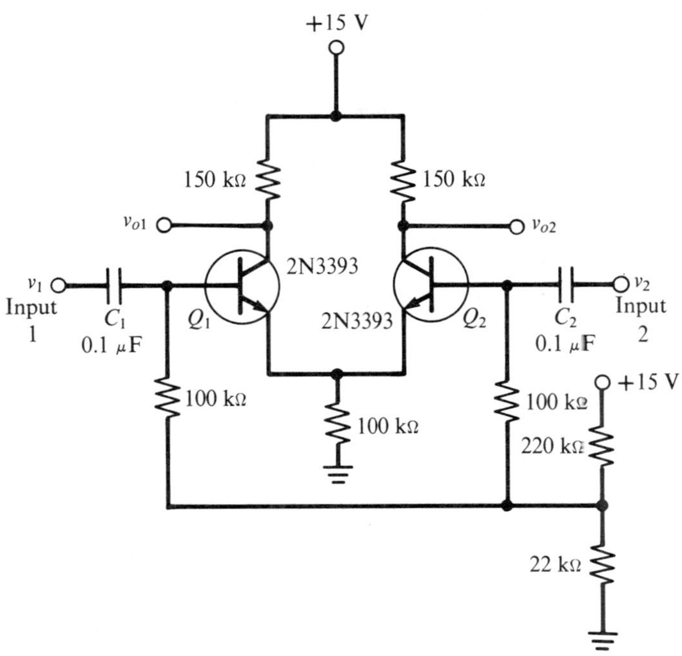

Fig. E2–62 BJT difference amplifier.

Set the VRS at exactly the same voltage as the +5 V supply, using the DVM to measure the voltages. Look at the outputs v_{o1} and v_{o2} with the scope. If the transistors are truly matched, the scope should display a dc signal at 0 V. *Note any deviations from this state below.*

Adjust the VRS until the output signal is just at 0 V. The difference between the VRS voltage and the +5 V supply voltage is the initial offset voltage of the amplifier and is a function of the difference in gain and in the conduction of the base-emitter junctions for the two transistors. *Record the initial offset value.*

The temperature of the semiconductor greatly affects its characteristics. If the temperature of one of the transistors, and not the other, should change, the balance of the amplifier will be greatly disturbed. Spray one transistor and not the other with freon to cool it. *What happens to the difference signal?*

Now spray both Q_1 and Q_2 with freon to bring them to the same temperature. *What happens to the difference signal?*

Insert capacitors C_1 and C_2 and the 100 kΩ base resistors, and allow Q_1 and Q_2 to reach ambient temperature. Connect input 2 to ground and the other input to a 1000 Hz sine wave source of about 100 mV amplitude. The amplifier is now in the "phase inverter" mode. Measure the output signal with the scope between each output v_{o1} and v_{o2} and ground.

Record the input and each output signal, noting relative magnitudes and time relationships.

As has been seen, a difference amplifier should respond only to signals at each input which are different. If the same signal is applied to both inputs, there should be no output. The ability of an amplifier to reject such a signal is called the common mode rejection ration (CMRR). It is given by

$$\text{CMRR} = \frac{\text{gain of amplifier to difference signal}}{\text{gain of amplifier to common signal}}.$$

Short out one of the 150 kΩ collector load resistors to make the circuit a single-ended difference amplifier. With one input grounded and with a 1000 Hz sine wave of about 40 mV amplitude applied to the other, measure the gain, v_o/v_{in}, taking the output from the collector that still has the 150 kΩ load. *Record the difference signal gain:* ________ .

Connect the same signal into both inputs at the same time and measure the gain. *Record the common mode gain:*

Calculate the common mode rejection ratio:

CMRR = ____________________

Construct the circuit of Fig. E2–63. The JFET's in the same package are a matched pair. *Verify this by characterizing this amplifier as the previous one was.*

Initial offset voltage = ____________________

Record the input and output signals for the phase inverter mode.

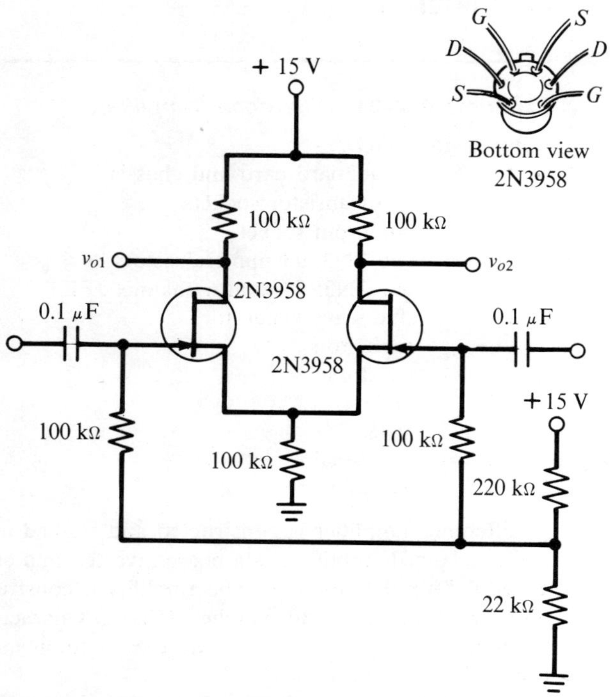

Fig. E2–63 Matched JFET difference amplifier.

Short one of the 100 kΩ collector load resistors. *Record the difference signal gain:* ____________ .

Record the common mode gain: ________ .

Calculate the CMRR:

CMRR = ____________________

Discuss why it is desirable for a difference amplifier to have a high CMRR. How might the CMRR be improved in the amplifiers constructed here?

Experiment 2–32 *Operational Amplifier Characteristics*

Equipment: ADD
741 type operational amplifier mounted on OA card
OA card
sine wave generator
oscilloscope
precision resistors
VRS
DVM

Two important characteristics of operational amplifiers which are easily measured are the common mode rejection ratio (CMRR) and the output resistance (R_o). In this experiment these values are measured for two different types of operational amplifiers in order to compare and contrast these characteristics. Other OA characteristics can be measured if desired.

Construct the circuit of Fig. E2–64 using the FET type OA. Connect the VRS and obtain v_o as a function of v_{in} as shown in the table. Measure v_o and v_{in} with a DVM. It was seen in Experiment 2–31 that a difference amplifier should respond only to signals at each input which are different. If the same signal is applied to both inputs, there should be no output voltage for a wide range of common mode signals. The CMRR is

$$\text{CMRR} = \frac{\text{gain of amplifier to difference signal}}{\text{gain of amplifier to common signal}}.$$

For the circuit of Fig. E2–64 the CMRR can be calculated by dividing the nominal voltage v_n by the voltage difference $(v_o - v_{in} - v_{off})$.

Nominal voltage, v_n	v_{in}	v_o	$v_o - v_{in}$	$v_o - v_{in} - v_{off}$
0.1				
1.0				
5.0				
10.0				

Note: Use $v_o - v_{in}$ at $v_{in} = 0.1$ V for v_{off} value.

CMRR = ________________

Next the OA output resistance will be measured. Set the VRS at about 1 V and connect it at v_{in}. Measure the output voltage v_o with a DVM. Now connect a 50 Ω, 1% resistor as the load across the output and again measure the output voltage v_{oL}.

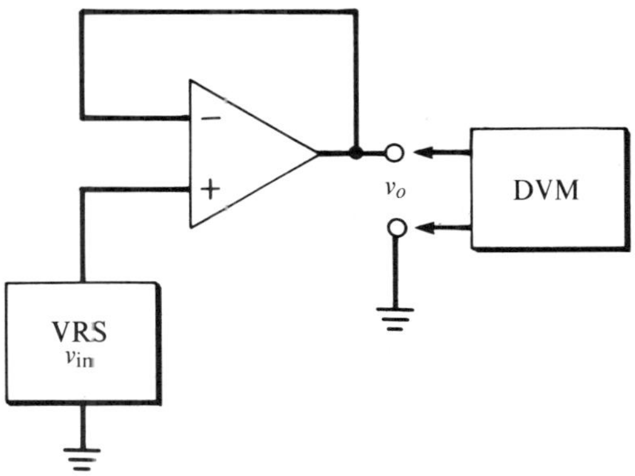

Fig. E2–64 Circuit to measure common mode gain.

Calculate R_o.

v_o = ________________

v_{oL} = ________________

$i_o = v_{oL}/50 =$ ________________

$R_o = (v_o - v_{oL})/i_o =$ ________________

Repeat the above measurements of CMRR and R_o for the 741 OA.

CMRR = ________________

R_o = ________________

Automatic Feedback Control

Section 2–6

The basic concepts of feedback control (servo) systems were described in Section 1–4 of Module 1, *Electronic Analog Measurements and Transducers.* The potentiometric recorder, potentiometric amplifier, and operational amplifier voltage and current followers have all been presented as part of a family of automated null comparison systems that are usually referred to as **feedback control systems.** The same basic automated null comparison concepts are inherent in all servo systems whether they are used in controlling a ship, spacecraft, manufacturing process, or laboratory experiment.

The applications of the feedback control concept are rather overwhelming, yet the underlying idea is simple and elegant. The servo system recognizes a specific command (such as a command to move an object to a certain position), responds to the command (by comparing the actual position to the desired position of an object), and then takes the necessary steps to correct any difference.

The amplifiers, power supplies, motors, gears, and other hardware can be important parts of the servo system, but it is the total system that must operate so as to recognize and fulfill a specific command by feedback control. Obviously, system design is very important, and as one looks at real servos certain practical system problems become apparent.

No system can respond instantaneously to a command, and unfortunately the more quickly the system reacts to obey the command, the greater are the chances of instability. Consider, for example, a position servo. No mass can be moved instantaneously to fulfill a command for change of position because this would require infinitely rapid acceleration and deceleration and therefore infinite force. Nor can any electronic circuit respond instantaneously, because of propagation delays in moving charge carriers and because of the capacitance and inductance that are inherent in the devices and leads used in circuit construction (as discussed under slew rate in the previous section).

All systems that detect a difference between the actual and the desired states of a controllable quantity (e.g., voltage, temperature, chemical composition, position) and then feed the difference information back to a controlling device, which then reacts and causes the difference to become essentially zero, are classified as **servo systems.** A jet airliner or spacecraft is controlled by many types of mechanical, electromechanical, and electronic servos. A human is controlled by many types of servos, and one type of human servo will be used in the next paragraph as an illustration for focusing on the functional blocks of all servo systems. Then some of the servo response characteristics will be considered and the section will conclude with a discussion of programmable power supplies operated in current-regulated and voltage-regulated modes.

2–6.1 BASIC SERVO COMPONENTS

An understanding of the basic components in a servo system can be gained by considering how a human operates in response to an order to position an object. As shown by the block diagram (Fig. 2–215a), the command is picked up by the eye, ear, or touch, and a signal is transmitted to the human brain, which controls the body muscles that exert a force on the load to move it toward the desired location. The eye or other sensor acts as an information feedback device that describes the actual position. The mind compares the command and feedback information and controls the body so as to change the load position in accordance with the error signal, i.e., the difference between the desired and actual positions of the object. The man-made servo operates in a similar way, as shown in Fig. 2–215b. The command v is picked up by a device that puts it in the form of an acceptable command signal r, and the information on the actual position of the load is put in the form of a feedback signal b. These are usually voltage signals that can be compared to give a difference voltage e that is fed to an electronic signal controller. The controller uses the error signal e to regulate the electrical power to the motor (or other positioner), and the motor operates to move the load to the desired position. When the load is at the desired position, the error (actuating) signal e is essentially zero. Note that the terms "error signal" and "actuating signal" have the same meaning here and are used interchangeably.

The AIEE (American Institute of Electrical Engineers) has recommended standard symbols for feedback control systems; these are shown with a general block diagram (Fig. 2–216) that contains the basic elements and variables. The definitions are as follows:

The **command** v is the input, which is established by some means external to and independent of the feedback control system.

The **reference input elements** a produce a **reference input** r which is proportional to the command.

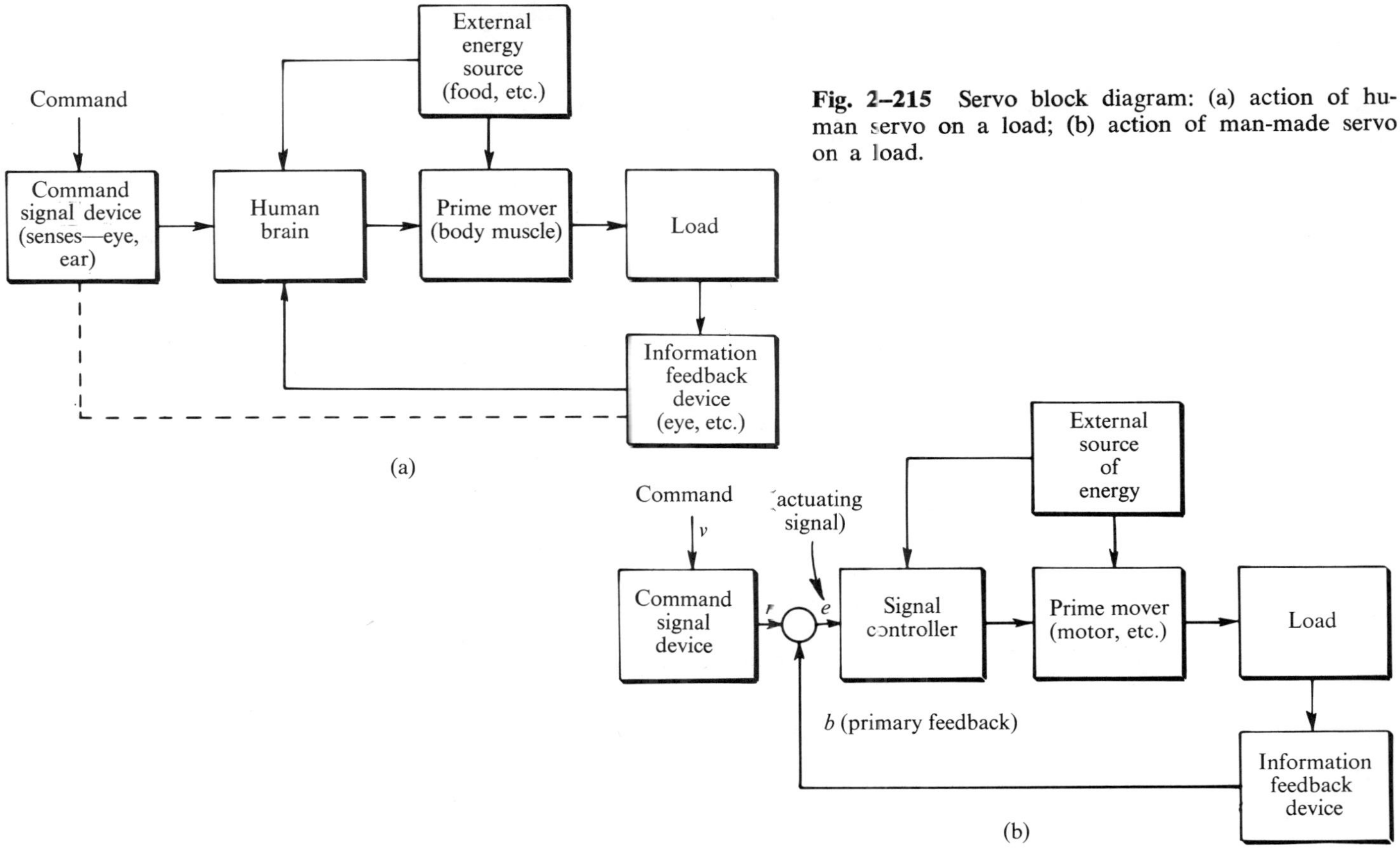

Fig. 2–215 Servo block diagram: (a) action of human servo on a load; (b) action of man-made servo on a load.

Fig. 2–216 General block diagram for the feedback control system.

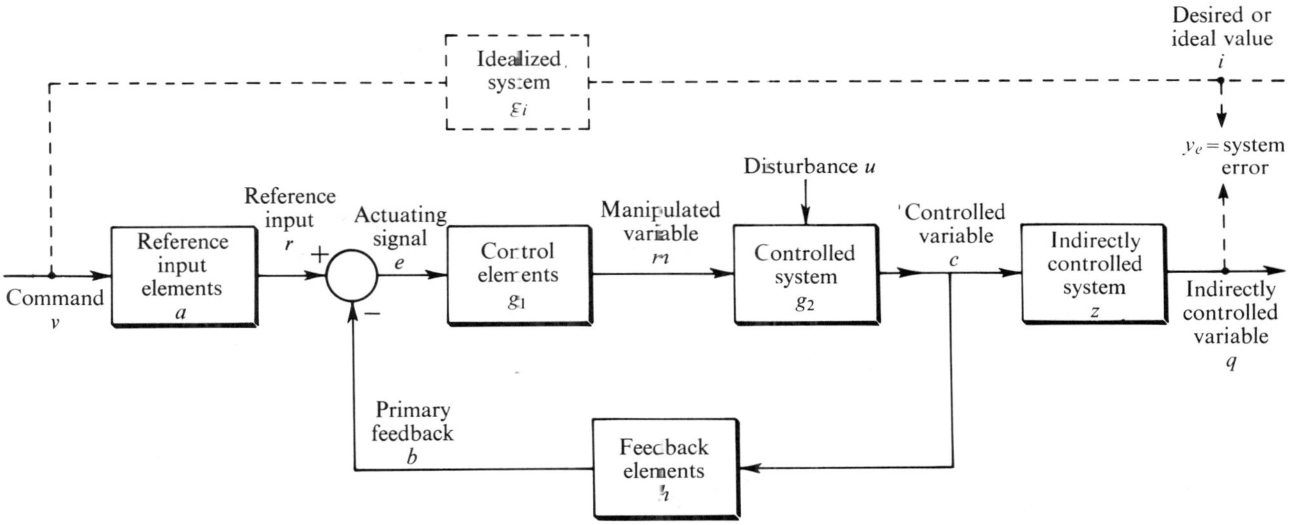

The **control elements** g_1 produce the **manipulated variable** m which is applied to the controlled system. The manipulated variable is generally at a higher energy level than the actuating signal and may also be modified in form.

The **controlled system** g_2 is the device that is to be controlled.

The **controlled variable** c is the quantity that is directly measured and controlled. It is the output of the controlled system.

The **feedback elements** h produce the **primary feedback** b, which is a function of the controlled variable and is compared with the reference input to obtain the **actuating signal** e. Therefore, $e = r - b$; this signal is usually at a low energy level and is fed to the control elements.

The **indirectly controlled variable** q is the output quantity and is related through the **indirectly controlled system** z to the controlled variable. It is outside the closed loop and is not directly measured for control.

The **ultimately controlled variable** is, in a general sense, the indirectly controlled variable. In the absence of the indirectly controlled variable, this term refers to the controlled variable.

The **idealized system** g_i is one whose performance is agreed upon to define the relationship between the **ideal value** i and the command.

The **ideal value** i is the value of the ultimately controlled variable that would result from an idealized system operating from the same command as the actual system.

The **system error** y_e is the ideal value minus the value of the ultimately controlled variable.

The **disturbance** u is the unwanted signal that tends to affect the controlled variable. The disturbance may be introduced into the system at many places.

2–6.2 SERVO RESPONSE CHARACTERISTICS

It has been mentioned that no servo system, whether mechanical, chemical, electromechanical, or all-electronic, can respond instantaneously to a command signal. The inherent characteristics of the devices used to perform the servo operations introduce certain time lags. The general response considerations are quite similar for all servos, and in this section the electromechanical position servo (servomechanism) is used as an illustrative example.

Force Balance in a Servomechanism

The motive force in the electromechanical servo system is the electromotive torque developed in the servo drive motor (prime mover). This torque acts to accelerate the load and to overcome the frictional and damping forces in the system. For most instrument servomotors the torque is roughly proportional to the voltage in the control winding. If the error-signal amplifier is linear, the output voltage of the amplifier is proportional to the error-signal input. Finally, if the system is completely linear, the error signal is proportional to the difference between the actual and the command positions of the load. Therefore the relationship between the load position c and the motor torque T_m is

$$T_m = K'(c_0 - c), \tag{2–185}$$

where c_0 is the command position of the load, or that load position which would give no error signal to the amplifier, and K' is a constant with units of torque per load displacement.

For Eq. (2–185) to be true at all times, it is necessary that the amplifier gain, which affects K', not be a function of frequency. For chopper amplifiers, then, the chopper frequency must be high compared with the highest frequency command that the servo is expected to follow. The signal delay in the amplifier (phase distortion) should also be negligible for signals of the fastest expected response speed.

In considering the linear translation of a load, such as the pen in a recording potentiometer, it is more convenient to use the unit "force" than "torque." The force is simply the torque divided by the lever arm. When the load is positioned by a drive pulley concentric to the motor shaft, the lever arm l is a constant and the force developed in the motor, F_m, as applied to the system, is

$$F_m = T_m/l = (K'/l)(c_0 - c) = K(c_0 - c). \tag{2–186}$$

The force that actually accelerates the load, F_a, is the motor force F_m minus the force lost to damping and frictional forces, F_d:

$$F_a = F_m - F_d. \tag{2–187}$$

The force F_a is related to the acceleration a of the load by

$$F_a = Ma, \tag{2–188}$$

where M is the mass of the load. The damping and frictional forces are generally proportional to the velocity u of the load. The constant of proportionality will be given the symbol D:

$$F_d = Du. \tag{2–189}$$

We assume that all frictional surfaces are lubricated so that the starting friction and unlubricated sliding friction are negligible. Substituting Eqs. (2–186), (2–188), and (2–189) into Eq. (2–187), we obtain

$$Ma = K(c_0 - c) - Du \tag{2–190}$$

or

$$Kc_0 = Ma + Du + Kc. \tag{2–191}$$

Expressing the acceleration and the velocity as the second and first time derivatives of c, we can write

$$Kc_0 = M\frac{d^2c}{dt^2} + D\frac{dc}{dt} + Kc. \tag{2–192}$$

Response to Step Command

To characterize the response of the system, one of the traditional tests is to apply a step signal at the input, commanding that the load, at standstill, be located at a new position instantly. From Eq. (2–190), it can be seen that, the greater the command displacement, the greater the accelerating force, and the greater the mass, the less the acceleration.

To consider the effect of the damping term, consider the step response of a system with no damping, i.e., one in which the damping and frictional forces are negligible ($D = 0$). The load accelerates at a rate of $K(c_0 - c)/M$. As c approaches c_0, the acceleration decreases, but until $c = c_0$, the load has continued to gain speed. At this point the load ceases to accelerate and would just maintain its acquired velocity except that, in going beyond c_0, an accelerating force in the opposite direction is developed by the system. Finally, at some position on the other side of c_0 from the initial position, the load is stopped by the accelerative force, which always acts toward c_0. Now the load, accelerated toward c_0, increases its velocity in that direction until it passes c_0 again. This sequence is shown as load position plotted against time in Fig. 2–217. If there are no damping or frictional forces, the load will continue to oscillate about c_0.

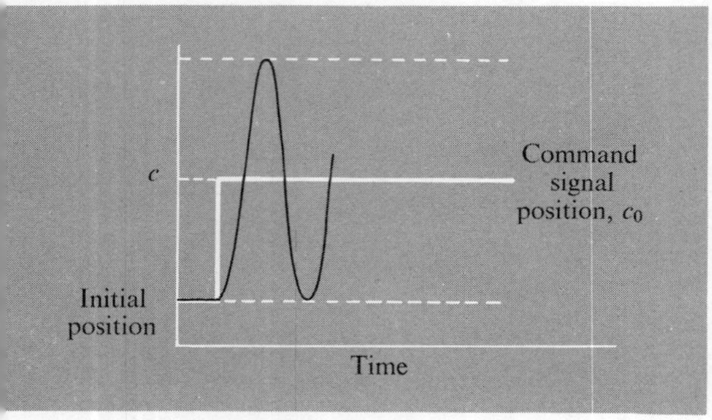

Fig. 2–217 Servo response to step command, no damping.

When damping or viscous-friction forces are present, the accelerative force is decreased by an amount proportional to the velocity. Again consider Eq. (2–190) as applied to a load approaching c_0. As the velocity u increases and the error $(c_0 - c)$ decreases, Du will, at some c, exceed $K(c_0 - c)$ and the net force will become decelerative (negative Ma). The deceleration of the load before c_0 is attained reduces the overshoot of the load. The overshoot will decrease each half cycle until the load comes to rest at c_0. This action is called a damped oscillation and is shown in Fig. 2–218. If the damping is increased still further, the overshoot can be eliminated altogether but the approach to c_0 is much slower. This is shown in Fig. 2–219.

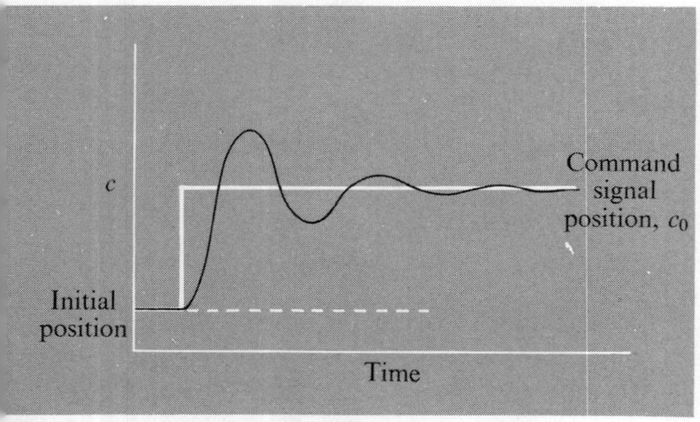

Fig. 2–218 Response to step command, some damping.

Both overshoot and slow response are distortions of the desired response to the command signal, and therefore they are important parameters of the servo system. The overshoot is often described as a fraction of the

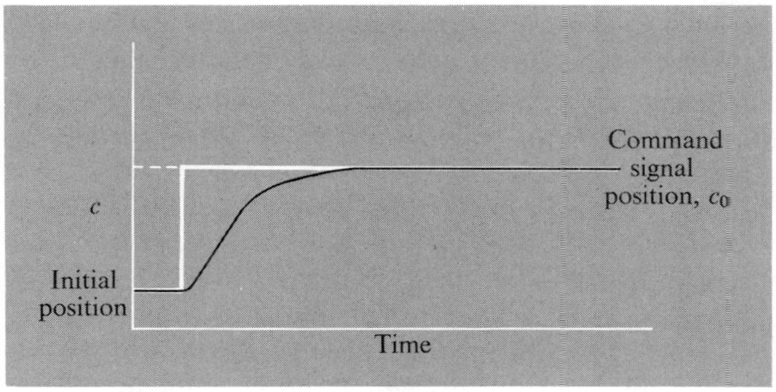

Fig. 2–219 Response to step command, damped to eliminate overshoot.

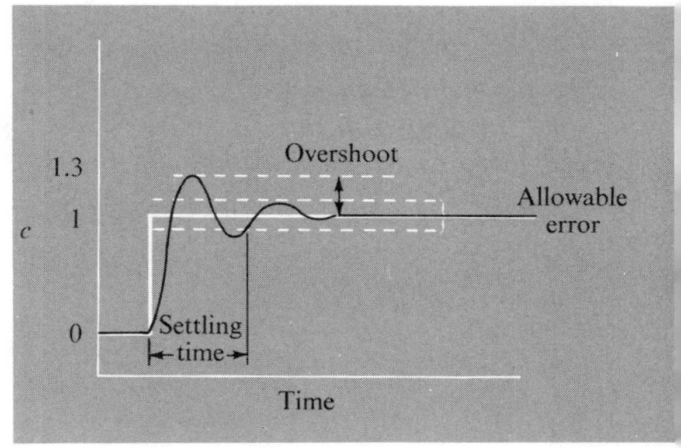

Fig. 2–220 Servo response characteristics.

command displacement. That is, if the command was to move the load 10 cm and the initial overshoot was 1 cm, the overshoot would be 0.1, or 10 percent. A measure of the response speed is the "settling time," the time required for the load to come to the command position within a prescribed error tolerance. The measure of overshoot and settling time is shown in Fig. 2–220. This is the same type of consideration as for the settling time of OA feedback circuits discussed in the previous section.

An *RLC* Analog of Servo Response

It can be shown that the undamped servo system is an oscillator. This is easiest to do by showing that the positioning of the load according to Eq. (2–192) is exactly analogous to the movement of charge in a series-resonant *LC* circuit. When the switch of Fig. 2–221 is closed, the command is to transfer charge to capacitor *C* through inductance *L* and resistance *R*. The sum of the voltages across the circuit elements is equal to the battery voltage. The voltage across the resistor is iR; across the conductance, $L(di/dt)$; across the capacitor, q/C, where q is the charge on the capacitor. The summation of the voltages in the loop of Fig. 2–221 may be written

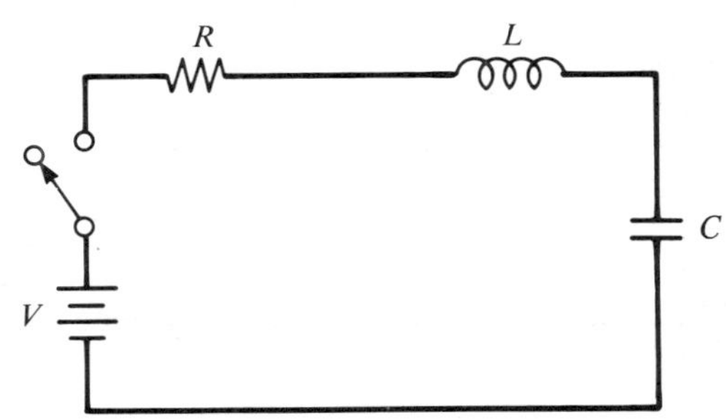

Fig. 2–221 Electrical analog of servo system.

$$V = L\frac{di}{dt} + iR + \frac{q}{C}$$

or, since $i = dq/dt$,

$$V = L\frac{d^2q}{dt^2} + R\frac{dq}{dt} + \frac{q}{C}. \tag{2–193}$$

Thus we see that L is analogous to M, R to D, and $1/C$ to K of Eq.

Note 10. Resonant Frequency.
The resonant frequency, f, of an RLC circuit occurs when the capacitive reactance, X_C, is equal to the inductive reactance, X_L, and is $f = 1/(2\pi\sqrt{LC})$.

(2–192). If there were no resistance, the *LC* circuit excited as in Fig. 2–217 would oscillate at its resonant frequency[10] theoretically forever. The introduction of resistance results in a damped oscillation, or **ringing.** The resonant frequency is $f_0 = 1/2\pi\sqrt{LC}$ (as will be explained in Module 3). It follows that the natural frequency of the undamped servo system will be

$$f_0 = \frac{1}{2\pi} \cdot \frac{\sqrt{K}}{M}. \qquad (2\text{–}194)$$

Damping Factor

It is advantageous to know the amount of damping necessary to reduce the overshoot of the system to a tolerable level. The amount of damping required will be larger for a larger mass, because a larger mass acquires less velocity in a given time for the same accelerating force. Also, in order to bring about a deceleration for a particular value of $c_0 - c$, D must be larger for larger values of the motive force K. The following formula for the damping factor ξ shows the relationship between D, K, and M for a constant damping effect:

$$\xi = \frac{D}{2\sqrt{KM}}. \qquad (2\text{–}195)$$

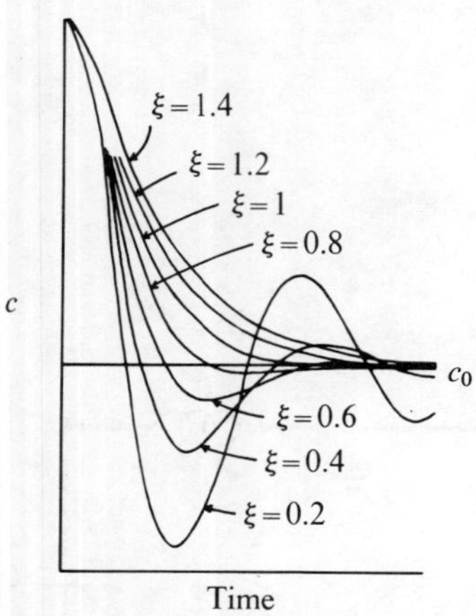

Fig. 2–222 Servo response as a function of damping factor.

Figure 2–222 shows the step response as a function of the damping factor. Note that, when the damping factor is less than unity, there are overshoot and damped oscillations. An unnecessarily long time is required to reach command position when the damping factor is greater than 1. A system with a damping factor of unity is said to be critically damped. Most servo systems are operated with a damping factor between 0.6 and 1.0. The lower damping factors give a quicker response and, in spite of the overshoot, a smaller settling time than critical damping. If the damping factor is much below 0.6, the damped oscillations take too long to subside and the settling time is again increased.

Note also from Fig. 2–222 that the damping factor has an effect on the period of the damped oscillation. The actual frequency is given by the following equation when $\xi < 1$:

$$f = f_0\sqrt{1-\xi^2} = \frac{1}{4\pi M}\sqrt{4KM - D^2}. \qquad (2\text{–}196)$$

Saturation. The assumption of the preceding arguments that the motor torque is proportional to the error in the load's position cannot be true for very large errors. There is a maximum torque for any given motor, and any amplifier has a maximum output capability. When one of these maxima is reached, the drive system is saturated and a further increase in error signal cannot increase the restoring torque. Today most servo systems are designed to saturate with an error which is only a small percentage of the span. For large displacement, therefore, the term $K(c_0 - c)$ in Eq. (2–190) has a maximum value regardless of c:

$$Ma = K(c_0 - c) - Du.$$

This maximum force will be called K_{max}. The initial acceleration of the load will be $a = K_{max}/M$. As the velocity increases, the acceleration decreases until finally the damping force is equal to the driving force and the velocity is constant and maximum. When this is true, $K_{max} = Du_{max}$. Solving for the maximum velocity, we obtain

$$u_{max} = \frac{K_{max}}{D}. \tag{2–197}$$

This demonstrates that the maximum velocity in saturation is inversely proportional to the damping; so, here again, excess damping must be avoided if response speed is an objective.

One of the advantages of operating the system in saturation is that the system always responds with its maximum response speed. When the load comes close enough to the command position to bring the system out of saturation, the driving torque decreases and the load decelerates as it approaches c_0. The response of the system operating for a time in the saturated region is shown in Fig. 2–223. In this case, the settling time is a linear function of the displacement, and *the overshoot is a constant fraction of the control-system span,* rather than a constant fraction of the input step.

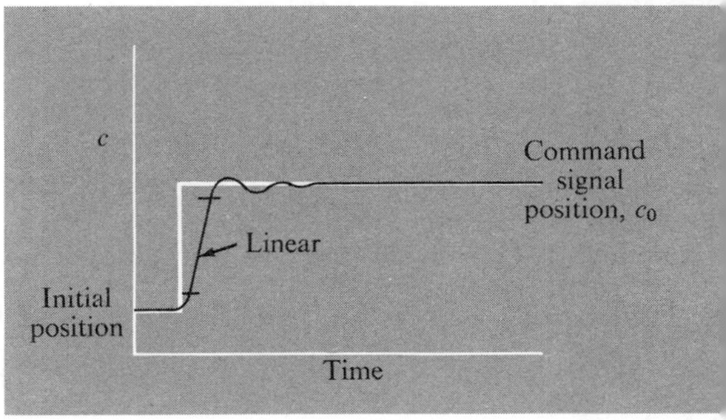

Fig. 2–223 Step response of saturated servo system.

Position Errors

Errors in servo positioning arise from several sources. One of these errors is that the load position does not immediately coincide with the command position. This difficulty has been discussed above. Since it cannot be overcome completely, it is certainly of concern when the load must accurately follow the command signal.

Another source of position error is the "dead zone" of the system. This is the position error for which the error signal is too small to result in a rotation of the servo motor. An obvious way to correct this difficulty is to increase the amplifier gain. This also has the effect of driving the system into saturation for smaller displacements. The limit of practical amplification is reached when the amplifier is so sensitive that it cannot distinguish between noise and the error signal.

Backlash, due to a looseness in the coupling between the load and the motor, leads to a kind of oscillation. This is due to the momentum built up by the motor rotating through the backlash region. Backlash between the load and the position indicator will cause a hysteresis in the positioning.

An oscillation can also occur as the result of a time delay in the feedback control loop. That frequency for which the delay is exactly a half cycle will be fed back in phase rather than out of phase.

The Ramp Input Test

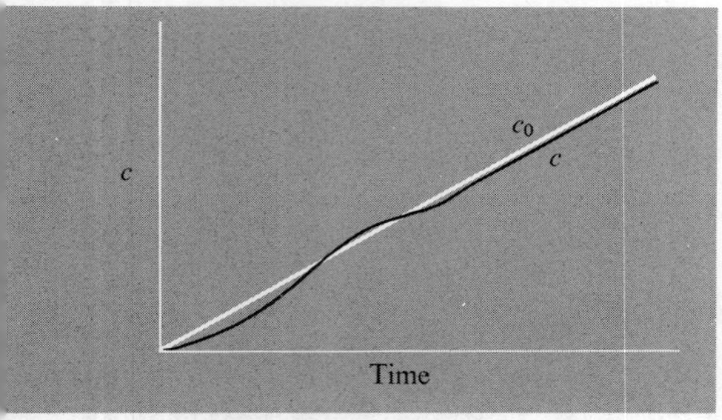

Fig. 2–224 Servo response to ramp command.

Another standard test command signal is the ramp, a linearly increasing potential. The usual response of the linear servo system to the ramp is shown in Fig. 2–224. The servo should move the load at a constant velocity from the initiation of the ramp. The load cannot be accelerated to that velocity instantly, and so it lags. As the load accelerates to catch up, its velocity exceeds the command velocity and overshoot occurs. The system settles down to follow the command velocity, but the command signal is always a little ahead of the load at any given time. The reason is that, to have the servo motor rotating at a constant velocity, there must be an error signal at the input. It can be seen from Eq. (2–190) that the steady-state displacement error will be proportional to D and u and inversely proportional to K. As the command signal approaches u_{max}, the servo loses the ability even to reproduce the slope of the command signal.

2–6.3 OPERATIONAL POWER SUPPLIES

Several power supplies are usually found in each modern measurement or control instrument. For example, a relatively simple instrument for color measurement requires a current-regulated supply at 2 A for a nominal 12 V light source, and voltage-regulated supplies of +5 V at 1 A for the TTL circuits, +15 V and −15 V at 100 mA for the OA's, and 1000 V at 1 mA for the photomultiplier transducer. Other instruments not only require regulated supplies at fixed current or voltage levels, but also require that specific supplies be programmable and respond rapidly to control signals provided either by an internal program or by remote signals from a computer or other signal controller.

These modern regulated and programmable power supplies are electronic servos. They can recognize a command to provide a specific current or voltage output, make a comparison between the actual and desired values, and take action either to make the actual value equal to a preset command value (regulated supply) or to respond rapidly and track accurately a remote command value (programmable supply).

Many types of discrete component circuits have been used for power supplies, but most of the newest supplies utilize integrated operational amplifier circuits to provide the feedback control. The regulated supply can be described in terms of OA analysis, and a supply with an OA controller is frequently called an **operational power supply.**

The great usefulness of accurate and precisely controlled voltage and current power supplies, and the high accuracy that can be obtained by electronic feedback control, make these supplies ideal examples to conclude this section on automatic feedback control.

Voltage-Regulated and Controlled Power Supply

When the ac line power is adjusted with a transformer, rectified, and filtered to provide a dc output voltage, the magnitude of the output from the raw dc supply is dependent on the line voltage and the attached load. The principles of a series regulator that will adjust and hold the output voltage at a command value, even with rather large changes in both line voltage and load, were introduced in Section 2–4.1. The discussion of these concepts is now enlarged in terms of feedback control, and a practical circuit is developed.

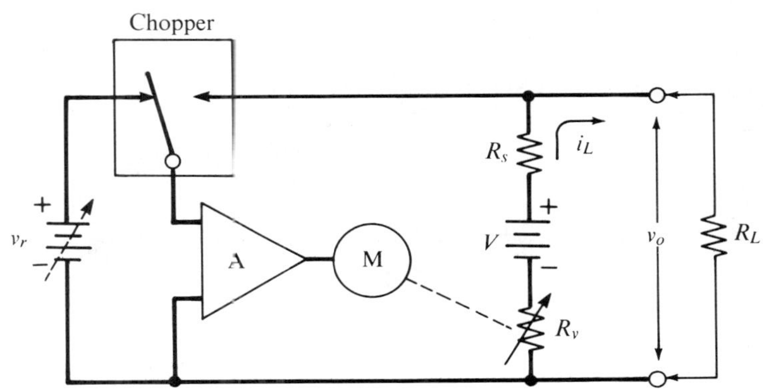

Fig. 2–225 Voltage-regulated supply with electromechanical feedback.

Electromechanical Feedback. The raw dc supply as obtained by a transformer, rectifier, and filter is represented simply by V with an internal resistance R_s in Fig. 2–225. In series with the raw supply is a variable power resistor, R_v, which is varied by mechanical coupling to a motor (M) which operates from the output of a servo amplifier. At the input of the servo amplifier the output voltage v_o is compared with the reference voltage v_r by using a chopper to look alternately at v_r and v_o. If there is any difference between v_r and v_o, the output of the high-gain amplifier provides the necessary power output for the motor shaft to turn, changing R_v and varying v_o until $v_o = v_r$. Since

$$v_o = V - i_L(R_s + R_v), \qquad (2\text{–}198)$$

it is readily seen that any change in the raw supply voltage V because of line voltage changes, or in the term $i_L(R_s + R_v)$ because of load current changes, will cause a change in v_o. However, this error (difference between v_o and v_r) is quickly sensed at the input to the servo amplifier, and the motor shaft rotates and changes R_v so as to bring v_o back to the stable preset reference voltage v_r. If v_o is greater than v_r, the motor shaft turns in the direction to increase R_v, and vice versa. Obviously, the servo amplifier must recognize whether $v_o > v_r$ or $v_o < v_r$ so that the motor shaft will turn in the correct direction to increase or decrease R_v.

This type of electromechanical servo voltage regulator is seldom used anymore because of its high cost and slow response, but it does illustrate the basic principles of all series voltage regulators.

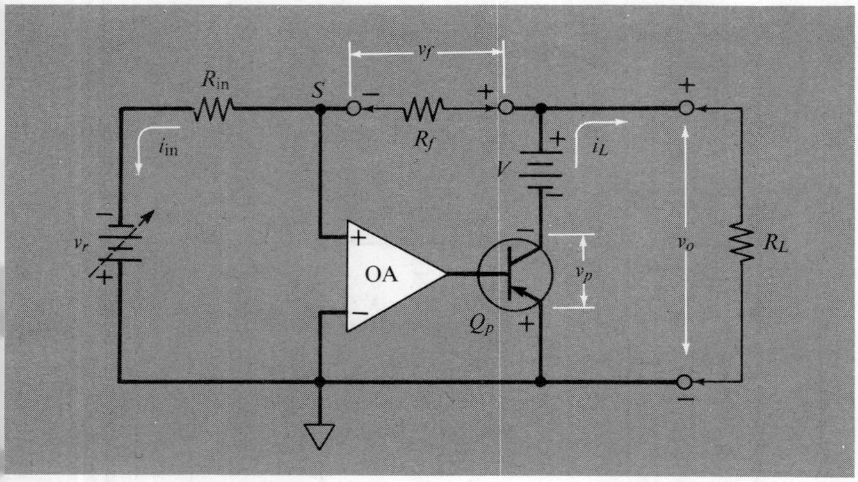

Fig. 2–226 Voltage-regulated supply with electronic feedback.

Electronic Feedback. An electronic servo voltage regulator is illustrated in Fig. 2–226. A pnp power transistor that is capable of carrying the load current i_L is connected in series with the raw supply V. The power transistor is called the **pass transistor** and acts essentially as a variable resistance similar to R_v of Fig. 2–225. It provides the variable voltage drop v_p which compensates for any fluctuations in V or i_L.

As in all OA feedback circuits, the sum of the voltages around the feedback loop from the summing point S to the noninverting input is zero. Therefore,

$$v_o = -v_f \tag{2–199}$$

and, since $v_f = i_f R_f$, $i_f \cong i_{in}$, and $i_{in} = v_r/R_{in}$, it follows that

$$v_o = -i_{in}R_f = -v_r\left(\frac{R_f}{R_{in}}\right). \tag{2–200}$$

The circuit in Fig. 2–226 to the right of the summing point, including the raw supply, control amplifier, and pass transistor (but not the feedback resistor R_f), can be considered as a high-powered OA. The combination can be represented by a triangular symbol and used as an OA for typical operational functions such as adding and integrating, but with the advantage of higher power output. Of course, the output voltage for the circuit shown in Fig. 2–226 will swing only in the positive direction. This is sufficient in many cases. It is necessary to add an npn power transistor and oppositely connected raw supply to the circuit of Fig. 2–226 if the OA output is to be bipolar. If this is done, the output v_o will swing either positive or negative depending on the direction of the input current.

Voltage Control and Programming. If the reference voltage v_r and R_{in} are fixed values, then R_f can be used as the **voltage control.** It can be seen from Eq. (2–200) that v_o is directly proportional to R_f, and the volts/ohm ratio for R_f is equal to v_r/R_{in}. For example, if a 5 V Zener supply is used for v_r and $R_{in} = 5000\ \Omega$, then v_o changes by 1 mV per ohm change of R_f. From Eq. (2–200) it is also apparent that a variable reference source could be used for linear control of the output v_o, and R_{in} could be used for an inverse relationship.

The output voltage can be electromechanically or electronically programmed. A timing motor with a cam selector switch can switch in different values of v_r, R_f, or R_{in}. Much higher-speed programming can be accomplished by using electronic switches driven by logic circuits.

Voltage Output Errors. The voltage-regulated power supply is evaluated and specified by its changes in output v_o versus changes in line fluctuations, load variations, temperature variations, and changes over long time periods. In Section 2–5 the input bias current (i_b) and offset voltage (v_{off}) of OA's were described and their variation with temperature was indicated. If the OA gain is very high and i_b and v_{off} are significant, the expression for output voltage is

$$v_o = -v_r\left(\frac{R_f}{R_{in}}\right) \pm i_b R_f \pm v_{off}\left(1 + \frac{R_f}{R_{in}}\right) \qquad (2\text{–}201)$$

and the variation in the output voltage is

$$\Delta v_o = -\Delta v_r\left(\frac{R_f}{R_{in}}\right) \pm \Delta i_b R_f \pm \Delta v_{off}\left(1 + \frac{R_f}{R_{in}}\right). \qquad (2\text{–}202)$$

If the accuracy of the voltage output is to be reliably specified, it is important to include data on Δv_r, Δi_b, and Δv_{off} as a function of temperature, time, line voltage, and load. Also, any changes in R_{in} or R_f should be considered, but it is assumed that resistors with low temperature coefficients are used and that the input and feedback currents are within bounds so that self-heating of the resistors does not cause significant change in the resistance ratio R_f/R_{in}. The value of v_o can be regarded as the output signal and the largest $|\Delta v_o|$ is the worst-case error. The maximum relative error for specified conditions is

$$\text{percent relative error} = 100 \frac{\Delta v_o}{v_o}. \tag{2–203}$$

A very well-regulated voltage supply would have a relative error of <0.01% for line variation of 105–125 V, a change from no load to full load, a change of ±5°C from ambient temperature, and over an eight-hour period.

Integrated Circuit Regulators. Inexpensive integrated circuit voltage regulators are now available for many applications that provide 0.01 to 1% regulation. These regulators contain both a control amplifier and a Zener supply on the semiconductor chip. If the pass transistor is included on the same semiconductor chip, it must dissipate considerable heat. Even with a good heat sink, the changes in V and i_L cause significant changes in the temperature-dependent parts of the circuit and cause regulator errors. Therefore, at present the high accuracy reference supplies do not include the pass transistor on the regulator chip.

Current-Regulated and Programmable Power Supply

There are many transducers, light sources, and circuits that require a constant current or a programmable sequence of accurate current signals for operation. Again OA feedback circuits can provide accurate and programmable current control. Two types of current-regulated supplies are shown in Fig. 2–227.

The current-regulated supply of Fig. 2–227a is basically the voltage-follower-with-gain configuration. The current i_L in the load R_L is supplied from the raw power supply and the power transistor Q_p. A current-sensing resistor R_{cs} is used to monitor the load current. If the amplifier bias current is negligible, the current in R_{cs} is equal to i_L. The amplifier maintains the voltage v_u at the inverting input equal to the reference voltage v_r. Since $v_u = i_L R_{cs}$, it follows that $i_L = v_r/R_{cs}$ within the limitations imposed by the current capabilities of the power supply. The load current can be adjusted very precisely by appropriate choices of v_r and R_{cs}. Also remote programming is possible by controlling v_r. This type of current-regulated supply is limited to rather low-voltage applications since the output voltage of the OA is approximately equal to the voltage drop across the load.

A second type of current-regulated supply that has been frequently used is illustrated in Fig. 2–227b. The current i_L through the load is equal to the

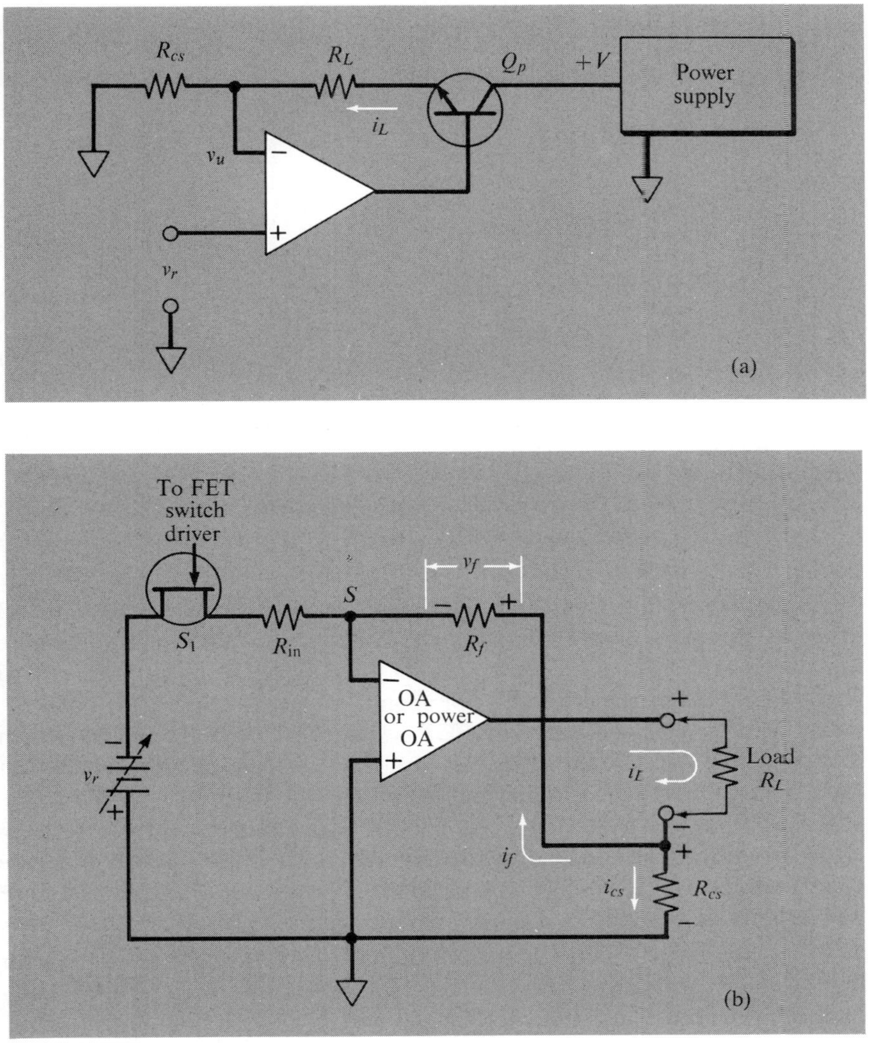

Fig. 2–227 Programmable current-regulated power supplies: (a) noninverting OA; (b) inverting OA.

sum of the current i_{cs} through the current-sensing resistor R_{cs} and the current i_f through the feedback resistor R_f. That is,

$$i_L = i_{cs} + i_f. \tag{2–204}$$

The voltage drop v_{cs} across the current-sensing resistor must equal the voltage drop v_f across R_f to maintain the sum of the voltages between common and the summing point S equal to zero; and since

$$v_{cs} = i_{cs}R_{cs} \qquad \text{and} \qquad v_f = i_f R_f,$$

it follows that

$$i_{cs}R_{cs} = i_f R_f \qquad \text{and} \qquad i_{cs} = i_f\left(\frac{R_f}{R_{cs}}\right). \tag{2–205}$$

If Eq. (2–205) is substituted in Eq. (2–204),

$$i_L = i_f\left(\frac{R_f}{R_{cs}}\right) + i_f = i_f\left(\frac{R_f}{R_{cs}} + 1\right). \tag{2–206}$$

The input current $i_{\text{in}} = i_f$ and $i_{\text{in}} = v_r/R_{\text{in}}$, which can be substituted for i_f in Eq. (2–206), so that

$$i_L = \frac{v_r}{R_{\text{in}}}\left(\frac{R_f}{R_{cs}} + 1\right). \tag{2–207}$$

It can be seen from Eq. (2–207) that the load current can be set at a desired value by choosing precise values for v_r, R_f, R_{in}, and R_{cs}. The current can be controlled over a range by varying any of these values. The current can also be programmed by remote control of v_r and/or the resistors R_f, R_{in}, and R_{cs} with logic level signals driving FET switches. The FET analog switch shown in Fig. 2–227 can be used to turn the load current ON and OFF at very high speeds. When the switch S_1 is ON, the load current will be as specified by Eq. (2–207), but when S_1 is OFF, the load current will be zero. The current through the load will be controlled accurately according to the parameters in Eq. (2–207) and can be programmed in a precise ON-OFF sequence as determined by logic level pulses to the FET driver.

This type of programmed current source can be used for low voltage or high voltage applications and the programmed current range will be determined by the power capability of the raw supply and the operational amplifier, and the magnitudes of the parameters shown in Eq. (2–207). For loads such as hollow-cathode lamps that require high voltage and high current when operated in a switched mode, the triangular symbol in Fig. 2–227 would have to be a power OA similar to the combination of OA, Q_p, and V in Fig. 2–226, but with R_f in the circuit as shown in Fig. 2–227.

Switching Regulators

Voltage regulators which utilize high-frequency ON-OFF switching can offer several advantages over conventional continuously controlled, series-

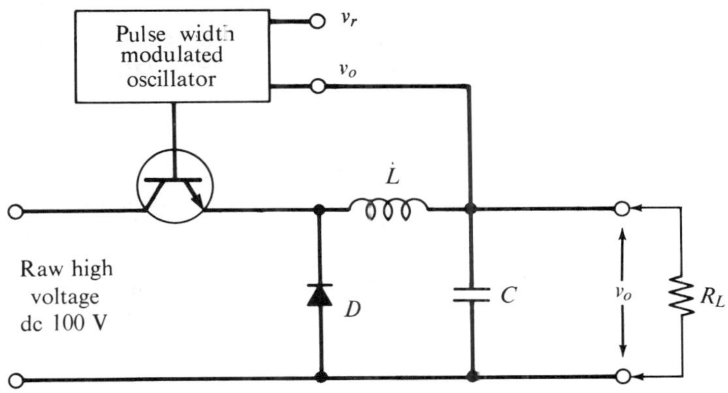

Fig. 2–228 Switching regulator.

type regulators. There are several different types of switching regulators, one of which is illustrated in Fig. 2–228. In this type of regulator the series transistor is turned ON and OFF by an oscillator whose pulse width is variable in proportion to the voltage difference between a reference voltage and the regulator output. The oscillator has a constant repetition rate (typically 20 kHz), but the ON time of the oscillator pulse is variable. The LC filter at the output smooths the pulsed output from the switching transistor. When the transistor conducts, current is delivered to the load from the raw supply. When the transistor is OFF, diode D conducts and current is supplied to the load from the energy stored in the inductor and capacitor.

The major advantage of the switching regulator is its higher efficiency compared to conventional series regulators. Efficiencies of a switching regulated supply can be as high as 80%, while typical series-pass regulated supplies have efficiencies of 50% or less.

Another type of switching-regulated supply makes use of direct rectification of the 60 Hz line voltage. The relatively high dc voltage is then converted to a high-frequency ac (20 kHz) where a stepdown in voltage can take place with a high-frequency transformer. The resulting high-frequency signal can then be rectified and applied to a switching regulator of the type shown in Fig. 2–228. This type of supply can offer considerable size reduction over conventional supplies because the high-frequency transformer can be much smaller than a 60 Hz transformer and less heat sink area is required because of higher efficiency. In addition, the size of the filter capacitor is reduced and its efficiency improved.

At present, because of the number of components needed for a switching-regulated supply, the costs are usually higher than for a conventional supply. However, this type of supply is expected to become less expensive as integrated circuit techniques provide cheaper components.

Experiment 2–33 *An OA Voltage Follower Source and Integrated Circuit Voltage Regulator*

The effectiveness of an OA voltage follower with gain for greatly decreasing the loading of a reference cell or electronic voltage reference source is first tested. The OA voltage follower circuit is also modified to include a booster amplifier in the OA feedback loop to show that the booster reduces drift of the input offset voltage with changes of output load and provides regulated output voltage over a wide range of loads. Then an integrated circuit voltage regulator that includes a reference source is tested with various loads. The drift of the regulator because of temperature change with change of load is measured and compared with the manufacturer's specifications.

Experiment 2–34 *A Programmable Current-Regulated Power Supply*

In many instrumentation systems it is important either to maintain the current in a device at a fixed value or to program the current accurately according to some function such as a precise rectangular pulse or a ramp signal. In this experiment an electronic OA feedback system is constructed which provides accurate programmable control of the current. A booster amplifier is included in the first part to provide bidirectional control of current. In an optional part a circuit can be constructed using a power transistor which provides only unidirectional current control.

Experiment 2–35 *An Electronic Servo for Control of Illuminance*

In this experiment the illuminance from a light source is regulated by an electronic servo system. The response characteristics of the feedback control element are first varied to illustrate their effect on the response of a servo system to a command signal when the controlled device (a light-emitting diode) responds very rapidly (10^{-8} sec). Then a tungsten source with a relatively slow response time is controlled, and it is shown how the servo system can cause continuous oscillation in the controlled device. It is then shown how suitable changes in the feedback control circuit can damp the oscillation and enable the controlled device to follow the command signal.

PROBLEMS

1. In operational amplifier feedback regulators, as in other OA applications, the input bias current can cause an output voltage error. Consider the OA inverting voltage amplifier shown in Fig. P–1. Assume that the voltage offset

$v_{off} = 0$. In the inverting amplifier only the bias current into or out of the inverting input $i_{b(-)}$ is important. (a) In which resistor will the bias current exist? (b) If $i_{b(-)} = 15$ nA, what will be the output voltage v_o due to $i_{b(-)}$?

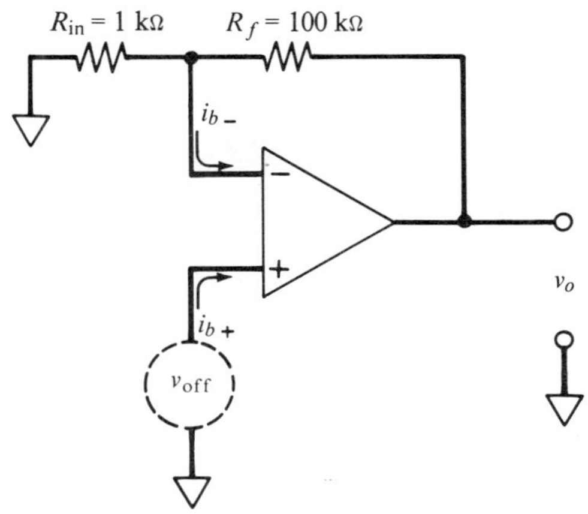

Fig. P–1

2. Input offset voltage in OA's can also cause output voltage errors. The offset voltage can be considered a voltage source in series with one of the inputs, as shown in Fig. P–1. Assume that the bias currents shown in Fig. P–1 are negligible and that $v_{off} = 1$ mV. What is the magnitude of the output voltage v_o due to v_{off}?
3. In the OA inverting amplifier of Fig. P–1 consider a 50 mV signal to be applied between R_{in} and common. (a) What is the maximum value of $i_{b(-)}$ which can be tolerated for a 1% or less output voltage error (assume ideal resistors and $v_{off} = 0$)? (b) What is the maximum value of v_{off} that can be tolerated for the same error (assume $i_b = 0$)?
4. The voltage-regulated supply of Fig. 2–226 is operated with $R_{in} = 5$ kΩ, $R_f = 10$ kΩ, and $v_r = -2$ V. If $i_{b(-)} = 1$ μA into the noninverting input and $v_{off} = +2$ mV into the inverting input, what is the output voltage? What is the percentage error due to $i_{b(-)}$ and v_{off}?
5. In the voltage-regulated supply of problem 4, v_{off} drifts with time at a rate of ±100 μV/day. The input bias current drift rate is ±1 μA/day and v_r drifts at a rate of ±1 mV/day. What is the maximum drift rate of the regulated output v_o?

Experiment 2–33 *An OA Voltage Follower Source and Integrated Circuit Voltage Regulator*

Equipment: 1.35 V mercury battery and holder
operational amplifiers (discrete and IC)
ADD
100 kΩ, 1% resistor
load resistors
DRB
DVM
IC voltage regulator (723)
booster OA
dual in-line card

The effectiveness of an OA voltage follower with gain for greatly decreasing the loading of a reference cell or electronic voltage reference source is first tested. The OA voltage follower circuit is also modified to include a booster amplifier in the OA feedback loop to show that the booster reduces drift of the input offset voltage with changes of output load and provides regulated output voltage over a wide range of loads. Then an integrated circuit voltage regulator that includes a reference source is tested with various loads. The drift of the regulator because of temperature change with change of load is measured and compared with the manufacturer's specifications.

E2–33.1 Application of an OA Follower for Unloading a Voltage Reference Source

Accurately measure, to 1 mV, the voltage of a mercury cell with a DVM. Connect a 1 kΩ resistor across the cell, repeat the measurement within a few seconds after connecting the load, then disconnect the load. Record the voltage values and determine the change in voltage of the cell under load. Also note how the battery voltage drifts over a few minutes after removing the load, as the battery voltage slowly recovers. Do not use heavier loads (smaller resistors); if you do, the cell voltage will be unstable for long periods.

Calculate the internal resistance of the mercury cell using the above data.

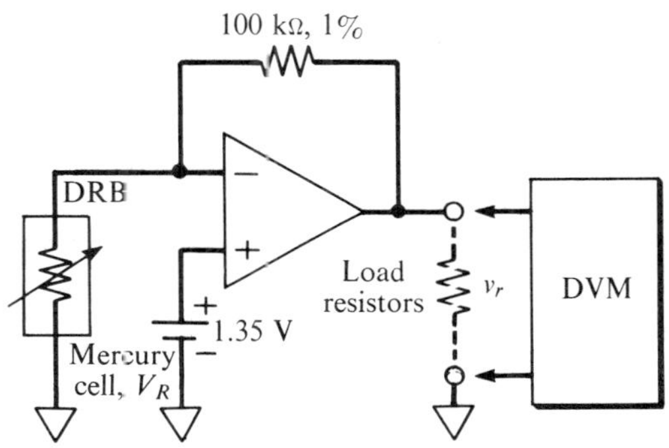

Fig. E2–65 An OA voltage follower reference source.

Connect the circuit shown in Fig. E2–65 using an IC operational amplifier card. Use the voltage value of the mercury cell V_R measured above and calculate the approximate setting for the DRB (and set it) to provide an output reference voltage $v_r = 5.0$ V. Measure v_r with the DVM and adjust the DRB if necessary to make $v_r = 5.000$ V, accurate to within 1 mV. Now connect load resistors of approximately the values shown in the table, record the values of v_r and $V_R - v_r$ for each load, and *calculate or measure the load current.*

R_{load}	v_r	$V_R - v_r$	Load current, mA
∞ (no load)			
50 kΩ			
5 kΩ			
2.5 kΩ			
1 kΩ			
500 Ω			
200 Ω			
150 Ω			
10 Ω			

Note the load current at which the output voltage begins to change significantly or any drift in readings.

Repeat the above test with a discrete OA card substituted for the IC OA card and record results in the same table as for the IC OA. Note any differences between the results obtained with the IC and with discrete OA's. Now modify the circuit of Fig. E2–65 to include a booster amplifier as shown in Fig. E2–66. *Repeat the measurements of v_r with various loads and record the results.*

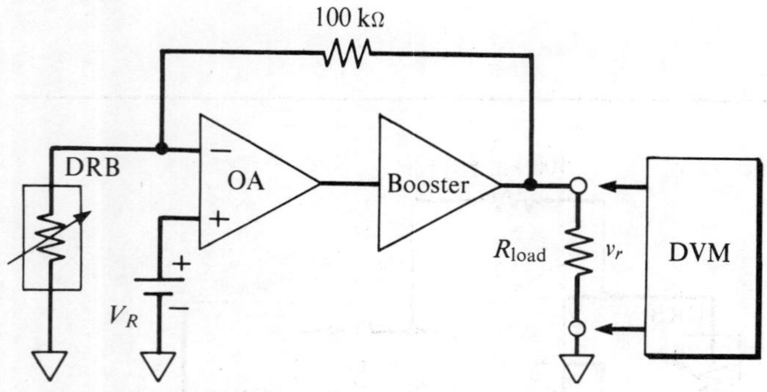

Fig. E2–66 A regulated voltage source with booster power output stage.

R_{load}	v_r	$V_R - v_r$	Load current, mA
∞ (no load)			
50 kΩ			
5 kΩ			
2.5 kΩ			
1 kΩ			
500 Ω			
200 Ω			
150 Ω			
10 Ω			

Make a plot of v_r versus load current. Connect patch wires (from connections 3 and 6 of the booster OA card to the output) to change the current limiting of the booster. Note how much the short-circuit current limit is changed by this modification.

Note that, regardless of the load used with the booster circuit of Fig. E2–66, no significant additional power is delivered by or dissipated in the control OA; consequently no drift is caused by heating effects in the control OA with changes of load. Also, in addition to decreasing the errors caused by the offset voltage drift, the booster can greatly increase the effective open-loop gain of the OA, and this can significantly improve regulation. *Do your data illustrate the advantages of the booster?*

E2–33.2 An Integrated Circuit Voltage Regulator

Obtain a type 723 integrated circuit voltage regulator and insert it into a dual in-line card. Be sure it is inserted in the correct orientation, and be careful not to damage the device leads (it may be necessary to bend them inward slightly with pliers before they will fit in the socket). Note that on the top of the dual in-line card, each patch wire connector has two numbers next to it; be sure to use the numbers that correspond to a 14-pin integrated circuit. Patch together the circuit shown in Fig. E2–67. You should always turn the ADD power off when assembling a circuit; double-check your wiring before applying power.

This integrated circuit contains an internal reference voltage source V_R of approximately 7 V. The voltage V_R is reduced by the voltage divider formed from R_1 and R_2 to a 5 V level that is applied to the noninverting input (N.I.) of a differential amplifier similar to an OA. The output is connected back to the inverting input (INV), forming a voltage follower. Thus the output voltage will be about 5 V. Any voltage from 2 to 7 V can be obtained at the output by varying the ratio of R_1 to R_2, and output of up to 37 V can be obtained with a slight circuit rearrangement. The current limit (CL) and current sense (CS) pins on the IC, and R_3, are used to set the maximum short circuit output current; and C_1, connected between the inverting input and the compensation terminal (COMP), prevents high-frequency oscillation.

Record in the table the output voltage for various loads, measured with a DVM. Comment on the degree of regulation observed.

R_L	v_r	i_L
∞		
10 kΩ		
1 kΩ		
500 Ω		
200 Ω		
100 Ω		
10 Ω		

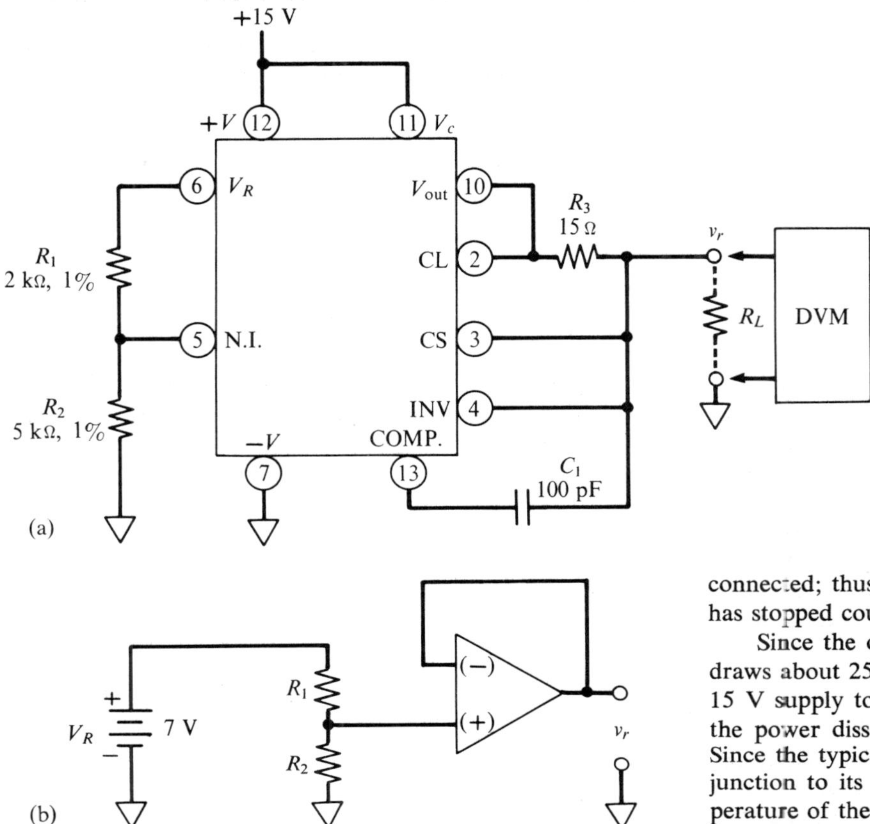

Fig. E2–67 Integrated circuit voltage regulator: (a) connection to IC chip; (b) simplified equivalent circuit.

Calculate or measure the current output for each load, and observe the current limiting characteristics.

Remove any load from the 723 IC and allow the output voltage to stabilize (1–2 min). Apply a 200 Ω load and measure and record the output voltage immediately (allowing about 2 sec for the DVM reading to stabilize). Record the voltage at 30 sec intervals until no further rapid drift is observed (1–2 min). The difference between the reading taken immediately after connecting the load, and the final reading, represents drift due to the increase in the IC temperature. Note that the direction of the drift may tend to cancel the drop in voltage produced at first when the load is connected; thus the voltage reading obtained when the drift has stopped could be close to the initial no-load voltage.

Since the output voltage is close to 5 V, a 200 Ω load draws about 25 mA. The voltage regulator has to reduce the 15 V supply to 5 V by dissipating power internally, and so the power dissipated is (15 − 5) V × 25 mA = 250 mW. Since the typical thermal resistance from the semiconductor junction to its surroundings is about 0.15°C/mW, the temperature of the IC will increase by about 0.15 × 250 mW = 37.5°C. Divide the voltage drift determined above by this temperature change to compute the drift per °C, and convert to a relative value by dividing by 5 V. A typical value is 0.002%/°C; the maximum value is not more than 0.015% per degree. Therefore, the overall regulation of this IC regulator is generally about 0.1% even for relatively large changes of load during use (i.e., 0.002% × 40°C = 0.08% ≈ 0.1%).

Experiment 2–34 *A Programmable Current-Regulated Power Supply*

Equipment: VRS
OA card
booster OA card
ADD
DVM
20 kΩ, 10 kΩ, and 100 Ω, 1% resistors
load resistors
2N3393 npn transistor
FET switch card

In many instrumentation systems it is important either to maintain the current in a device at a fixed value or to program the current accurately according to some function such as a precise rectangular pulse or a ramp signal. In this experiment an electronic OA feedback system is constructed which provides accurate programmable control of the current. A booster amplifier is included in the first part to provide bidirectional control of current. In an optional part a circuit can be constructed using a power transistor which provides only unidirectional current control.

E2–34.1 A Current-Regulated Power Supply

Connect the circuit shown in Fig. E2–68. Do not include any jumpers on the OA booster card which would increase the current limit beyond 30 mA. Set the VRS at 2 V, and start with a short circuit (wire jumper) for the load R_L. Measure the voltage V_{cs} across the current-sensing resistor, R_{cs}, using a DVM, and record it in the table. Increase the value of R_L using the approximate values shown in the table, and *record the value of V_{cs} for each load.*

R_L, Ω	V_{cs}, V	I_{load}, mA
0		
100		
500		
1000		
2000		

Determine the current in each load from the measured values of V_{cs} and compare with the theoretical value.

Explain why the load current drops off for loads above 1 kΩ. *Why is it often desirable to use a relatively small value of R_{cs} so that V_{cs} is only about* 1 V *or less?*

Insert a load resistance of about 10 Ω into the circuit. Vary the VRS voltage from 0 to 10 V as indicated in the table, and *record your measurements of V_{cs}.*

V_{VRS}, V	V_{cs}, V	I_{load}, mA	$I_{load\,(theor.)}$, mA
0.00			
2.00			
4.00			
6.00			
8.00			
10.00			

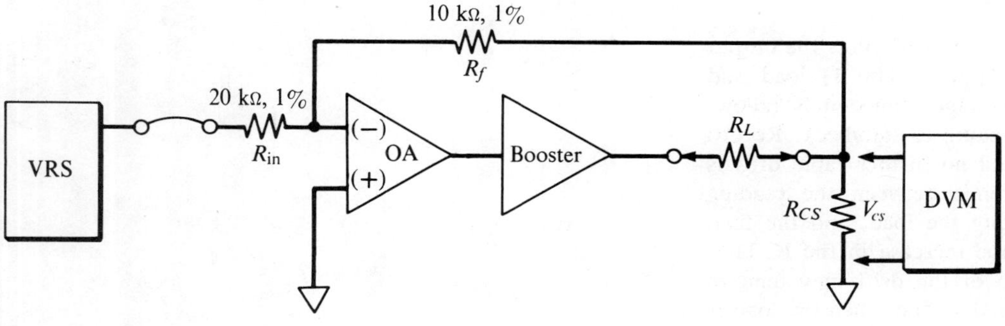

Fig. E2–68 A programable current-regulated power supply.

Calculate the load current for each VRS setting and compare with the expected current.

Why does the current drop below the theoretical value when the VRS voltage reaches 6–8 V?

Observe that the current control is bidirectional.

E2–34.2 Programmable Current-Regulated Power Supply

Modify the circuit of Fig. E2–68 by inserting a FET switch (from the FET switch card) between the VRS terminal and R_{in}. Use the DTM square wave output to control the FET switch. Set the VRS at 2 V, and connect a 500 Ω load. Observe the DTM square wave and the voltage V_{cs}, using both channels of your oscilloscope. Note the response of the programmable current source, and *determine (and record) the limits of useful performance by varying the DTM frequency*. If the FET and R_{in} positions in the circuit are reversed, spikes will be produced in the load current. *Observe this effect.*

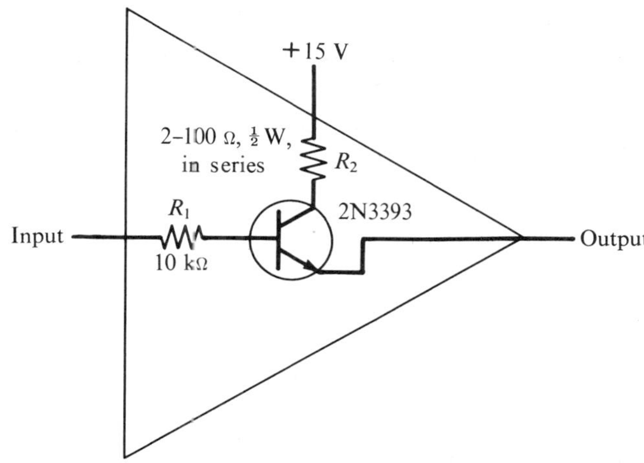

Fig. E2–69 A power transistor booster amplifier.

E2–34.3 (Optional) Unidirectional Current-Regulated Supply

When it is desirable to boost the output current capability of an OA, and only current sourcing is required, a simple transistor emitter follower will serve as an OA booster. In the circuit of Fig. E2–69, R_1 limits the maximum current that will be drawn from the amplifier, and R_2 provides current limiting to protect the pass transistor. ***Build this circuit and substitute it for the booster used in the previous experiment and determine its performance.***

Experiment 2–35 *An Electronic Servo for Control of Illuminance*

Equipment: three OA's (OA2 and OA3 are FET OA's)
one booster OA
ADD
light-emitting diode (LED)
phototransistor
oscilloscope
tungsten light bulb
assorted resistors

In this experiment the illuminance from a light source is regulated by an electronic servo system. The response characteristics of the feedback control element are first varied to illustrate their effect on the response of a servo system to a command signal when the controlled device (a light-emitting diode) responds very rapidly (10^{-8} sec). Then a tungsten source with a relatively slow response time is controlled, and it is shown how the servo system can cause continuous oscillation in the controlled device. It is then shown how suitable changes in the feedback control circuit can damp the oscillation and enable the controlled device to follow the command signal.

E2–35.1 Programmed Control of Illuminance from a Light-Emitting Diode

In this experiment an electronic servo system is used to regulate the illuminance from a light-emitting diode (LED). A phototransistor serves as the light sensor or feedback element. This is an instance where the response of the controlled system, the LED, is much faster than that of the sensor. The light output from the LED follows the current through it with a response time of about 10 nsec (in fact, the limitation on the effective response speed of the LED in this experiment is the slewing rate of the driving circuitry). The phototransistor has a response speed of the order of 30 μsec at the low current levels used here.

Construct the circuit shown in Fig. E2–70. Set the DTM frequency to 5 kHz. When the circuit is turned ON, the LED should not be emitting light until the phototransistor is shielded from the room light by the hand or another opaque object. Now the LED and phototransistor should be placed as close together as possible, and covered with a short piece of opaque tubing to exclude room light. The diode D_1 clamps the OA booster output by preventing it from swinging to negative saturation when the control signal from the DTM turns the LED OFF. The diode D_2 further protects the LED by preventing reverse current through it. The 270 Ω resistor from the cathode of the LED to ground protects the LED from excessive forward current, and provides, at point B in the schematic, a voltage proportional to the current. The LED should always be used with a series resistor to limit the current, to less than 50 mA in this case.

Observe the DTM control signal and the voltage at point B. The waveform at B should appear approximately as a square wave, with some initial overshoot followed by a damped oscillation. When the oscillation has damped out, the voltage should be about 0.5–3 V. If, because of normal variations in the characteristics of the LED or phototransistor, the

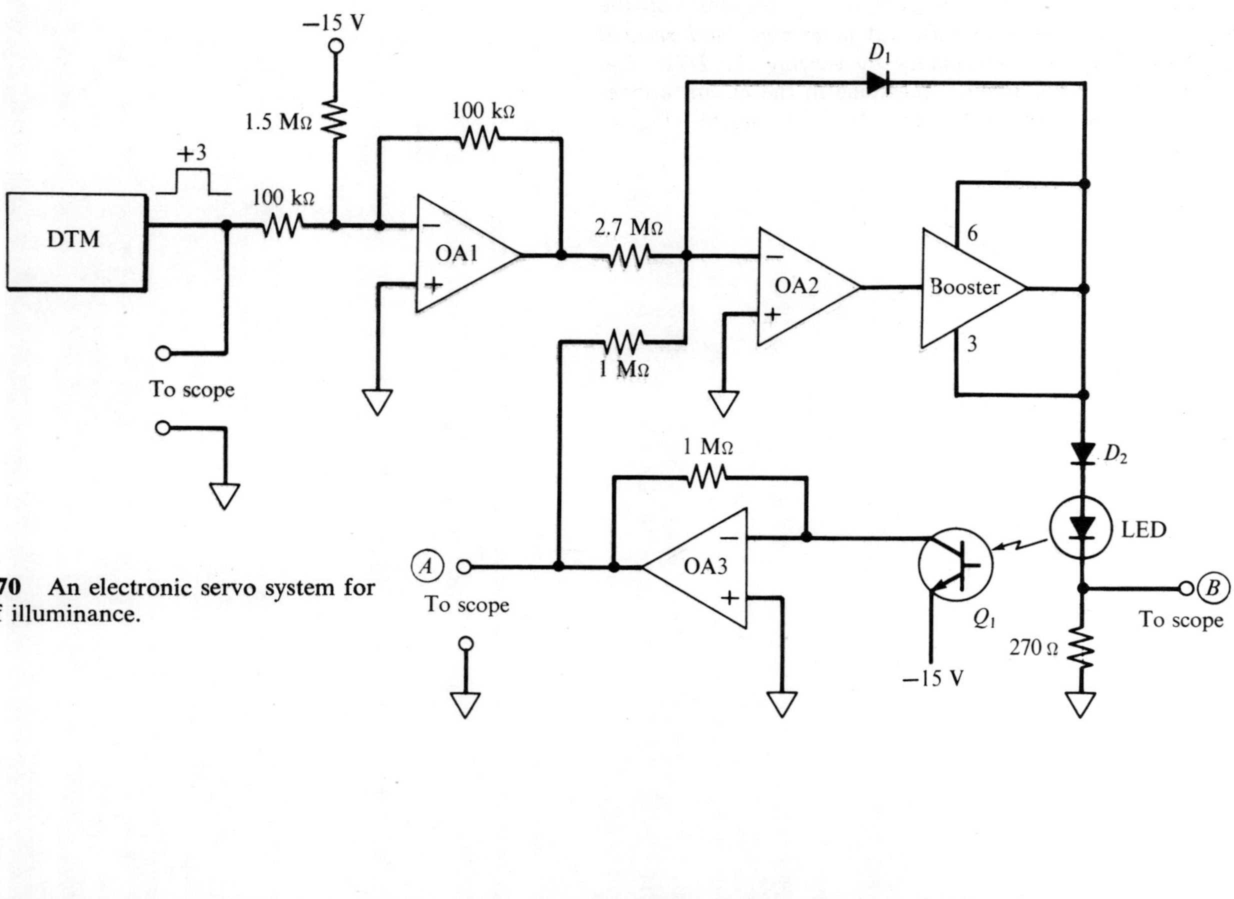

Fig. E2–70 An electronic servo system for control of illuminance.

voltage is higher, increase the value of the feedback resistor for OA3. If the voltage is less, separate the LED and phototransistor slightly.

Record the waveforms at points B and A (a voltage proportional to the phototransistor current) versus the DTM square wave control signal.

Insert a 1 MΩ resistor between the collector of the phototransistor and the summing point of OA3. This greatly slows down the response speed of the phototransistor. It will be necessary to reduce the DTM frequency to about 500 Hz in order to obtain an approximate square wave output demonstrating overshoot and damped oscillation. Momentarily place a 0.01 μF cap across the series resistor; explain why the waveforms appear as they do with the phototransistor directly feeding the OA summing point. Remove the capacitor. Note that as the response of the feedback network is slowed down, the frequency of the damped oscillation decreases. *Record and compare the waveforms at B and A.*

Insert a 0.001 μF capacitor across the phototransistor. This adds to the capacitance inherent in the phototransistor, and further slows its response. Reduce the DTM frequency again, to about 50 Hz, and *record the waveforms at points A and B.* Note that the period of the damped oscillation is about 1 msec. This is the same as the time constant of the 1 MΩ resistor in series with the phototransistor and the 0.001 μF capacitor across it. From the period of the damped oscillation with no added capacitor across the phototransistor *calculate the approximate effective shunt capacitance inherent in the phototransistor.* Using this value of shunt capacitance and the oscillation period for the case where no series resistor or shunt capacitor is used with the phototransistor, *calculate the effective resistance of the phototransistor in series with its capacitance.*

Note also that the phase relationship between the current through the LED (the controlled system) and the phototransistor current sensed at point *A* is such that the LED current leads the sensor current *B* by 90 degrees. You will recall that 90 degrees is the maximum phase shift that can be produced by a simple one-stage *RC* circuit, which in this case is the resistor in series with the phototransistor and the shunt capacitor. Note that it has been assumed that the basic responses of the phototransistor and the LED do not alter the simple *RC* behavior of the feedback network. This is true because, to a first approximation, the light output from the LED is directly proportional to the current through it, which in turn is proportional to the OA booster output voltage, and the phototransistor current is proportional to the light intensity striking it. Since the source and detector are both linear, they do not affect the response characteristic.

The overshoot and damped oscillation observed can be eliminated by adjusting the response speed of the feedback amplifier to match that of the feedback network. Insert progressively larger capacitors between the summing point of OA2 and the booster output. Cover a range of capacitors from 10 pF to 0.005 μF. Note that the larger the capacitor, the longer the rise time of the signals at *A* and *B*, and the greater the damping effect. Also, the period of the damped oscillation increases. *Record your observations below.*

E2–35.2 Electronic Servo Control of Illuminance from a Tungsten Lamp

For this experiment, the light output from a tungsten filament lamp will be controlled, using the phototransistor as a sensor. The response speed of the lamp is much slower than that of the phototransistor. Modify the circuit of Fig. E2–70 by replacing the LED and 270 Ω resistor with a CM–47 lamp. Make D_2 a rectifier diode, and remove D_1. Observe the voltage across the lamp so as to monitor the controlled system (call the junction of the cathode of D_2 and the lamp point B).

Start with the DTM frequency set at about 0.5 Hz. The lamp should be off until the phototransistor is shielded from room light. Place the phototransistor and lamp as close together as possible, and cover them with a piece of light-excluding tubing. Observe that when the control signal from the DTM commands current into the lamp, a steady oscillation is observed at point B, between the limits of about 0 and +2 V. The positive limit is set by the current limiting of the OA booster. Note that the period of the oscillation is about 3 msec. *What does this tell you about the response speed of a tungsten filament lamp? Compare the controlled system and sensor signals at points B and A respectively, and note that the phototransistor output signal appears to be nearly steady while the lamp is turned* ON.

The servo system is operating similarly to the common ON-OFF control systems often used for heating control. When the phototransistor senses that the light output from the lamp is below the controlled value, the control amplifier turns fully ON (the OA booster undergoes current limiting). When the lamp output goes above the desired value, power to the lamp is cut off. If the phototransistor output is observed carefully, it will be seen that there is some ripple in the form of a triangular waveform that rises when power is applied to the lamp and falls when power is removed.

Insert increasingly larger capacitors, ranging from 10 pF up to 0.1 μF, between the booster output and the control OA2 summing point. Note that, while a sufficient capacitance in the feedback loop exerts a damping effect that prevents continuous oscillation, the chief effect of adding capacitance beyond the value that just damps the oscillation is merely to lengthen the period of the damped oscillation. Further, the light output from the lamp, as measured by the phototransistor output, varies much more with the added damping (until the oscillation finally dies out) than it did with the simple ON-OFF control system. With a more sophisticated means of damping the feedback amplifier response, the response could be matched to that of the lamp better than is possible with simple first-order RC damping, and thus the overshoot could be eliminated. *Record your observations below.*

Derivation of the Delta-Y, Y-Delta Transformation

Appendix A

Delta to Wye

From Fig. 2–22a the resistance from A to B is $R_{AY} + R_{BY}$, and from Fig. 2–22b the resistance is $R_{AB}(R_{AC} + R_{BC})/(R_{AB} + R_{BC} - R_{AC})$. Therefore,

$$R_{AY} + R_{BY} = \frac{R_{AB}R_{AC} + R_{AB}R_{BC}}{R_{AB} + R_{BC} + R_{AC}}. \tag{A1}$$

In the same way,

$$R_{BY} + R_{CY} = \frac{R_{BC}R_{AC} + R_{BC}R_{AB}}{R_{AB} + R_{BC} + R_{AC}} \tag{A2}$$

and

$$R_{AY} + R_{CY} = \frac{R_{AC}R_{AB} + R_{AC}R_{BC}}{R_{AB} + R_{BC} + R_{AC}}. \tag{A3}$$

Subtract (A2) from (A1) to obtain

$$R_{AY} - R_{CY} = \frac{R_{AB}R_{AC} - R_{BC}R_{AC}}{R_{AB} + R_{BC} + R_{AC}}. \tag{A4}$$

Now subtract (A4) from (A3) to get

$$R_{AY} = \frac{R_{AB}R_{AC}}{R_{AB} + R_{BC} + R_{AC}}. \tag{A5}$$

The expressions for R_{BY} and R_{CY} are obtained similarly.

Wye to Delta

From the delta-Y transformation equations,

$$\frac{R_{AY}}{R_{BY}} = \frac{R_{AC}}{R_{BC}} \quad \text{and} \quad R_{AC} = \frac{R_{BC}R_{AY}}{R_{BY}}. \tag{A6}$$

Similarly,

$$\frac{R_{AY}}{R_{CY}} = \frac{R_{AB}}{R_{BC}} \quad \text{and} \quad R_{AB} = \frac{R_{BC}R_{AY}}{R_{CY}}. \tag{A7}$$

Now substitute (A6) and (A7) for R_{AC} and R_{AB} in (A5) to get

$$R_{BC} = R_{BY} + R_{CY} + \frac{R_{BY}R_{CY}}{R_{AY}} \tag{A8}$$

and obtain the expressions for R_{AB} and R_{AC} similarly.

Fabrication of Transistors

Appendix B

It was indicated in the previous discussions that the method of fabrication of the transistor influences its properties. To obtain the desired properties, the location and concentration of the n- and p-type impurities must be carefully controlled. The most popular ways of forming the desired transistor pn junction are by the alloy-junction and diffused-base (mesa and planar) techniques.

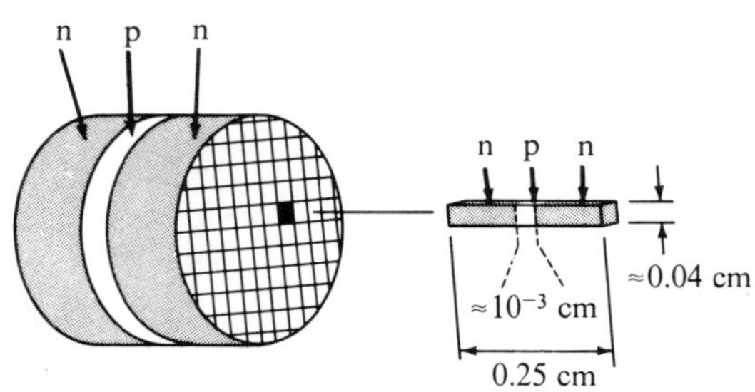

Fig. B–1 Sectioning of grown-junction doped semiconductor to obtain small bars of npn material for transistors.

Grown-Junction Transistors

The first commercial transistors were produced by the grown-junction technique. With this method a rather large semiconductor crystal is grown as it is continuously pulled from the molten semiconductor material, e.g., silicon. As the crystal grows the composition of the silicon melt is changed from n-type to p-type and back to n-type by doping the melt while continuously turning and pulling the crystal from the melt. By cutting the doped crystal into short cylinders and then into small bars, many segments of npn material are obtained as illustrated in Fig. B–1. Leads are connected to the two n- and the p-sections of the semiconductor bar, which is then mounted on a header as shown in Fig. B–2.

When using the grown-junction technique it is impossible to control the base-layer thickness very well, and the base thickness is relatively large, about 10^{-3} cm. Because of the base thickness, the transistors made by this process do not have good high-frequency response, and the process is quite complicated. Therefore, the grown-junction process has generally been superseded by other fabrication techniques.

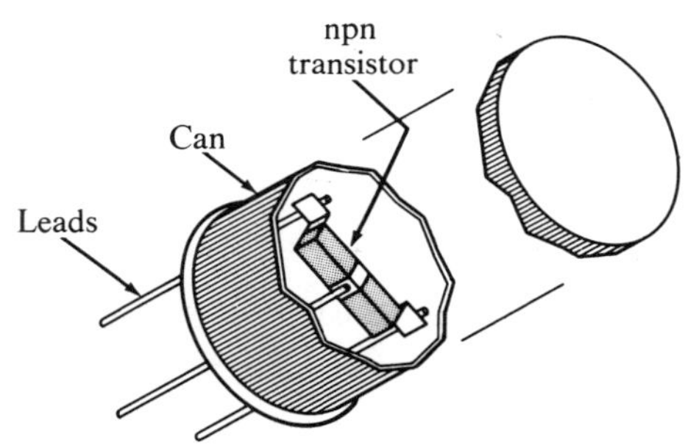

Fig. B–2 Mounting for a grown-junction transistor.

Alloy-Junction Transistors

A thin slice (about 20 mils) of semiconductor material, e.g., germanium, can have emitter and collector dopants (indium) alloyed to either side of the wafer, as shown in Fig. B–3. When the indium melts, it dissolves some of the germanium and during the subsequent cooling the dissolved germanium recrystallizes on the base with many indium atoms incorporated in the crystal structure. Therefore, the recrystallized germanium is now doped as p-type, and a pnp transistor results.

The larger size of the collector pellet as compared to the emitter enables the carriers injected at the emitter to be collected more efficiently. The dimensions of the junction regions are also hard to control by the alloy process, so that the high-frequency response is generally not good.

Diffused-Base Transistors

It is possible to control the junction fabrication process precisely by diffusion. By a suitable sequence of masking and exposing silicon semiconductor material to a hot gas of antimony or boron dopants and etching, it is possible to produce the diffused silicon mesa or planar transistors. Cross sections of these two types are shown in Fig. B–4.

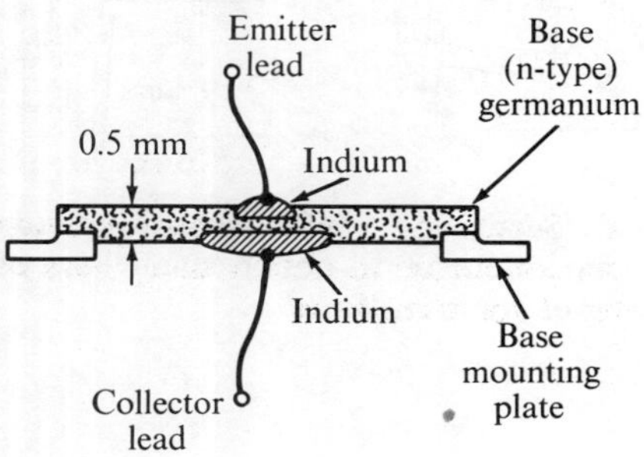

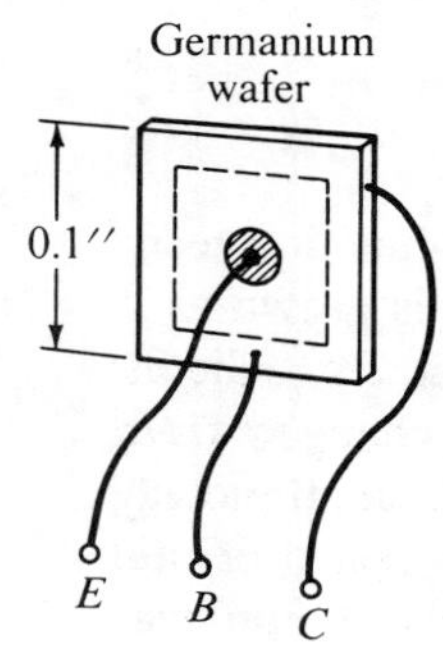

Fig. B–3 A pnp alloy transistor.

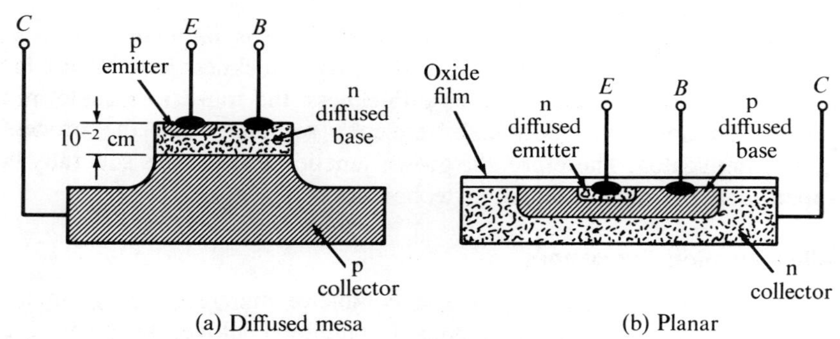

Fig. B–4 Cross-sectional view of diffused silicon transistors: (a) mesa; (b) planar.

Measures of Sine Wave Amplitude

Appendix C

Sine wave voltage and current amplitudes are commonly described in four ways: the peak value, the peak-to-peak value, the average value, and the rms (or *root-mean-square*) value. The particular description used depends on the application. Measuring instruments such as vacuum-tube voltmeters and oscilloscopes *respond* to the peak-to-peak value, which for a sine wave is merely twice the peak value. Moving-coil meters *respond* to the average value (effective rectified dc value), and the rms value is mostly used for the measurement of *power* in an ac circuit. Figure C–1 graphically summarizes the four commonly used measures.

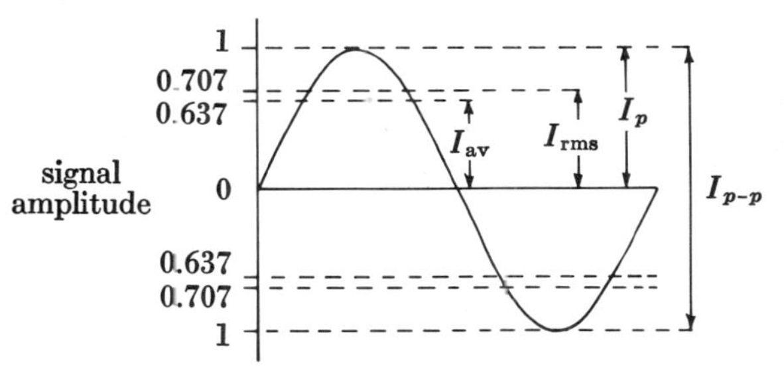

$I_{p-p} = 2I_p$
$I_{rms} = 0.707I_p$
$I_{av} = 0.637I_p$

Fig. C–1 Common measures of sine wave amplitude.

Average Value

Moving-coil meters respond to the average value of the current through them. The mechanism by which the meters operate is discussed in Module 1. The average value is that current above and below which the current-time areas of a half-cycle are equal. This is illustrated graphically in Fig. C–2.

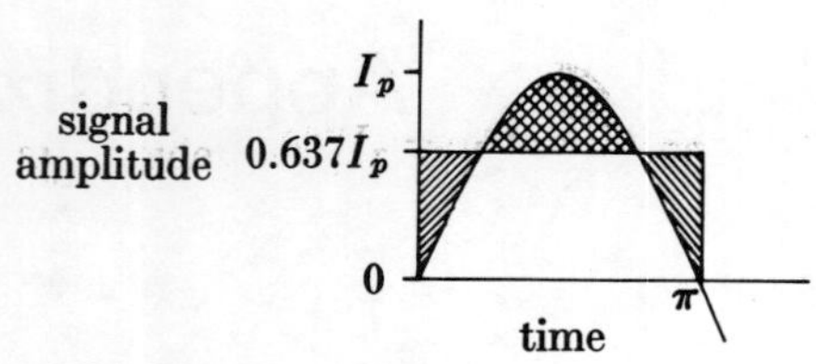

Fig. C–2 Determination of the average value of a sinusoidal signal.

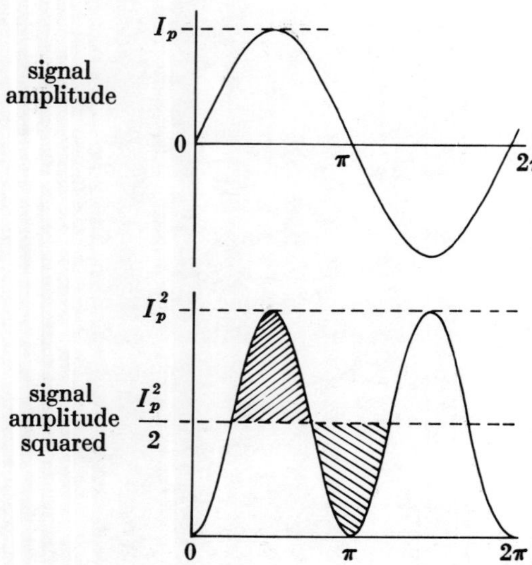

Fig. C–3 Determination of the rms value of a sinusoidal signal.

The average value may be obtained by integrating the area over a half-cycle:

$$I_{\text{av}} = \frac{1}{\pi}\int_0^{\pi} I_p \sin \omega t \, dt$$

$$= \frac{2I_p}{\pi} = 0.637 I_p. \qquad \text{(C1)}$$

In practice, the sine wave must be rectified (full-wave) to obtain meter readings corresponding to Eq. (C1).

RMS Value

The rms value is the *effective* value needed in power calculations. The rms value of a sine wave current will produce the same heating (power) in a resistor as an identical dc current. In other words, 1 rms ac ampere produces the same amount of heat in a resistor in a given time as 1 ampere of dc current. The rms value is obtained by taking the *square root* of the *mean* (average) of the *squared* values of the current. This operation is shown graphically in Fig. C–3. The square of the sine wave ($I_p^2 \sin^2 \omega t$) is plotted directly below the sine wave, because the heating effect of a current is directly proportional to the square of the current. The result is another "sine wave" which is completely above zero and twice the frequency. The mean or average value of this "sine wave" is clearly $I_p^2/2$. On taking the square root,

$$I_{\text{rms}} = \sqrt{\frac{I_p^2}{2}} = \frac{I_p}{2}\sqrt{2} = 0.707 I_p. \qquad \text{(C2)}$$

The relation between the effective emf and the peak emf is the same as for effective and peak current. Thus, $V_{\text{rms}} = 0.707 V_p$. Calculations of voltages and currents in ac circuits provide self-consistent values; e.g., if rms voltages are used, the currents will have rms values, etc. Note that peak values are obtained by multiplying the rms values by the reciprocal of 0.707 (1.414), so that

$$I_p = 1.414 I_{\text{rms}},$$

$$V_p = 1.414 V_{\text{rms}}.$$

Equipment Appendixes

Permission was obtained from Heath Company, Benton Harbor, Michigan, for the use of the illustrations and some of the copy which appear in the Equipment Appendixes.

1. THE BINARY INFORMATION MODULE

The binary information module (BIM) shown in Fig. A2–1 serves a dual purpose in the analog-digital designer (ADD).

First, the front panel switches and lamps can be used to provide and display binary information. Second, the BIM can hold and provide power for up to seven plug-in cards that are used in the ADD.

On the front panel there are eight single-pole, double-throw switches marked A through H. In addition, two spring-return pushbutton switches, PB1 and PB2, are available. Each of the ten switches has complementary logical level outputs available at the top of the BIM. The circuit for each switch is illustrated in Fig. A2–2. The switch outputs are labeled A and $\bar{A}$. $\bar{A}$ is interpreted as "not A" and is the complement of the level at the A output. When switch A is placed in the "0" position, a voltage of 0 V appears at the A output and +4–5 V at the $\bar{A}$ output. When switch A is placed in the "1" position, as shown in Fig. A2–2, +4–5 V appears at the A output and 0 V at the $\bar{A}$ output. The same circuits and output arrangements are used for all ten switches, except that PB1 and PB2 spring-return to the "0" position after they are pushed.

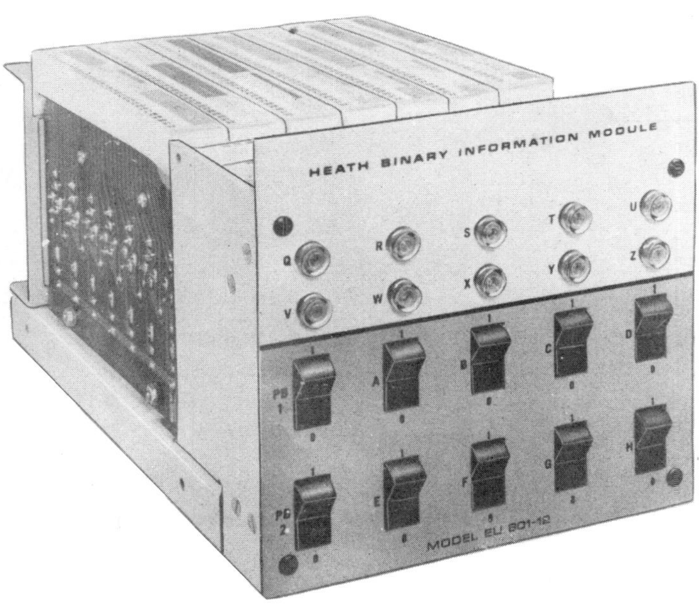

Fig. A2–1 Binary information module.

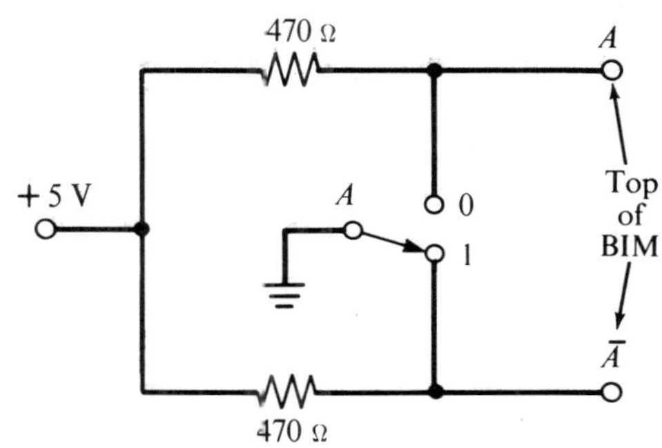

Fig. A2–2 Circuit diagram of BIM switches.

The ten neon lamps marked Q through Z on the front panel are used to read logic levels. The lamp inputs are on the top of the BIM and are labeled Q through Z. The lamps glow when logic "1" signals (+2.4–5 V) are applied and are unlit when logic "0" signals are applied.

2. THE RELAY CARD

A pictorial and card top diagram of the relay card are shown in Fig. A2–3.

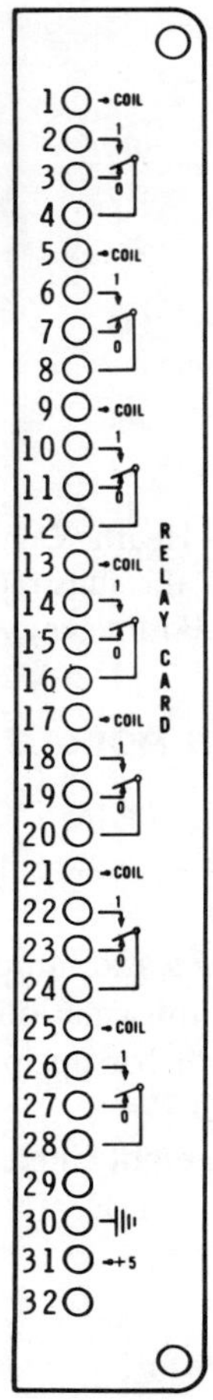

Fig. A2–3 Picture and card top diagram of relay card.

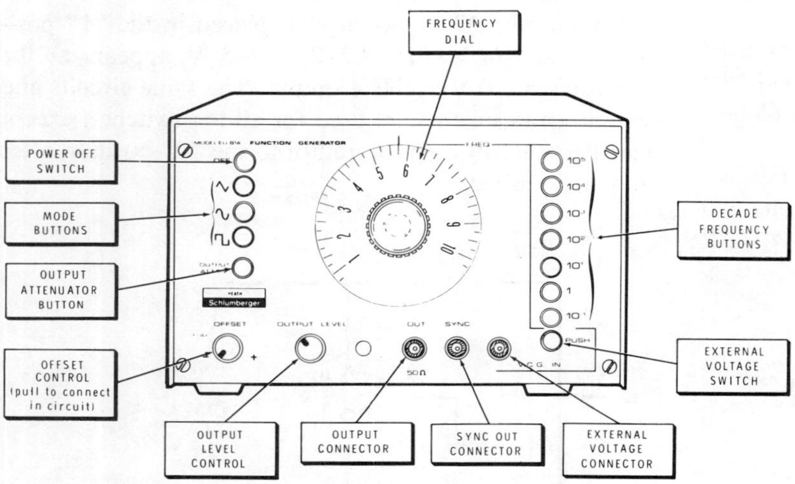

Fig. A2–4 EU-81A function generator.

Seven single-pole, double-throw relays are mounted on the card. Each relay can be activated by a +2.4–5 V logic level signal applied to inputs marked COIL. By looking at the card top diagram of Fig. A2–3, you can easily understand the operation of one relay. Locate the first relay whose contacts are brought to connectors 1–4 on the top of the relay card. Connector 1 is the coil input. Connectors 2 and 3 are the fixed contacts of the relay, and connector 4 is the movable contact whose position depends on the signal applied to the COIL input. When a logic "0" (0 V) is applied to the coil, the movable contact is connected to the fixed contact marked 0. Thus connectors 3 and 4 on the card are connected together. When a logic "1" (+2.4–5 V) is applied to the coil, the movable contact is changed to the fixed contact marked 1. Thus connectors 2 and 4 are shorted together. Note the seven identical relays on the card schematic.

3. WAVEFORM GENERATORS

In this appendix the EU-81A function generator and the EUW-27 sine-square generator are described.

3.1 Function Generator

The EU-81A function generator is pictured in Fig. A2–4. This unit can provide sine, square, and triangular waveforms from 0.1 Hz to 1 MHz. The frequency is continuously adjustable between the seven decade ranges.

Refer to Fig. A2–4 to locate the specific controls, switches, and connectors described below.

POWER OFF. This pushbutton switch turns the power to the generator off when depressed.

MODE BUTTONS. Pressing a mode button turns the power on. Triangular waves, sine waves, or square waves can be selected.

DECADE FREQUENCY BUTTONS. These buttons select the frequency decade within which the exact frequency is set by the continuously variable FREQUENCY DIAL. The decade buttons are thus multipliers for the 1–10 frequency dial setting. If the 10^5 button were pressed in Fig. A2–4, the output frequency would be 5.75×10^5 Hz.

OUTPUT LEVEL CONTROL. This control adjusts the amplitude of the output continuously from 200 mV peak-to-peak to 20 V peak-to-peak.

OFFSET CONTROL. When this control is pulled out, a dc offset can be applied to the waveform. Positive offset is provided by a clockwise rotation of the control, while a counterclockwise rotation provides negative offset.

OUTPUT ATTENUATOR. When this button is depressed, the output signal is attenuated by approximately 20 dB.

EXTERNAL VOLTAGE CONNECTOR. By means of this connector and the EXTERNAL VOLTAGE SWITCH, the output frequency of the generator can be controlled by an external voltage. With the external VCG button depressed, the frequency dial is disconnected, and the frequency is determined entirely by the voltage (1–10 V) applied at the external voltage connector. When the external VCG button is OUT, the frequency that appears at the output connector is the sum of the dial frequency and the voltage-controlled frequency. Thus an external voltage can be used to raise or lower the frequency set on the dials.

SYNC OUT CONNECTOR. This connector provides a 0–5 V square wave of the same frequency as the output for synchronizing external circuits.

3.2 The Sine-Square Generator

The sine-square wave generator (SSG) of Fig. A2–5 provides simultaneous sine and square waves from 20 Hz to 1 MHz frequency. The output voltages can range from 0 to 10 V rms.

The SSG is an oscillator whose frequency is determined by the position of the FREQUENCY MULTIPLIER and FREQUENCY controls. The SSG output voltage is divided, and the fraction of the total voltage available at the output is selected by a RANGE switch and the AMPLITUDE control. These controls are roughly calibrated in volts rms for the sine wave output and volts peak-to-peak for the square wave. The sine and square wave outputs are electrically independent but necessarily of the same frequency. The sine wave output is available at the right terminals, while the square wave is available at the left terminals.

4. DIGITAL TIMING MODULE

The digital timing module (DTM) shown in Fig. A2–6 provides timing pulses and other repetitive waveforms for the digital and analog circuits of an ADD unit. Logic one and logic zero pulses, square waves, and comple-

Fig. A2–5 The sine-square generator.

Fig. A2–6 Digital timing module.

mentary square waves and ramps are all available simultaneously.

The rate of the output signals can be varied in five decade steps from less than 0.1 Hz to 10 kHz, or adjusted to any frequency within that range. When the VAR/CAL switch is in the CAL position, the variable control is disabled, and the output signals are fixed and calibrated for each position of the multiplier switch. Use of an external capacitor will extend the output frequency range to 100 kHz.

The module also holds up to seven plug-in cards and provides power connections for them. Three 100 kΩ controls on the front panel can be patch-wired to the circuit cards for use as timing and/or level controls. A connector panel at the top of the module provides patch connections for all the generator output functions and to the control potentiometers.

5. BREADBOARD CONNECTION SYSTEM

The three-card breadboard chassis used in many experiments is shown in Fig. A2–7 with three cards inserted in the card guides. A power patch card with a power cable and octal plug is used to supply voltages from the digital power module in the ADD or other power supplies to the patch cards. These voltages are then readily connected by No. 22 wire to the cards in the other two slots. A multiple connector card is used to provide multiple common connections, and the breadboard card is inserted so that components can be readily interconnected as shown in Fig. A2–7.

At the end of the breadboard chassis are three slots for connecting switches and potentiometers.

Again, when the breadboard card is used, *never connect more than one wire in each hole.*

6. LOW VOLTAGE POWER SUPPLIES

In this appendix the Heath EU-41A low voltage power supply (LVPS) and the EUW-17 transistorized power supply (TPS) are described.

6.1 Low Voltage Power Supply

A picture of the LVPS is shown in Fig. A2–8 along with a summary of the operating controls. The LVPS provides regulated dc voltages in the range of 0–15 V at currents up to 750 mA. The output voltage is regulated within ±0.02% and is continuously adjustable over the entire range. The output is also protected by a current limiting circuit that allows the maximum short circuit current to be adjusted from 35 to 750 mA. A current limiting light on the front panel indicates an overcurrent condition.

6.2 Transistorized Power Supply

Figure A2–9 shows a picture of the EUW-17 transistorized power supply, designed to supply low voltages and currents. A variable voltage adjust on the front panel allows adjustments of the ac voltage from 0 to 35 V with no output current or 0 to 25 V at maximum output current. The supply has an output current capability

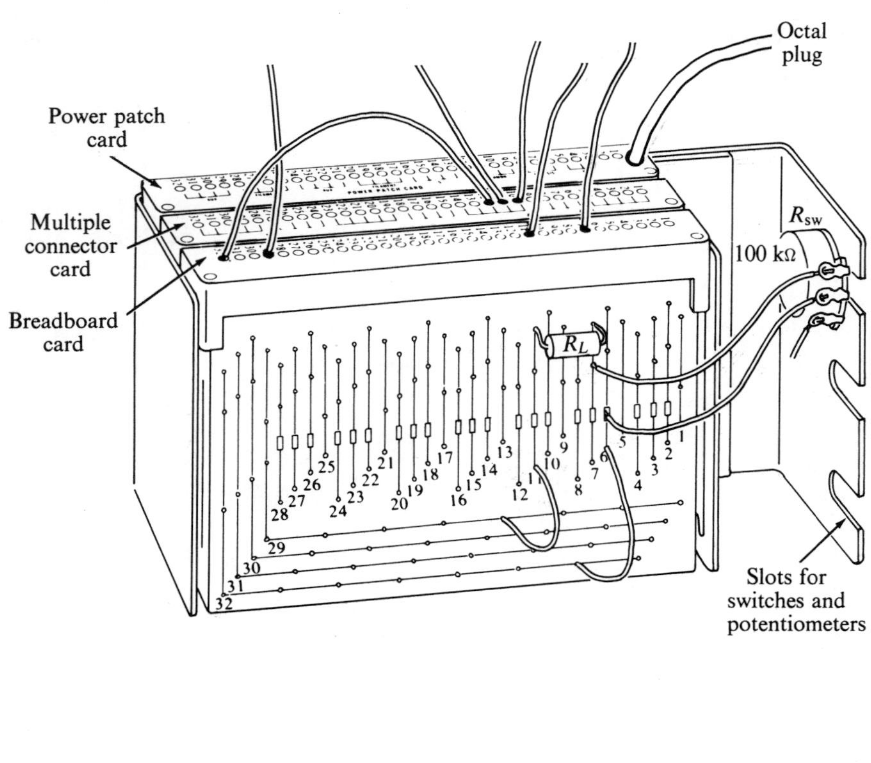

Fig. A2–7 Breadboard connection system.

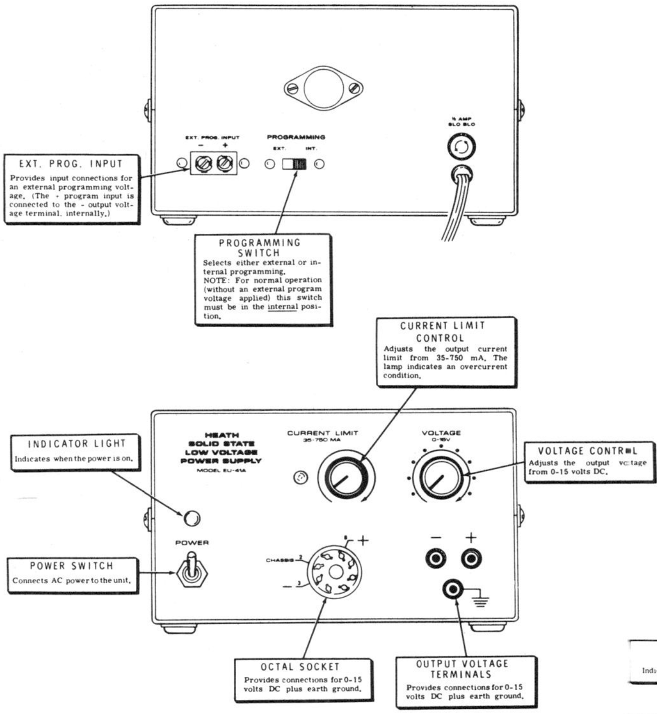

Fig. A2–8 Low-voltage power supply controls and connectors.

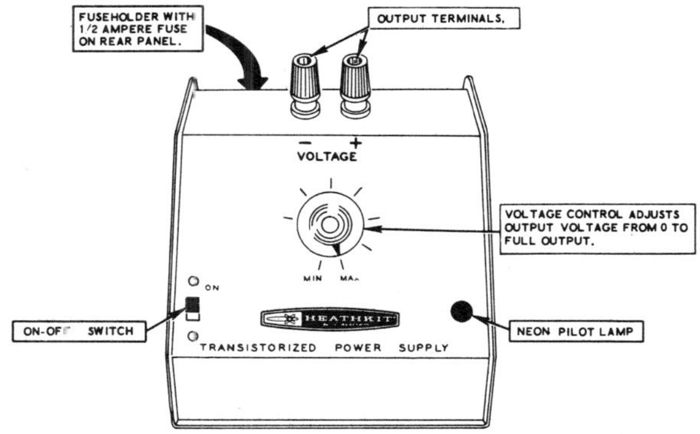

Fig. A2–9 Transistorized power supply.

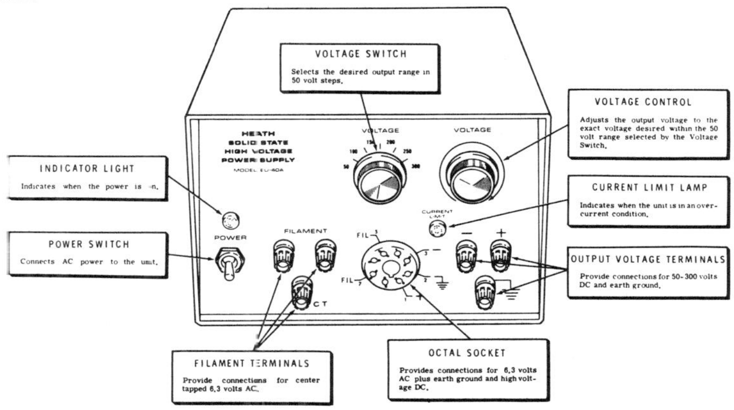

Fig. A2–10 High-voltage power supply controls and connectors.

which varies with voltage from a low of ≈60 mA at 2 V to a high of 200 mA at voltages greater than 18 V. The output resistance of the supply is less than 50 Ω.

7. HIGH VOLTAGE POWER SUPPLIES

In this appendix the Heath EU-40A high voltage power supply (HVPS) and the EUW-15 universal power supply (UPS) are described.

7.1 High Voltage Power Supply

Figure A2–10 illustrates the operating controls of the HVPS. The supply provides regulated dc voltages from 50 to 300 V with output currents up to 20 mA. A supply of 6.3 V ac is also provided. The dc output is continuously variable over its entire range and is protected by a current limiting circuit that causes the maximum short-circuit output current to be about 25 mA. A current limiting lamp indicates an overcurrent condition. All power output connections can be made to 5-way binding posts or an octal socket.

7.2 Universal Power Supply

The EUW-15 UPS is pictured in Fig. A2–11. Each of the front panel controls and connectors is identified in the figure.

The output terminals in the center of the UPS are 5-way binding posts, which make available 6.3 V ac, chassis ground, —dc, and +dc. A connecting terminal is also provided so that a milliammeter (MA) can be easily connected in series with the +dc output.

The power switch at the lower left turns on the ac power to the UPS, while the high-voltage switch at the lower right connects the dc high voltage to the panel connectors.

There are two octal socket connectors in the lower center of the UPS; 6.3 V ac and —dc are always available at these sockets, and +dc is available only when

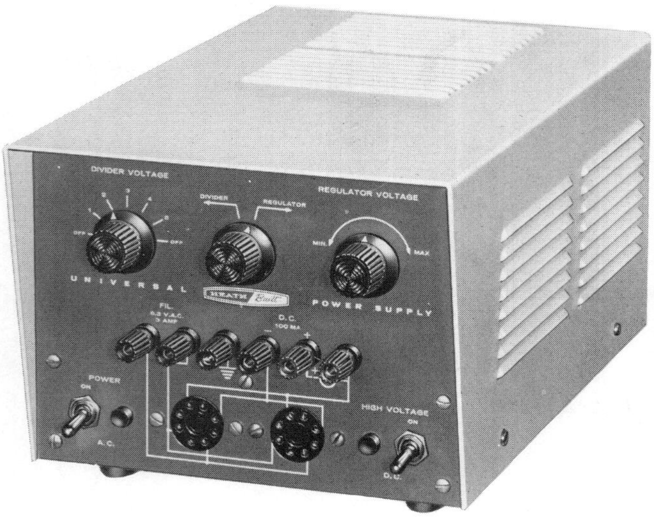

Fig. A2–11 Universal power supply.

the +dc and MA terminals are shorted together or connected by a milliammeter.

The divider-regulator switch at the top center of the supply switches the unit so that it operates either as a regulated power supply or as an unregulated power supply with a voltage divider output. In the divider mode, the five voltages available are selected by the divider switch at the top left and are 500, 400, 300, 200, and 100 V. In the regulator position, the regulator voltage control on the top right allows continuous adjustment of the output voltage from 200 to 350 V.

8. EIGHT-BIT ANALOG SWITCH CARD

A block diagram and card top diagram of the EU-900-JA analog switch card are shown in Fig. A2–12. The card permits switching of up to eight analog signals simultaneously without relays. The FET switches give no "contact bounce" and have a high ON-to-OFF resistance ratio. Six of the switch circuits (connectors 1–3, 7–9,

Fig. A2–12 Analog switch card.

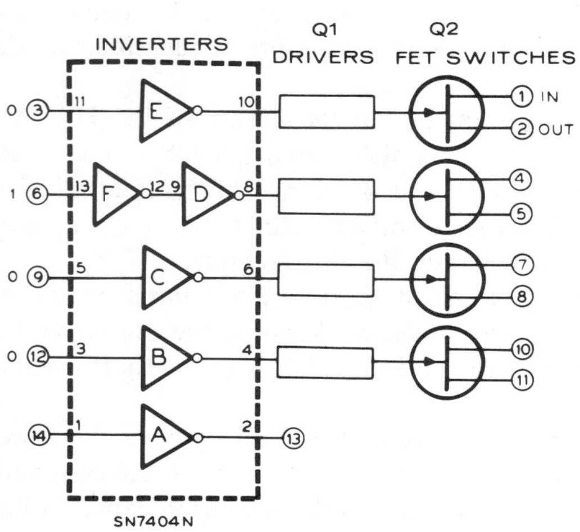

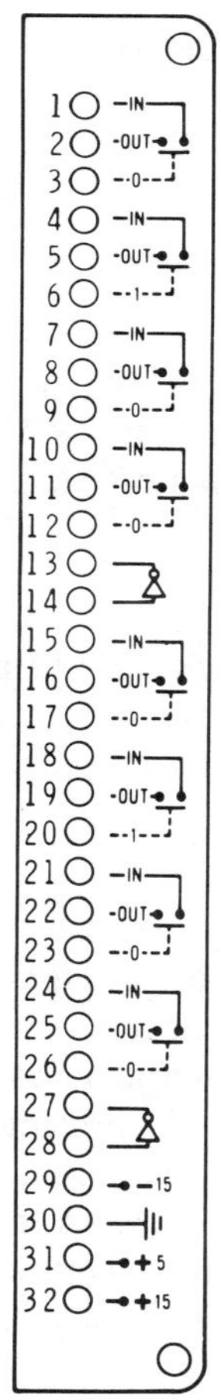

10–12, 15–17, 21–23, and 24–26) require +5 V to close the switch and two switches (connectors 4–6 and 18–20) require ≈0 V to close the switch. An open control input acts as a logic 1 (+5 V).

Also contained on the card are two inverters (connectors 13–14 and 27–28) that can be patch-wired for signal inversion. The FET switches can accept analog input voltages from −5 V to +30 V. The ON resistance is typically 10 Ω, while the OFF resistance is typically 10^{10} Ω. The maximum switching current is 10 mA, while the maximum switching rate is 100 kHz.

9. NAND GATE CARD

A card top diagram of the EU-800-JC NAND gate card is shown in Fig. A2–13. The card contains two TTL (transistor-transistor logic) integrated circuits. All the input and output connections of the gates are brought to the top of the card for patch wiring.

One of the NAND gate integrated circuits is a quadruple two-input gate, while the other is a dual four-input NAND gate. Each of the six gates can be used independently to perform NAND gate functions.

The NAND gate is a negative- or inverted-output AND gate. As such, the NAND gate output is low (≈0 V) only when all inputs are high (≈5 V).

An unused input of a NAND gate acts like a high input and has no effect on the gate function. If only one input is used, the NAND gate output is always the opposite, in logic level, of the input. Thus if only one input is used and it is high (≈5 V), the NAND gate output is low (≈0 V), and vice versa. A single-input NAND gate is therefore a logic level inverter.

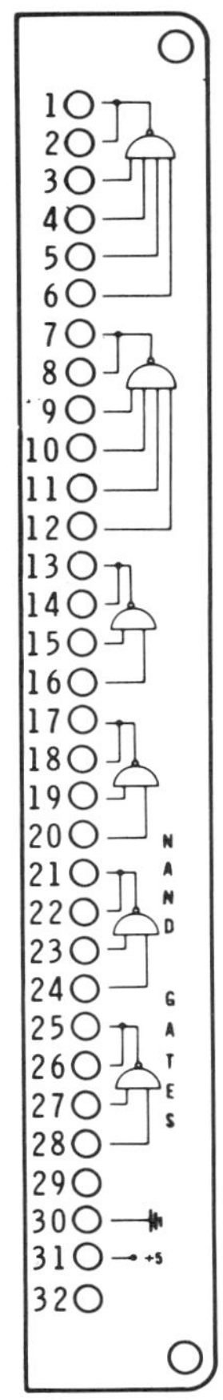

Fig. A2–13 NAND gate card top diagram.

Bibliography

Angelo, E. J., Jr., *Electronics: BJT's, FET's and Microcircuits,* McGraw-Hill, New York, N.Y., 1969.

A book intended for a first course in electronics for electrical engineering students. It contains an excellent basic description of amplifiers.

De Waard, H., and D. Lazarus, *Modern Electronics,* Addison-Wesley, Reading, Mass., 1966.

The first chapter of this book is an excellent introduction to the mechanisms of conduction in solids.

Haase, R., *Thermodynamics of Irreversible Processes,* Addison-Wesley, Reading, Mass., 1969.

A helpful reference for mathematical treatment of thermoelectric phenomena.

Halliday, D., and R. Resnick, *Physics, Parts I and II,* Wiley, New York, N.Y., 1966.

A standard college physics text that includes good discussions of basic electrical quantities.

Lion, K. S., *Instrumentation in Scientific Research—Electrical Input Transducers,* McGraw-Hill, New York, N.Y., 1959.

Many basic discussions of transducers are included in this classic text.

Malmstadt, H. V., C. J. Delaney, and E. A. Cordos, "Instruments for Rate Determinations," *Analytical Chemistry 44,* 1972, p. 79A.

A description of the principles and circuits for quantitative measurement of rate of change of electrical signals, with special emphasis on integrating types of differentiators.

Malmstadt, H. V., and C. G. Enke, *Digital Electronics for Scientists,* W. A. Benjamin, Menlo Park, Calif., 1969.

The integrated presentation of electronic switching devices and applications first presented in Chapters 2 and 3 of this book has been greatly modified and updated and introduced in Section 2–3 of this new text.

Malmstadt, H. V., C. G. Enke, and E. C. Toren, Jr., *Electronics for Scientists,* W. A. Benjamin, Menlo Park, Calif., 1962.

Some of the basic amplification and servo control concepts presented in this earlier text have been modified and presented in the present text in a new perspective, with an emphasis on the overall control of electrical quantities in instrumentation.

Millman, J., and H. Taub, *Pulse, Digital and Switching Waveforms,* McGraw-Hill, New York, N.Y., 1965.

A standard source book for switching and wave shaping. Contains detailed discussions of diode and transistor switches, including gating.

Motorola, *The Semiconductor Data Book,* 1969.

Contains specifications for many transistors, FET's, IC's, OA's, diodes, etc., and also descriptive material on FET's, phototransistors, unijunction transistors, and their applications.

Schmid, H., *Electronic Analog/Digital Conversions,* Van Nostrand Reinhold Co., 1970.

Includes an excellent section on electronic switches.

SCR Manual, 5th ed., General Electric Company, Syracuse, N.Y., 1972.

Principles and applications of pnpn devices, including light-activated devices. Contains many practical circuits for use of thyristors.

Seely, S., *Electronic Circuits,* Holt, Rinehart and Winston, New York, N.Y., 1968.

A well-known text in which electronic devices and basic semiconductor and tube-driven circuits are presented.

Smith, R. J., *Circuits, Devices and Systems,* 2nd ed., Wiley and Sons, New York, N.Y., 1971.

A basic text that provides an understanding of the fundamental principles of electrical engineering, a familiarity with the viewpoint and terminology of electrical engineers, and an introduction to some modern techniques and many good illustrations and problems.

Sze, S. M., *Physics of Semiconductor Devices,* Wiley-Interscience, New York, N.Y., 1969.

The electrical characteristics of modern semiconductor devices are developed from basic semiconductor theories in a thorough mathematical treatment.

Texas Instruments Inc., *Transistor Circuit Design,* McGraw-Hill, New York, N.Y., 1963.

A good basic description of transistors and their applications.

Tobey, G. E., J. G. Graeme, and L. P. Huelsman (Eds.), *Operational Amplifiers,* McGraw-Hill, New York, N.Y., 1971.

A modern book on operational amplifiers that contains many applications of OA's as well as basic design considerations of the amplifiers.

Solutions to Problems

SOLUTIONS TO PROBLEMS IN SECTION 2–1

1. a) If $R_m = 0$, the measured thermistor resistance R'_T is given by

$$R'_T = 5\ \text{V}/0.99 \times 10^{-3}\ \text{A} = 5.05\ \text{k}\Omega.$$

b) $R_m = 250\ \Omega$.

c) With the 1 mA meter the true resistance is

$$R_T = \frac{V}{I} - R_m = 4800\ \Omega,$$

and

$$\text{RPE} = \frac{R_T - R'_T}{R_T} = 0.052, \qquad \text{or } 5.2\%.$$

2. a) For the current meter a series circuit consisting of the 1 V battery, the thermistor, and the current meter will be considered. The theoretical current is 10^{-4} A $= 100\ \mu$A. The measured current is

$$I = \frac{1\ \text{V}}{R_T + R_m} = \frac{1}{10^4 + 2.5 \times 10^3} = 0.8 \times 10^{-4}\ \text{A},$$

and

$$\text{RPE} = 0.20, \qquad \text{or } 20\%.$$

b) For voltage measurements, a series circuit consisting of the 1 V battery, a 10 kΩ resistor, and the thermistor is considered. The voltmeter is in parallel with the thermistor. The theoretical voltage V_T may be calculated from

$$V_T = \frac{VR_T}{R_T + 10\ \text{k}\Omega} = 0.5\ \text{V}.$$

The voltage measured may be calculated in the same manner, substituting R_{eq} for R_T, where

$$R_{eq} = \frac{R_m R_T}{R_m + R_T} = 9.9 \times 10^3 \ \Omega.$$

Then

$$V_m = 0.497 \text{ V}$$

and

$$\text{RPE} = 0.006, \qquad \text{or } 0.6\%.$$

3. The total current in the circuit may be written as

$$I_t = \frac{V}{R_p + R_r + R_u}.$$

If $R_u = 0$, $I_t = 0.107$ A. When R_u is present, $I = \frac{2}{3} I_t = 0.0715$ A. Hence

$$R_u = \frac{V}{I} - (R_p + R_r) = 7 \ \Omega.$$

4. From Eq. (2–20),

$$R_u = R(R_A/R_B) = 6.28 \text{ k}\Omega.$$

5. By successive reductions, the current through $R_7 = 7.05$ mA.
6. $V = 0.71$ V.
7. Use the Kirchhoff junction and loop equations and solve by determinants. $I_1 = 6.22$ mA, $I_2 = 1.87$ mA, $I_3 = 4.35$ mA, $I_4 = -0.15$ mA, $I_5 = 1.72$ mA, $I_6 = 4.50$ mA.
8. The Thévenin equivalent voltage is 0.279 V, $R_{Th} = 925 \ \Omega$. If $V_m = V_{Th}$, IR_{Th} must be

$$IR_{Th} \leqslant 0.0001(V_{Th}) = 2.79 \times 10^{-5} \text{ V}.$$

Then

$$I = 3.02 \times 10^{-8} \text{ A},$$

and

$$R_{meter} = \frac{V_{Th}}{I} - R_{Th} \cong 9.25 \text{ M}\Omega.$$

9. Analyzing by superposition, we obtain:

	1 kΩ	2.5 kΩ	5 kΩ
I_1, mA	3.75	2.5	−1.25
I_2, mA	−1.87	0.75	2.72
I_t, mA	1.88	3.25	1.37

10. If $IR_3 = 0$, then at the junction of the three resistors,

$$I_1 = -I_2 \qquad \text{or} \qquad \frac{V_1}{R_1} = \frac{-V_2}{R_2},$$

$$V_2 = -V_1 \frac{R_2}{R_1}.$$

The OA circuit is the inverting voltage amplifier.

11. Since R_3 is to be varied to determine the maximum power P_{max}, it is best to remove R_3 and form a Thévenin equivalent circuit of the rest of the network. When R_3 is removed, the Thévenin equivalent voltage between the two terminals created, V_{Th}, is

$$V_{Th} = 14\text{ V} - 20I = 9\text{ V} + 5I.$$

Since

$$I = \frac{5\text{ V}}{25\ \Omega} = 0.2\text{ A},$$

$$V_{Th} = 14\text{ V} - 4\text{ V} = 10\text{ V}.$$

R_{Th} is the parallel combination of the 20 Ω and 5 Ω resistors, or

$$R_{Th} = \frac{20 \times 5}{20 + 5} = 4\ \Omega.$$

The simple series circuit after reinserting R_3 is thus a 10 V source, a 4 Ω resistance, and R_3. The power in the circuit is given by

$$P = I^2R_3 = \left(\frac{10\text{ V}}{4\ \Omega + R_3}\right)^2 R_3.$$

The maximum power occurs when $dP/dR = 0$,

$$\frac{dP}{dR_3} = \frac{(4 + R_3)^2 100 - 100R_3 \times 2(4 + R_3)}{(4 + R_3)^4} = 0,$$

from which $R_3 = 4\ \Omega$.

$$P_{max} = \left(\frac{10\text{ V}}{8\ \Omega}\right)^2 \times 4\ \Omega = 6.25\text{ W}.$$

12. The Thévenin equivalent voltage is

$$V_{Th} = 30\text{ V},$$

and

$$R_{Th} = 21\ \Omega.$$

Thus the current through the ammeter, I_m, is

$$I_m = \frac{V_{\text{Th}}}{R_{\text{Th}} + 9\ \Omega} = \frac{30\text{ V}}{30\ \Omega} = 9\text{ A}$$

13. a) R_s must equal the setting on ΔR_3 times the multiplier:

$$R_u = R_s = \Delta_1 R_3\ (R_2/R_1),$$

$$R_s + \Delta R_s = (\Delta_1 R_3 + \Delta_2 R_3)(R_2/R_1),$$

or

$$\Delta R_s = \Delta_2 R_3 (R_2/R_1),$$

where $\Delta_2 R_3$ is the change in resistor ΔR_3 required to rebalance the bridge.

b) If $\Delta R_3 = 0.1\ \Omega$, $\Delta R_s = 0.1\ \Omega$, then

$$\Delta l/l = \frac{\Delta R_s/R_s}{F} = \frac{0.1/120}{2} = 0.000416.$$

14. a) R_{Th} should be no greater than 0.01% of 100 kΩ, or 10 Ω. $R_{\text{Th}} = R_1R_2/(R_1 + R_2) = 10\ \Omega$ and since this is a 4/1 divider, $R_1 = 4R_2$. Therefore, $R_1 = 50\ \Omega$, $R_2 = 12.5\ \Omega$.

b) $I_N = 5\text{ V}/50\ \Omega = 0.1\text{ A}$, $R_N = 50\ \Omega \times 12.5/62.5 = 10\ \Omega$.

15. $R_N = 21\ \Omega$, $I_N = I_{30} - I_{20} = -(I_{90} - I_{10}) = 1.45\text{ A}$, $I_{\text{meter}} = I_NR_N/(R_N + R_m) = 1\text{ A}$.

16. a) $v_o = v_x - v_y = V\left(\dfrac{R_u}{R_A + R_u} - \dfrac{R}{R_B + R}\right).$

b) $$\frac{dv_o}{dR_u} = V\left[\frac{(R_A + R_u) - R_u}{(R_A + R_u)^2}\right] = V\left[\frac{R_A}{(R_A + R_u)^2}\right].$$

c) The bridge output will be linear if the derivative in part (b) is a constant. This will occur if $R_A \gg R_u$. Thus

$$\frac{dv_o}{dR_u} = V\frac{R_A}{R_A^2} = \frac{V}{R_A}.$$

d) $$\frac{d^2v_o}{dR_u^2} = V\left[\frac{-R_A \times 2(R_A + R_u)}{(R_A + R_u)^4}\right] = 0.$$

The above condition will hold only if $R_A \ll R_u$.

e) No, since when $R_A \ll R_u$ the slope determined in part (b) is not constant.

SOLUTIONS TO PROBLEMS IN SECTION 2–2

1. Substituting into Eq. (2–28), $W_d = CV^2/2 = 10^{-9}\text{ F} \times (8.32)^2/2 = 3.46 \times 10^{-8}$ J.

2. a) Since the voltage on each capacitor is 5 V and $q = CV$, $q_1 = 0.5 \times 10^{-6}$ C, $q_2 = 0.25 \times 10^{-6}$ C, $q_3 = 0.1 \times 10^{-6}$ C.

   b) Since the charges on parallel capacitors add, $q_T = 0.85 \times 10^{-6}$ C.

   c) $C_p = 0.17$ μF.

3. Substituting into Eq. (2–33), $C_s = 0.0125$ μF,

$$\frac{1}{C_s} = \frac{1}{C_1} + \frac{1}{C_2} + \frac{1}{C_3}.$$

   The charge on each capacitor $= q_T = CV = 6.25 \times 10^{-8}$ C.

4. a) $C_{\text{eq}} = 0.083$ μF.

   b) V_1 and $V_2 = 5$ V. The charges on C_3, C_4, and the parallel combination of C_5 and C_6 are equal. The equivalent capacitance of that branch of the network is 0.023 μF. The charge on C_3 and $C_4 = 1.13 \times 10^{-7}$ C. $V_{C3} = q_3/C_3 = 1.13$ V, $V_{C4} = 2.26$ V, $V_{C5} = V_{C6} = 1.61$ V.

   c) The charge on each capacitor is:

	C_1	C_2	C_3	C_4	C_5	C_6
q	5	25	11.3	11.3	3.2	8.1 $\times 10^{-8}$ C.

5. Using Eq. (2–37), $i = (V/R)e^{-t/RC}$ or $V_R = Ve^{-t/RC}$, $V_C = V - V_R$.

t, msec	V_R, V	V_C, V
0.5	0.95	0.05
1	0.90	0.1
5	0.61	0.39
10	0.37	0.63
50	0.0067	0.9933
100	4×10^{-5}	0.99996

6. a) The charging time constant is

$$\tau_c = \frac{CR_1R_2}{R_1 + R_2} = 1.76 \text{ msec.}$$

The discharge time constant is

$$\tau_D = \frac{CR_1R_3}{R_1 + R_3} = 2.18\ \mu\text{sec}.$$

For practical purposes, 5τ is sufficient time for charging or discharging, so $5\tau_c = 8.8$ msec, $5\tau_D = 10.9$ μsec.

7. a)

t, msec	V_R, V	V_C, V
0	10	0
0.1	3.68	6.32
0.3	0.50	9.50
0.5	0.067	9.933
1.0	0.0045	9.9955
1.3	−0.50	0.50
1.5	−0.067	0.067

b)

t, msec	V_R, V	V_C, V
0	10	0
0.5	6.06	3.94
1.0	3.68	6.32
1.5	−3.83	3.83

8. $v_o = -\dfrac{1}{RC}\displaystyle\int v_{\text{in}}\,dt,$

$$\frac{dv_o}{dt} = -\frac{v_{\text{in}}}{RC} = -\frac{5\text{ V}}{2.5\text{ sec}} = -2\text{ V/sec}.$$

The output changes by −1 V in 0.5 sec.

9. Using Eqs. (2–27) and (2–45),

$$C = \frac{Vt}{V_oR} = \frac{q}{V} = \frac{0.096\text{ C}}{5\text{ V}}, \qquad C = 0.019\text{ F}.$$

The resistor value is unimportant.

10. a) This is a follower with gain.

$$v_o = v_{\text{in}}\left(\frac{R_f}{R_{\text{in}}} + 1\right) = 5 \times 10^{-5}\text{ V}\left(\frac{1}{10^5} + 1\right), \qquad v_o \approx 50\ \mu\text{V}.$$

b) $v_o = -\dfrac{1}{RC}\displaystyle\int v_{\text{in}}\,dt,$

$$\Delta t = -\frac{v_oRC}{v_{\text{in}}} = -\left(\frac{+12}{-5 \times 10^{-5}}\right)(10^5)(10^{-7}),$$

$$\Delta t = 2.4 \times 10^3\text{ sec}.$$

c) It must be at least 100×50 μV or 5 mV.

d) 24 sec ±1%.

e) For $v_{\text{in}} = 0$, Δt must be greater than 10^5 sec.

$$-v_{\text{in}} = \frac{v_o}{\Delta t} RC = -10^{-6}\ \text{V} = 1\ \mu\text{V offset.}$$

11. a) $v_o = -RC \dfrac{dv_{\text{in}}}{dt}$, $\qquad RC = -\dfrac{(-5.0)\ \text{V}}{8\ \text{V/sec}} = 0.625$ sec.

Any R and C with a product of 0.625 sec may be used. Example: $R = 10^6\ \Omega$, $C = 0.625\ \mu$F.

b) $RC = 0.067$ sec. Example: $R = 1\ \text{M}\Omega$, $C = 0.067\ \mu$F. The differentiator can be followed with an inverting unity-gain OA.

12. a) Using Eq. (2–62),

$$v_o = k \log \frac{v_r}{v_s} K = K_1 K_2,$$

$$K_1 = \frac{i_{\text{in}}}{C_2} = \frac{v_{\text{in}}}{R_{\text{in}} C_2}, \qquad K_2 = RC(2.3),$$

$$v_o = \frac{0.5}{(10^5)(10^{-6})} [(2.3)(10^6)(0.5 \times 10^6)] \log 10,$$

$$v_o = 5.75\ \text{V}.$$

b) $v_o = 5.75 \log 5 = 1.82$ V.

c) $v_o = 0.91$ V.

13. Substituting into Eq. (2–66),

$$L = \frac{V}{di/dt} = \frac{2}{15} = 0.133\ \text{H}.$$

14. $V = 1.5$ V.

15. From Eqs. (2–69) and (2–70),

$$V_L = Ve^{-tR/L}, \qquad V_R = V(1 - e^{-tR/L}).$$

t, msec	V_L, V	V_R, V
0.5	1.17	0.33
1	0.91	0.59
2	0.55	0.95
10	0.010	1.49
100	~0	1.50

SOLUTIONS TO PROBLEMS IN SECTION 2–3

1. a) From the equations preceding Eq. (2–98) the voltage drops across the closed switch and open switch, V_{sc} and V_{so}, can be calculated:

$$V_{sc} = 5\left(\frac{80}{50 \times 10^3 + 80}\right) \text{V} = 0.008 \text{ V},$$

$$\text{V}_{so} = 5\left(\frac{900 \times 10^3}{50 \times 10^3 + 900 \times 10^3}\right) \text{V} = 4.737 \text{ V}.$$

The voltage drop across the load is the source voltage minus the voltage drop across the switch. For the closed switch, $V_L = 5 - 0.008 = 4.992$ V, and for the open switch $V_L = 5 - 4.737 = 0.263$ V.

b) For τ_o use Eq. (2–98),

$$\tau_o = \frac{400 \times 10^{-12} \times 50 \times 10^3 \times 900 \times 10^3}{50 \times 10^3 + 900 \times 10^3} = 18.9 \ \mu\text{sec},$$

and for τ_c use Eq. (2–99),

$$\tau_c = \frac{400 \times 10^{-12} \times 50 \times 10^3 \times 80}{50 \times 10^3 + 80} = 0.032 \ \mu\text{sec}.$$

2. At 300°K, $Q_e/kT = 39$, so that Eq. (2–95) can be written

$$I = I_i(e^{39V} - 1) = 10^{-10}(e^{39V} - 1).$$

For $V = 0.30$, $I = 10^{-10}(e^{11.7} - 1) = 15 \ \mu\text{A}$.

For $V = 0.40$, $I = 10^{-10}(e^{15.6} - 1) = 0.6$ mA.

For $V = 0.50$, $I - 10^{-10}(e^{19.5} - 1) = 30$ mA.

For $V = 0.60$, $I = 10^{-10}(e^{23.4} - 1) = 1.45$ A.

Note that the final value of 1.45 A would not be reached in practice because of the ohmic resistance of the junction.

3. For the source at +10 V, the voltage across the load

$$V_L = +10 \times \frac{1 \text{ k}\Omega}{1 \text{ k}\Omega + 65\Omega} = 9.39 \text{ V}.$$

For the source at —10 V,

$$V_L = -10 \times \frac{1 \text{ k}\Omega}{1 \text{ k}\Omega + 5 \text{ M}\Omega} = -0.002 \text{ V}.$$

4. Substituting in Eq. (2–103), $I_m = 0.8$ mA, we find

$$R_{fd} = \frac{26}{0.8} = 32.5\ \Omega.$$

5. a) From Eq. (2–118),

$$I_{B1} = \frac{1\text{ V} - 0.6\text{ V}}{4\text{ k}\Omega} = 100\ \mu\text{A}.$$

The transistor is saturated if the condition of Eq. (2–116) holds. Since $h_{FE}I_{B1} = 100 \times 100\ \mu\text{A} = 10$ mA, and

$$I_{C(\max)} = \frac{5\text{ V} - 0.2\text{ V}}{1\text{ k}\Omega} = 4.8\text{ mA},$$

the transistor is in saturation.

b) As calculated above, $I_{C(\text{sat})} = 4.8$ mA.

c) $R_{\text{ON}} = \dfrac{V_{CE(\text{sat})}}{I_{C(\text{sat})}} = \dfrac{0.2\text{ V}}{4.8\text{ mA}} = 42\ \Omega.$

6. a) The minimum voltage across the coil for turn-on, $V_{\text{coil}(\min)} = 200\ \Omega \times 20\text{ mA} = 4$ V. Hence V_{CE} must be less than 1 V for turn-on, and the transistor need not be in saturation.

b) The source current $I_s = (4\text{ V} - V_{BE(\text{ON})})/10\text{ k}\Omega = 3.4\text{ V}/10\text{ k}\Omega = 0.34$ mA.

c) $h_{FE(\min)} \geqslant I_{C(\min)}/I_B$ or $h_{FE(\min)} \geqslant 20\text{ mA}/0.34\text{ mA} = 59.$

7. The ON resistance R_{ON} is $R_{\text{ON}} = V_{CE(\text{sat})}/I_{C(\text{sat})} = 0.4\text{ V}/50\text{ mA} = 8\ \Omega$. When the transistor is in cutoff, the circuit can be considered to be R_L in series with the transistor OFF resistance, R_{OFF}. Hence,

$$I_{CO} = \frac{V_{CC}}{R_L + R_{\text{OFF}}}$$

and

$$R_{\text{OFF}} = \frac{15\text{ V} - 20\ \mu\text{A} \times 1\text{ k}\Omega}{20\ \mu\text{A}} = 749\text{ k}\Omega.$$

8. Substituting in Eq. (2–121), $V_p = 15$ V, $V_\gamma = 0.6$ V, and $V_{BB} = 19$ V, we have

$$\eta = \frac{15 - 0.6}{19} = 0.758.$$

9. a) At the peak point, $V_B = I_p R_2 + V_{BB}$, where I_p is the peak-point current. From Eq. (2–120),

$$V_p = V_\gamma + \eta V_{BB} = V_\gamma + \eta V_B - \eta I_p R_2.$$

The peak-point current I_p is

$$I_p = \frac{V_B}{R_{B1} + R_{B2} + R_2} \approx \frac{V_B}{R_{BB}}.$$

Hence

$$V_p = V_\gamma + \eta V_B - \frac{\eta V_B R_2}{R_{BB}}.$$

b) If R_2 is chosen such that $R_2 \approx V_\gamma R_{BB}/\eta V_B$, then $V_p = \eta V_B$ and will be highly independent of temperature.

10. Use $v_L = v_u R_L/(R_L + R_e + R_s)$, where R_s is the switch resistance. For the closed switch,

$$v_L = \frac{5\ \text{V} \times 1\ \text{k}\Omega}{1\ \text{k}\Omega + 100\ \Omega + 50\ \Omega} = 4.35\ \text{V}.$$

For the open switch,

$$v_L = \frac{5\ \text{V} \times 1\ \text{k}\Omega}{1\ \text{k}\Omega + 100\ \Omega + 100\ \text{k}\Omega} = 0.049\ \text{V}.$$

11. The current in each diode is composed of two components, one due to the supply voltage $\pm V_c$ and the other due to the input signal v_u. If each diode is considered to be ideal (zero resistance), the current due to the supply voltage in Fig. 2–132 is $V_c/2R_c$. This current is in the forward direction in each diode. The current due to v_u is in the reverse direction in diodes D_1 and D_4. The reverse current in D_1 is $v_u/R_c + v_u/2R_L$, while that in D_4 is $v_u/2R_L$. Hence the larger reverse current is in D_1. The forward current $V_c/2R_c$ must thus be greater than $v_u/R_c + v_u/2R_L$. Hence

$$(V_c)_{\min} = \frac{v_u}{R_c} \times 2R_c + \frac{v_u \times 2R_c}{2R_L} = v_u \left(2 + \frac{R_c}{R_L}\right).$$

12. a) Substituting into the equation preceding Eq. (2–127), $b_\lambda = 0.3$, $Q_e = 1.603 \times 10^{-19}$ C, and $P_k = 10^{12}$ photons/sec, we obtain

$$-I_p = 0.3 \times 1.603 \times 10^{-19} \times 10^{12} = 4.81 \times 10^{-8}\ \text{A}.$$

b) Substituting $I_p = -4.81 \times 10^{-8}$ A, and $h_{FE} = 50$ into Eq. (2–127), we obtain

$$I_E = -4.81 \times 10^{-8}(51) = -2.45 \times 10^{-6}\ \text{A}.$$

13. a) For $v_{in} = +5\text{ V}$, $v_s = +5\text{ V} \times \dfrac{5\text{ k}\Omega}{5\text{ k}\Omega + 50\ \Omega} = 4.95\text{ V}$.

For $v_{in} = -5\text{ V}$,

$$v_o = -5\text{ V} \times \frac{5\text{ k}\Omega}{5\text{ k}\Omega + 500\text{ k}\Omega} = -0.049\text{ V}.$$

b) For $v_{in} = +5\text{ V}$,

$$v_o = +5\text{ V} \times \frac{500\text{ k}\Omega}{500\text{ k}\Omega + 5\text{ k}\Omega} = 4.95\text{ V}.$$

For $v_{in} = -5\text{ V}$,

$$v_o = -5\text{ V} \times \frac{50\ \Omega}{5\text{ k}\Omega + 50\ \Omega} = -0.049\text{ V}.$$

14. See Fig. S–14 for three clipping circuits which will solve the problem.

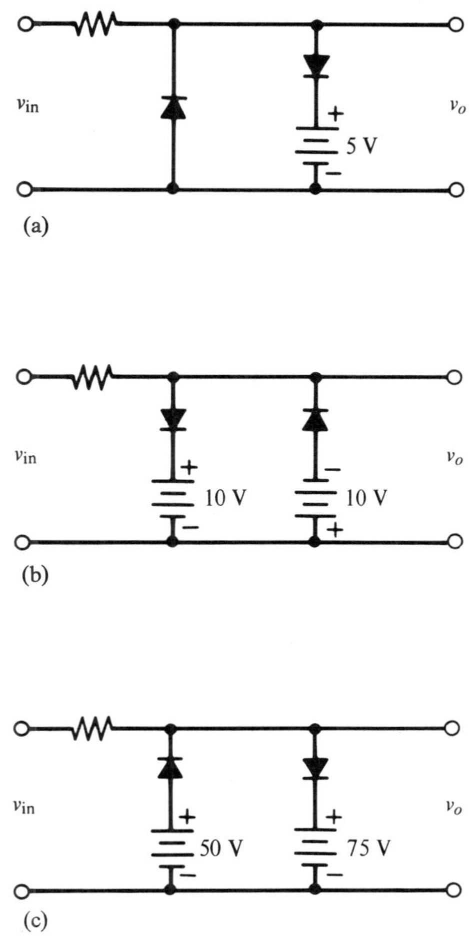

Fig. S–14

15. When $v_{in} = +5\text{ V}$, the diode is forward biased. The current I in the series circuit is

$$I = \frac{5\text{ V} + 2\text{ V}}{1\text{ k}\Omega + 50\ \Omega} = 6.66\text{ mA}.$$

The output voltage v_o is

$$v_o = 6.66\text{ mA} \times 50\ \Omega - 2\text{ V} = -1.67\text{ V}.$$

When $v_{in} = -5\text{ V}$, the diode is reverse biased. Now

$$I = \frac{5\text{ V} - 2\text{ V}}{5\text{ M}\Omega + 1\text{ k}\Omega} = 6 \times 10^{-7}\text{ A}$$

and $v_o = -2\text{ V} - 5\text{ M}\Omega \times 6 \times 10^{-7}\text{ A} = -5\text{ V}$.

16. a) Substituting in Eq. (2–130) $v_{in} = 1.5\text{ V}$, $R = 250\text{ k}\Omega$, and $C = 0.5\ \mu\text{F}$, we have

$$\frac{dv_o}{dt} = -\frac{1.5\text{ V}}{250\text{ k}\Omega \times 0.5\ \mu\text{F}} = -12\text{ V/sec}.$$

b) 1 sec.

SOLUTIONS TO PROBLEMS IN SECTION 2–4

1. Since by Eq. (2–133) the secondary voltage is directly proportional to the primary voltage, it will decrease by the same percentage, or $(115 - 95)/115 = 17\%$.

2. Substituting in Eq. (2–133) $v_s = 220$ V, $v_p = 4400$ V, we have $N_s/N_p = 220/4400 = 1/20$.

3. When the capacitor is present, it charges to $+1.4 \times 10 = 14$ V during the transformer's positive half-cycle. During the negative half-cycle, the diode must withstand the sum of the negative transformer voltage and the positive voltage on the capacitor. Thus $\text{PIV} = 2 \times 1.4 \times 10 = 28$ V.

4. a) By the same reasoning as in problem 3, $\text{PIV} = 2 \times 1.4 \times 5 = 14$ V.

   b) During the positive half-cycle the capacitor charges to $5 \times 1.4 = 7$ V. Two diodes conduct, while the other two must each withstand 7 V of reverse voltage. Likewise during the negative half-cycle the other two diodes must withstand 7 V. Hence the PIV rating is 7 V. The transformer required for the bridge circuit must have a 5 V rms secondary, whereas the transformer in part (a) must have a 10 V rms secondary with a center tap.

5. a) $V_{dc} = 1.4 \times 25 \text{ V} = 35$ V.

   b) 35 V.

   c) $2 \times 35 \text{ V} = 70$ V.

6. a) From the discussion preceding Eq. (2–135), it can be shown that

$$V_{dc} = 1.4\, V_{rms}[1 - 1/(2fCR_L)].$$

   Hence

$$V_{dc} = 1.4 \times 40\, [1 - 1/(2 \times 60 \times 10 \times 10^{-6} \times 5 \times 10^3)] = 46.7 \text{ V}.$$

   b) Since $I_{dc} = V_{dc}/R_L$ and $V_{dc} = 48$ and $R_L = 5$ kΩ, we have $I_{dc} \approx 47/5000 = 9.3$ mA.

   c) Substituting in Eq. (2–137) $f = 60$ Hz, $C = 10\ \mu$F, $R_L = 5$ kΩ, we find for the ripple factor

$$r = \frac{1}{2 \times 1.75 \times 60 \times 10 \times 10^{-6} \times 5 \times 10^3} = 0.095.$$

From Eq. (2–138)

$$V_{\text{ripple}} = rV_{\text{dc}} = 0.095 \times 47 = 4.4\ \text{V}_{\text{rms}}.$$

7. For V_{dc} use Eq. (2–135). With $C = 100\ \mu\text{F}$,

$$V_{\text{dc}} = 1.4 \times 18\,[1 - 1/(2 \times 120 \times 100 \times 10^{-6} \times 100)] = 14.7\ \text{V}.$$

Note that Eq. (2–135) is not reliable here because of the heavy load. With $C = 500\ \mu\text{F}$,

$$V_{\text{dc}} = 1.4 \times 18\,[1 + 1/(2 \times 120 \times 500 \times 10^{-6} \times 100)] = 23.1\ \text{V}.$$

With $C = 1000\ \mu\text{F}$, $V_{\text{dc}} = 24$ V. For the ripple voltages use Eqs. (2–137) and (2–138). For $C = 500\ \mu\text{F}$,

$$r = \frac{1}{2 \times 1.75 \times 120 \times 500 \times 10^{-6} \times 100} = 0.048,$$

$$V_{\text{ripple}} = 0.048 \times 23.1 = 1.11\ \text{V}_{\text{rms}}.$$

For $C = 1000\ \mu\text{F}$, $r = 0.024$, $V_{\text{ripple}} = 0.6\ \text{V}_{\text{rms}}$.

8. The current through the Zener diode $i_Z = P/v_Z = 100\ \text{mW}/5.1\ \text{V} \approx 20$ mA. The load current $i_L = v_Z/R_L = 5.1\ \text{V}/100 = 51$ mA. Hence the maximum supply current $i = i_Z + i_L = 71$ mA. To stay within this limit, $R_s = V/i = 10\ \text{V}/71\ \text{mA} = 141\ \Omega$.

9. From Eq. (2–143) the maximum current is

$$I = \sqrt{P/R} = \sqrt{0.5\ \text{W}/50} = 0.1\ \text{A}.$$

From Eq. (2–144) the maximum voltage is

$$V = \sqrt{PR} = \sqrt{0.5 \times 50} = 5\ \text{V}.$$

SOLUTIONS TO PROBLEMS IN SECTION 2–5

1. a) For the amplifier of Fig. P–1a, $v_o = 40v_{\text{in}}$, $A_V = 40$, while for the amplifier of Fig. P–1b, $i_o = 50i_{\text{in}}$, $A_i = 50$.

   b) The power gain A_P is the ratio of the output power to the input power, $A_P = P_o/P_{\text{in}}$. From Eq. (2–146) for the voltage amplifier, $P_o = (\mu v_{\text{in}})^2/R_L$ and $P_{\text{in}} = v_{\text{in}}^2/R_{\text{in}}$. Thus

$$A_P = \frac{(40v_{\text{in}})^2(10^6)}{v_{\text{in}}^2(10^6)} = 1600.$$

For the current amplifier of Fig. P–1b, $P_o = (\beta i_{\text{in}})^2 R_L$, and $P_{\text{in}} = (i_{\text{in}})^2 R_{\text{in}}$. Thus

$$A_P = \frac{(50 i_{\text{in}})^2 \times 10\ \text{k}\Omega}{i_{\text{in}}^2 \times 1\ \text{k}\Omega} = 25{,}000.$$

c) $v_o = i_o R_L = 50 i_{\text{in}} \times 10\ \text{k}\Omega, \qquad v_{\text{in}} = i_{\text{in}} R_{\text{in}} = i_{\text{in}} \times 1\ \text{k}\Omega,$

$$A_v = \frac{v_o}{v_{\text{in}}} = \frac{50 i_{\text{in}} \times 10\ \text{k}\Omega}{i_{\text{in}} \times 1\ \text{k}\Omega} = 500.$$

2. a) Substituting $g_m = 2 \times 10^{-3}\ \Omega^{-1}$ and $R_d = 10\ \text{k}\Omega$ into Eq. (2–152), we have $A_v = (2 \times 10^{-3})(10 \times 10^3) = 20$.

   b) Since $v_{ds} \approx 20 v_{gs}$, $v_{ds} \approx 20$ V. The current $v_{ds}/r_{ds} \approx 20\ \text{V}/100\ \text{k}\Omega = 2 \times 10^{-3}$ A. Hence v_{ds}/r_{ds} is 10% of the main current.

3. The small-signal model would be as in Fig. 2–201b, with $v_{gs} = 1$ V, $g_m = 2 \times 10^{-3}\ \Omega^{-1}$, $r_{ds} = 100\ \text{k}\Omega$, and $R_d = 10\ \text{k}\Omega$. This model differs from the ideal voltage amplifier in that the input resistance is not infinite and the output resistance is not zero. However, the JFET amplifier is a fairly good approximation to the ideal voltage amplifier at low signal frequencies.

4. From the slopes of the characteristic curves in Fig. 2–203, it is found that $h_{fe} \approx 25$, $h_{ie} \approx 700\ \Omega$, $h_{re} \approx 6 \times 10^{-4}$, $h_{oe} \approx 50 \times 10^{-6}\ \Omega^{-1}$.

5. a) For A_v, use Eq. (2–159):

$$A_v = \frac{-25 \times 1.5 \times 10^3\ \Omega}{(700\ \Omega \times 50 \times 10^{-6}\ \Omega^{-1} - 25 \times 6 \times 10^{-4}) \times 1.5 \times 10^3\ \Omega + 700\ \Omega}$$

$$A_v = \frac{-3.75 \times 10^4}{(0.035 - 0.015) \times 1.5 \times 10^3 + 700} = -51.$$

For A_i, use Eq. (2–161):

$$A_i = \frac{-25}{50 \times 10^{-6}\ \Omega^{-1} \times 1.5 \times 10^3\ \Omega + 1} = -23.2.$$

For A_p, use $A_p = A_v \times A_i = 51 \times 23.2 = 1183$.

   b) To find R_{in}, use Eq. (2–164):

$$R_{\text{in}} = \frac{700\ \Omega + (0.02) \times 1.5 \times 10^3\ \Omega}{1 + 50 \times 10^{-6}\ \Omega^{-1} \times 1.5 \times 10^3\ \Omega} = 679\ \Omega.$$

To find R_o use Eq. (2–166):

$$R_o = \frac{700\ \Omega + 2 \times 10^3\ \Omega}{0.02 + 50 \times 10^{-6}\ \Omega^{-1} \times 2 \times 10^3\ \Omega} = 22.5\ \text{k}\Omega.$$

6. The small-signal model would be as in Fig. 2–207 with the appropriate values. This differs in several ways from the ideal current amplifier. First, the input resistance is not zero (see problem 5b above). Second, the output resistance is finite and dependent on h_{oe} or $h_{ie}/\Delta h_e$ (see Eqs. 2–167 and 2–168). Third, the current gain is not strictly h_{fe}. Also, the output of the BJT influences the input through the voltage generator, $h_{re}v_{ce}$. However, the BJT can be considered an ideal current amplifier for many purposes.

7. a) For A_v, use Eq. (2–170):

$$A_v = 1 - \frac{dv_{BE}}{dv_u} = 1 - 0.05 = 0.95.$$

For A_i, use Eq. (2–173):

$$A_i = h_{fe} + 1 = 60 + 1 = 61.$$

For A_p, use $A_p = A_i \times A_v = 61 \times 0.95 = 58.$

b) R_{in} can be found from Eq. (2–177) with $R_E = 5\ \text{k}\Omega$ and $h_{fe} = 60$:

$$R_{\text{in}} = 5 \times 10^3\ (60 + 1) = 305\ \text{k}\Omega.$$

R_o can be found from Eq. (2–178) with $R_u = 100\ \Omega$ and $h_{fe} = 60$:

$$R_o = \frac{100\ \Omega}{61} = 1.6\ \Omega.$$

8. a) Substitute into Eq. (2–180)

$$\frac{\Delta v_o}{\Delta v_d} = \frac{5\ \text{V}}{250\ \mu\text{V}} = 20{,}000 \quad \text{and} \quad \frac{\Delta v_o}{\Delta v_{\text{cm}}} = \frac{1\ \text{V}}{500\ \text{mV}} = 2.$$

Then

$$\text{CMRR} = \frac{20{,}000}{2} = 10{,}000.$$

b) Use Eq. (2–181) to find CMRR in dB $= 20 \log 10^4 = 80$

9. To measure a 50 mV signal to 1% accuracy, we need to measure to 0.5 mV. With a CMRR of 10^4 the maximum common mode voltage which can be tolerated is $0.5\ \text{mV} \times 10^4 = 5\ \text{V}$.

10. Since R_L and R_o form a voltage divider for μv_{u1},

$$v_o = \mu v_{u1} \frac{R_L}{R_L + R_o} = 1000 v_{u1} \times \frac{2\ \text{k}\Omega}{2.2\ \text{k}\Omega} = 909 v_{u1}.$$

Hence $A_v = v_o/v_{u1} = 909$. To find A_i, we need i_{in} and i_o. The input current $i_{\text{in}} = v_{u1}/R_{\text{in}} = v_{u1}/50\ \text{k}\Omega$. The output current i_o is

$$i_o = \frac{\mu v_{u1}}{R_o + R_L} = \frac{1000 v_{u1}}{2.2\ \text{k}\Omega}.$$

Hence

$$A_i = \frac{i_o}{i_{\text{in}}} = \frac{1000 v_{u1}/2.2\ \text{k}\Omega}{v_{u1}/50\ \text{k}\Omega} = 22{,}727.$$

Since $A_p = A_v \times A_i$, $A_p = 909 \times 22{,}727 = 20.6 \times 10^6$.

11. a) The output power $P_o = v_o^2/R_L$, or since $v_o = \mu v_{u1}[R_L/(R_L + R_o)]$,

$$P_o = \frac{\mu^2 v_{u1}^2 R_L^2}{(R_L + R_o)^2 R_L} = \frac{\mu^2 v_{u1}^2 R_L}{(R_L + R_o)^2}.$$

The maximum power occurs when $dP_o/dR_L = 0$. This is found where $R_L = R_o$. The maximum power is then

$$P_{o\,(\text{max})} = \frac{1000^2 \times (5 \times 10^{-3})^2}{4 \times 200} = 31\ \text{mW}.$$

b) The input power $P_{\text{in}} = v_{u1}^2/R_{\text{in}} = (5 \times 10^{-3})^2/5 \times 10^4 = 5 \times 10^{-10}$ W.

$$A_p = \frac{P_o}{P_{\text{in}}} = \frac{31 \times 10^{-3}\ \text{W}}{5 \times 10^{-10}\ \text{W}} = 6.2 \times 10^7.$$

SOLUTIONS TO PROBLEMS IN SECTION 2–6

1. a) The bias current will exist in the feedback resistor R_f, since the input resistor is connected between the circuit common and virtual common.

   b) Since there is no current in R_{in}, $v_o = i_{b(-)} R_f = 15 \times 10^{-9}\ \text{A} \times 100 \times 10^3\ \Omega = 1.5$ mV. The sign of v_o will depend on the direction of $i_{b(-)}$.

2. The circuit as shown is a follower with gain to the offset voltage. Hence $v_o = v_{\text{off}}(1 + R_f/R_{\text{in}}) = 10^{-3}\ \text{V}\ (1 + 100) = 101$ mV.

3. a) The output voltage should be

$$v_o = \frac{-R_f}{R_{\text{in}}} v_{\text{in}} = -100 \times 50 \text{ mV} = -5 \text{ V}.$$

For 1% or less output error, 50 mV at the output due to $i_{b(-)}$ is all that can be tolerated. Hence,

$$i_{b(-)} = \frac{v_o}{R_f} = \frac{50 \text{ mV}}{100 \text{ k}\Omega} = 5 \times 10^{-7} \text{ A}.$$

b) Again 50 mV error is all that can be tolerated. So

$$v_{\text{off}} = \frac{v_o}{\left(1 + \dfrac{R_f}{R_{\text{in}}}\right)} = \frac{50 \text{ mV}}{101 \text{ k}\Omega} = 495 \ \mu\text{V}.$$

4. Substituting into Eq. (2–201), we have

$$v_o = -\left(-2 \text{ V} \times \frac{10 \text{ k}\Omega}{5 \text{ k}\Omega}\right) \pm 10^{-6} \text{ A} \times 10 \text{ k}\Omega$$

$$\pm 2 \times 10^{-3} \text{ V} \left(1 + \frac{10 \text{ k}\Omega}{5 \text{ k}\Omega}\right)$$

$$= +4 \text{ V} \pm 10^{-2} \text{ V} \pm 6 \times 10^{-3} \text{ V} = +4 \text{ V} \pm 16 \text{ mV}.$$

So the percentage error is 0.4%.

5. Use Eq. (2–202) as follows:

$$v_o = \pm 1 \text{ mV/day} \times \frac{10 \text{ k}\Omega}{5 \text{ k}\Omega} \pm 1 \ \mu\text{A} \times 10 \text{ k}\Omega$$

$$\pm 1 \text{ mV} \left(1 + \frac{10 \text{ k}\Omega}{5 \text{ k}\Omega}\right)$$

$$= \pm 2 \text{ mV/day} \pm 10 \text{ mV/day} \pm 0.3 \text{ mV/day}.$$

So the maximum drift rate is 12.3 mV/day.

Index

SELECTED PHYSICAL CONSTANTS

Quantity	Symbol	Value
Electron charge	Q_e	1.603×10^{-19} C
Faraday's constant	$\mathscr{F}$	96487.0 C $equiv^{-1}$
Gas constant	R	$\begin{cases} 8.32 \text{ J mole}^{-1}\ {}^{\circ}\text{K}^{-1} \\ 0.0821 \text{ liter atm mole}^{-1}\ {}^{\circ}\text{K}^{-1} \end{cases}$
Planck's constant	h	6.63×10^{-34} J-sec
Boltzmann's constant	k	1.38×10^{-23} J $°K^{-1}$
Ice point	T_0	273.15°K

TABLE OF PREFIXES FOR UNITS

Order	Prefix	Symbol
10^{12}	tera	T
10^{9}	giga	G
10^{6}	mega	M
10^{3}	kilo	k
10^{2}	hecto	h
10	deka	da
10^{-1}	deci	d
10^{-2}	centi	c
10^{-3}	milli	m
10^{-6}	micro	μ
10^{-9}	nano	n
10^{-12}	pico	p
10^{-15}	femto	f
10^{-18}	atto	a